RISK, SOCIETY, AND POLICY SERIES
Edited by Ragnar E Löfstedt

Risk, Media, and Stigma

Understanding Public Challenges to Modern Science and Technology

Edited by

James Flynn, Paul Slovic, and Howard Kunreuther

earthscan
from Routledge

First published in the UK and USA in 2001
by Earthscan Publications Ltd

For a full list of publications please contact:

Earthscan
2 Park Square, Milton Park, Abingdon, Oxfordshire OX14 4RN
711 Third Avenue, New York, NY 10017

First issued in hardback 2016

Earthscan is an imprint of the Taylor & Francis Group, an informa business

A catalogue record for this book is available from the British Library

ISBN 13: 978-1-138-15438-4 (hbk)
ISBN 13: 978-1-85383-700-5 (pbk)

Typesetting by Connie Kudura, Western Printers, Eugene, Oregon, USA
Cover design by Yvonne Booth

Contents

Nuclear Stigma

Place, Product, and Industry Stigma

Risk, Media, and Stigma

Coping with Stigma

List of Tables

List of Figures

Acronyms and Abbreviations

AAFC	Agriculture and Agrifood Canada
ADEQ	Arizona Department of Environmental Quality
AIDS	Acquired immune deficiency syndrome
AIS	Abbreviated Injury Scale
ANEC	American Nuclear Energy Council
AP	Associated Press
APA	American Psychological Association
ATSDR	Agency for Toxic Substances and Disease Registry
BGH	Bovine growth hormone
BSE	Bovine spongiform encephalopathy
BST	Bovine somatotrophin
CAST	Council for Agricultural Science and Technology
CBS	Columbia Broadcasting System Inc.
CENTED	Center for Technology, Environment, and Development
CEO	Chief Executive Officer
CJD	Creutzfeldt-Jakob disease
CRSP	Concerned Residents of South Phoenix
DDT	Dichlorodiphenyltrichloroethane
DEQ	Department of Environmental Quality (Georgia)
DNA	Deoxyribonucleic acid
DOE	U.S. Department of Energy
EMF	Electromagnetic field
EPA	U.S. Environmental Protection Agency
EPRI	Electric Power Research Institute
EU	European Union
FBI	Federal Bureau of Investigation
FDA	U.S. Food and Drug Administration
FMI	Food Marketing Institute
GAO	U.S. General Accounting Office
GIS	Geographic Information System
HCRA	Harvard Center for Risk Analysis
HIV	Human immunodeficiency virus
HLNW	High-level nuclear waste
IAEA	International Atomic Energy Agency
IFST	U.K. Institute of Food Science & Technology
J&J	Johnson & Johnson
LNG	Liquified natural gas
LQG	Large quantity generators
MAFF	U.K. Ministry of Agriculture, Food, and Fisheries
mph	Miles per hour
MLS	Multiple Listing Service
MRS	Monitored retrieval system
NHTSA	National Highway Traffic Safety Administration
NIDA	National Institute of Drug Abuse
NOAEL	No observed adverse effect level

NSF	National Science Foundation
NANP	Nevada Agency for Nuclear Projects
NRC	National Research Council
NTP	National Toxicology Program
NTSB	National Transportation Safety Board
nvCJD	new variant-CJD
NWPA	Nuclear Waste Policy Act
NWPO	Nuclear Waste Project Office
OII	Operating Industries, Inc.
ONWN	Office of the Nuclear Waste Negotiator
PBB	Polybrominated biphenyl
PCBs	Polychlorinated biphenyls
PCDD/Fs	Dioxins and furans
PCDDs	Polychlorinated dibenzodioxins
PCDFs	Polychlorinated dibenzofurans
ppm	parts per million
PR	Public relations
QPC	Quality printed circuits
RDD	Random digit dialing
RNA	Ribonucleic acid
TCDD	2,3,7,8-TCDD or Tetrachlorodibenzo-*p*-dioxin
TSEs	Transmissible spongiform encephalopathies
UCHSC	University of Colorado Health Science Center
UPI	United Press International
USDA	U.S. Department of Agriculture
WIPP	Waste Isolation Pilot Project

Preface

The emergence of technological stigma imposes important new demands on societal decision-makers and managers. This form of stigma expresses the modern world's concerns about human health and ecological risks—concerns that are amplified by the vast power of communications media to "spread the word" about risks. Since the industrial revolution, technological and industrial activities have often carried the potential for harm to people or the environment. What has changed in recent years is how technological risk is evaluated by the public. Many conditions are known to be hazardous; stigma refers to something that is to be shunned or avoided not just because it is dangerous but because it overturns or destroys a positive condition, signaling that what was or should be something good is now marked as blemished or tainted. As a result, technological stigmatization is a powerful component of public opposition to many technologies, products, and facilities and an increasingly significant factor in the acceptance of scientific and technological innovations.

Within social science, a framework for examining this phenomenon was outlined more than a decade ago with development of the concept of the Social Amplification of Risk. This conceptual framework linked earlier work on perception of risk to stigmatization by positing that social responses in a complex social and cultural context were important and could be described. A dominant part of the amplification process was the role of modern communications, especially the news media. Thus, the Social Amplification of Risk connects risk, media, and stigma in a logical conceptual framework.

Work on the issues of technological risk, the role of the media, and stigma requires the contributions of a wide range of social science disciplines. The people who have made many of the important contributions to these subjects come from a variety of geographical areas, have various institutional affiliations, and possess many different levels of management and research experience. A conference held at the Annenberg School for Communication of the University of Pennsylvania in March, 1997 invited a number of researchers with expertise in the areas of risk, media, and stigma. Kathleen Hall Jamieson, Dean of the Annenberg School, sponsored the conference, which was co-hosted by Howard Kunreuther from the Wharton Risk Management and Decision Processes Center (University of Pennsylvania) and Paul Slovic, president of

Decision Research and Professor of Psychology at the University of Oregon. This book is one major result of that conference.

The conference addressed the growing stigmatization of places, products, and technologies that arises from the association of these entities with an abnormal or unnatural degree of risk. Examples are stigmatized **places** such as contaminated communities or transport routes for nuclear wastes; **products** such as British beef, the blood supply (fear of AIDS or hepatitis), or apples treated with Alar; and **technologies** such as those developed in the nuclear, chemical, and biotechnology industries. Papers presented at the conference and included in this book characterize the phenomenon of stigma through numerous case studies. Based on our understanding of stigma, several papers examine strategies to help society better manage this phenomenon. The challenge is to reduce the vulnerability that stigma imposes on important products, industries, and institutions without suppressing the proper communication of risk information to the public.

The book contains six sections. The first section, an introduction, begins with the article by Gregory et al. entitled, "Technological Stigma." This article defines stigma and was sent to attendees prior to the conference. It was intended to initiate discussion at the conference and to motivate discussion on the importance, source, and range of this phenomenon. This is followed by the Kasperson et al. chapter, which considers the Social Amplification of Risk framework as an analytic aid to understanding technological stigma and its societal implications.

The next section, "Contamination Stigma," deals with cases where the quality of a thing or place is spoiled by contact with something that degrades and spoils it. Chapter 3, by Paul Rozin, presents results from innovative experiments demonstrating the power of contamination through touch and mental association. Once an object is seen as contaminated by a loathsome source the adverse reactions are long lasting and persist despite explanations intended to remove the stigma effects. Michael Edelstein recounts his conceptualization of "environmental stigma" from a 1984 case that has not been previously published. Environmental stigma is defined as a "discrediting [of] settings, places, objects, nonhuman lifeforms, and surroundings, as well as people associated with these environments." Anthropologist Theresa Satterfield presents a case study of a southern Georgia neighborhood that was contaminated over a number of years by an adjacent pesticide plant. The chapter recounts how residents were affected when the area was listed as a national priority for the U.S. Environmental Protection Agency's Superfund. The case study of stigmatization of place and people together produced a moving account of the experience of stigma at the level of individuals and their community.

Nuclear technologies have come to be the most feared and opposed of any industrial activities. Public reactions to nuclear risks serve as a prototype for technological stigma and offer researchers the clearest examples of stigma formation and effects. The section on "Nuclear Stigma" examines the potential for stigmatization of places associated with existing and potential radioactive contamination. Slovic and his colleagues report, in Chapter 6, their development of methods and measurements to gauge the potential for stigma effects. This research elicited images as a record of positive and negative evaluations of places. These images predict intended and actual behaviors and their potential economic impacts. Jenkins-Smith, in Chapter 7, reviews the work of

Slovic et al. and proposes a revised model of imagery and stigma. The revisions focus on predispositions to evaluating information with stigma potential. These predispositions include cultural perspectives, such as ideological guidance, and the existing set of place images that can modify the impact of new information.

Easterling, in Chapter 8, reviews the stigma studies undertaken in connection with the Yucca Mountain High-Level Radioactive Waste Repository not far from Las Vegas in Nevada. He finds that studies sponsored by the State suggest the potential for adverse, stigma-related visitor impacts but do not confirm that these effects will actually occur. The practical problems of measuring and placing a cost on stigmatized property are addressed in the Hunsperger paper (Chapter 9). In this case of property contaminated by radioactive plutonium from a nuclear weapons production facility, a number of methods were coordinated to understand and evaluate the stigma effects and public responses. The results of a public survey and media study described by Flynn et al. in Chapter 20, were obtained as the result of research commissioned by Hunsperger and were used to estimate the range of property value loss in the case of the facility at Rocky Flats, Colorado.

The potential for stigmatization exists for cases other than radioactive contamination. In the section entitled "Place, Product, and Industry Stigma," eight chapters address a broader range of risk, media, and stigma; two chapters deal with property, four chapters deal with products, and the other two chapters deal with stigma at the industry level. In Chapter 10 Adams and Cantor evaluate property stigma based upon litigation cases they have examined. They emphasize the special nature of home ownership as a place for family life and as an investment subject to public regard, both of which are threatened by property stigma. Pijawka and his colleagues present the case study of a major fire that caused health and environmental damages in an inner city neighborhood inhabited mainly by low-income, minority residents. They conclude that a hazardous event that took place in the context of existing environmental inequities produced disproportionate property losses to the residents.

The four chapters on products begin with Mitchell's classic essay on the economic costs of the 1982 Tylenol poisoning case. The company that produces Tylenol has been praised for its prompt, complete and under the circumstances successful response. Nonetheless, Mitchell finds that even this model of response to stigmatization cost the company about $1.2 billion. Powell, in Chapter 13, writes about the Mad Cow Disease case in Britain (BSE, bovine spongiform encephalopathy). His analysis focuses on the performance of management and regulation with reference to other cases of food contamination. Powell finds that the threat of stigma comes from lack of consumer confidence in the food supply, a condition that needs to be better understood and addressed by risk-management strategies.

Chan writes about the Canadian blood supply and its exposure to viral contamination. She notes the strong images involved with blood, the development of blood as an essential health product, and the development of stigmatization primarily from contamination from AIDS but also from other sources such as hepatitis. As in other cases, stigmatization was the result of a confluence of several factors: inadequate recognition and management of the threats to the blood supply, ineffective communication due to concern about public confidence, and lack of public information and risk evaluations. Graham addresses the risk of using airbags as a safety device in automobiles. He finds that although injuries caused by airbags meet many of the established criteria for "outrage" and stigmatization, this type of response failed to develop. He suggests some possible reasons for this and asks whether the risks from airbags have been presented properly to the public.

Two chapters address issues related to the chemical industry. In Chapter 16, Leiss examines the process by which chemicals containing dioxin came to be stigmatized and how these compounds came to represent the potentially hazardous products of the chemical industry. He argues that the stigmatizing process results from an information vacuum arising from a systematic failure of effective risk communication and goes on to illustrate this communication failure in the case of dioxins. Long, in Chapter 17, provides an insider's account of how a major chemical manufacturer struggled and changed in response to rapidly changing public attitudes toward their products over a 20-year period beginning in the 1960s. The result of this experience fundamentally changed the company as it struggled to modify purely scientific and engineering goals to incorporate public values. This was a process that involved great costs in resources and in organizational adjustments.

The section entitled "Risk, Media, and Stigma" begins with a study of images in the media presentation of risk events. Three subsequent chapters present case studies that employ the social amplification of risk framework to analyze the role of the media in creating and maintaining stigma. Ferreira, Boholm, and Löfstedt, in Chapter 18, observe that the images of hazards are pervasive in television and the news media, often presenting a different emphasis than the text or audio commentary. The authors argue that stigma is primarily created and represented by a visual mark, that photographs and visual presentations have a primary capability of creating negative imagery and thereby creating stigma.

Chapter 19, by Flynn et al., is a case study of a risk-communication program intended to obtain support for a radioactive waste facility in a region where the project was highly controversial. The failure to recognize the nature and the depth of the stigmatization of the waste disposal program led to widespread attacks on the risk-communication effort and the failure of the attempt to gain support. In Chapter 20, Flynn et al. present the results of a metropolitan area survey and an analysis of newspaper accounts of the operations and problems at a nuclear weapons facility near Denver, Colorado. The survey interviewed residents in the metropolitan area who were active in the home-buying market. The approach to this case study examines the roles of risk, media, and stigma as influences on residential property values.

The definition of technological stigma and the analyses that social scientists present in this volume inevitably leads to the final section, "Coping With Stigma." In Chapter 21, Kunreuther and Slovic present their view of the challenges and opportunities in managing stigma. After noting there are cases where stigma performs a useful social function, Kunreuther and Slovic address those cases where there are significant adverse impacts. The authors identify four strategies for coping with adverse effects of stigma: 1) prevent stigmatizing events, 2) reduce perceived risk, 3) reduce the social amplification of stigmatizing messages, and 4) reduce the impacts of stigma. Each of these strategies is discussed with the clear understanding that for the most vulnerable situations it may be necessary to employ all four. Walker, in Chapter 22, begins with a response to Kunreuther and Slovic, calling into question the assumption that researchers ought to adopt the normative values of reducing stigma by recommending a variety of strategies and techniques. He then contrasts stigma with unstigmatized (or "real") risk and raises questions regarding how the concept of stigma might be operationalized by researchers.

In the final chapter, Fischhoff takes on the challenge of reconsidering the definitions of stigma. He draws a number of inferences from this line of thought, not the least being his examination of what it means when "stigma reflects a moral statement of what constitutes acceptable behaviors." When this is the case, attempts to change or adjust perceptions, evaluations, or behaviors that stigma produces, presume a moral superiority on the part of those applying strategies to cope with stigma. As Fischhoff comments, this "may not be a comfortable or appropriate role for social scientists."

Altogether, the book provides a wide-ranging perspective on the impacts that stigma has on how we deal with risk and the challenges in managing these problems in the future. We hope this book stimulates future research in this area and look forward to being a part of this process.

James Flynn
Eugene, Oregon

Paul Slovic
Eugene, Oregon

Howard Kunreuther
Philadelphia, Pennsylvania

June 21, 2000

Acknowledgments

The conference and this book would not have been possible without the hospitality, encouragement, and financial support of Dean Kathleen Hall Jamieson of the Annenberg Public Policy Center and the Annenberg School for Communication of the University of Pennsylvania. Additional support was provided by the Wharton Risk Management and Decision Processes Center of the University of Pennsylvania and by the National Science Foundation grant CMS-9415730 to the University of Pennsylvania. Decision Research gratefully acknowledges support from the National Science Foundation under grants SES-9876587, SES-9876581, and SBR-9631635. Any opinions, findings, and conclusions or recommendations expressed in this material are those of the author(s) and do not necessarily reflect the views of the National Science Foundation.

We are grateful for the administrative assistance provided in conjunction with the conference by Debra Williams and Maxine Biederman of the Annenberg Center, by Anne Stamer of the Risk Management and Decision Processes Center, and by C. K. Mertz of Decision Research. Leisha Mullican and Connie Kudura did a superb job of preparing the book manuscript for publication with the assistance of Jan Kershner. We appreciate, as well, the support for the project provided by Ragnar Löfstedt and by Jonathan Sinclair Wilson and Frances MacDermott at Earthscan.

Introduction

This section contains two chapters. The first is a reprint of the Gregory, Flynn, and Slovic (1995) article "Technological Stigma" published in *American Scientist*. This article was provided to people attending the Risk, Media, and Stigma conference and served as a working definition of stigma to initiate discussion. The second chapter, by Kasperson, Jhaveri, and Kasperson, specifically addresses stigma as an outcome of the Social Amplification of Risk, a conceptual framework first defined by Kasperson et al. (1988). This framework initiated a large number of comments and produced dozens of articles during the subsequent decade. Therefore the Kasperson, Jhaveri, and Kasperson chapter also served as a review and introductory piece.

1 Technological Stigma[1]

Robin Gregory, James Flynn, and Paul Slovic

The word "stigma" was used by the ancient Greeks to refer to a mark placed on an individual to signify infamy or disgrace. A person thus marked was perceived to pose a risk to society. Within the social sciences, there exists an extensive literature on the topic of stigma as it applies to people. By means of its association with risk, the concept of stigma recently has been generalized to technologies, places, and products that are perceived to be unduly dangerous.

Stigma plays out socially in opposition to many technological activities, particularly those involving the use of chemicals and radiation, and in the large and rapidly growing number of lawsuits claiming that one's property has been devalued by perceptions of risk. For example, in 1993 in *Criscuola et al. v. New York Power Authority*, the New York State Court of Appeals ruled that landowners whose property is taken for construction of high-voltage power lines can collect damages if the value of the rest of their property falls because of public fears about safety, regardless of whether that fear is reasonable. In *City of Santa Fe v. Komis*, the Supreme Court of New Mexico in 1992 upheld the award of $337,815 to a Santa Fe couple for diminished property value resulting from the proximity of their land to a proposed transportation route for transuranic wastes. These and other similar cases have received national attention because they explicitly link public perceptions of a technological hazard with monetary compensation for a possible future decline in economic value.

This form of stigma has risen to prominence as a result of increasing concern about the human and ecological health risks associated with the use of technology. But stigma goes beyond conceptions of hazard. It refers to something that is to be shunned or avoided not just because it is dangerous but because it overturns or destroys a positive condition; what was or should be something good is now marked as blemished or tainted. As a reporter for *Time Magazine* commented in reference to a food contamination scare, "The most deep-seated fears are engendered when the benign suddenly turns menac-

---

[1]Reprinted from *American Scientist, 83*(3), 1995, pp. 220–223. Copyright 1995 by Plenum Publishing. Reprinted with permission.

ing." As a result, technological stigmatization is a powerful component of public opposition to many proposed new technologies, products, and facilities. It represents an increasingly significant factor influencing the development and acceptance of scientific and technological innovations and, therefore, presents a serious challenge to policymakers.

Stigma reminds us that technology, like the Roman god Janus, offers two faces: One shows the potential for benefit, the other shows the potential for risk. The existence of stigma reflects a widespread social concern about the risks from technology and provides evidence of an expectation that has not been met or a reputation that has become tarnished.

In this sense, scientific and technological prestige might be viewed as the opposite of stigma. For example, the fierce nationwide competition for the superconducting supercollider project was motivated in part by the prestige of hosting one of the world's most advanced scientific and technological research facilities. This shows that stigma is not an inevitable outcome of technological advance; it exists only when something has gone awry, so that prestige is replaced by fear and disappointment.

Characteristics of Stigma

The impetus for stigmatization is often some critical event, accident or report of a hazardous condition. This initial event sends a strong signal of abnormal risk. Roger E. Kasperson and colleagues have shown that the perceived risks of certain places, products and technologies are amplified by the reporting power of the mass media. Negative imagery and negative emotional reactions become closely linked with the mere thought of the product, place, or technology, motivating avoidance behavior. A theoretical model developed by one of us (Slovic) and his colleagues demonstrates how images of a place or product affect its desirability or undesirability, consistent with models used in the marketing and advertising world.

Stigmatized places, products, and technologies tend to share several features. The source of the stigma is a hazard with characteristics, such as dread consequences and involuntary exposure, that typically contribute to high perceptions of risk. Its impacts are perceived to be inequitably distributed across groups (e.g., children or pregnant women are affected disproportionately) or geographical areas (one city bears the risks of hazardous waste storage for an entire state). Often the impacts are unbounded, in the sense that their magnitude or persistence over time is not well known. A critical aspect of stigma is that a standard of what is right and natural has been violated or overturned because of the abnormal nature of the precipitating event (crude oil on pristine beaches and the destruction of valued wildlife) or the discrediting nature of the consequences (innocent people are injured or killed). As a result, management of the hazard is brought into question with concerns about competence, conflicts of interest or a failure to apply proper values and precautions. Specific examples of technological stigmatization demonstrate the importance of these features and the sometimes devastating effects of stigma.

Public evaluations of advanced technologies tend to be ambiguous, are often inaccurate and can, as such, contribute to the stigmatization of the technologies. It is not surprising that nuclear energy—touted so highly in the 1950s for its promise of cheap, safe power—is today subject to severe stigmatization,

"I think we're beginning to chip away at the stigma."

Figure 1.1 An improbable approach to managing stigma is depicted in this 1988 *New Yorker Magazine* cartoon.

reflecting public perceptions of abnormally great risk, distrust of management, and the disappointment of failed promises. Certain products of biotechnology also have been rejected in part because of perceptions of risk; milk produced with the aid of bovine growth hormone (BGH, or bovine somatotrophin, BST) is one example, with many supermarket chains refusing to buy milk products from BGH-treated cows. Startling evidence of stigmatization of one of the modern world's most important classes of technology comes from studies by Slovic and others asking people to indicate what comes to mind when they hear or read the word "chemicals." The most frequent response tends to be "dangerous" or some closely related term such as "toxic," "hazardous," "poison," or "deadly."

Stigmatization of places has resulted from the extensive media coverage of contamination at sites such as Times Beach, Missouri, and Love Canal, New York. Other well-known examples of environmental stigmatization include Seveso, Italy, where dioxin contamination following an industrial accident at a chemical plant resulted in local economic disruptions estimated to be in excess of $100 million, and portions of the French Riviera and Alaskan coastline in the aftermath of the *Amoco Cadiz* and *Exxon Valdez* oil spills.

Because stigmatization is based on perceptions of risk, places can suffer stigma in advance of or in the absence of any demonstrated physical impacts. More than a dozen surveys we conducted in Nevada during the past decade show that a majority of Nevadans are worried that the Las Vegas tourist industry might suffer from negative imagery associated with plans to construct the nation's first geologic repository for high-level nuclear wastes at nearby Yucca Mountain. In Tennessee, opposition by the governor and state legislature to a locally supported monitored retrievable storage (MRS) facility for high-level nuclear wastes was based on their fear that announcement of the facility would impose a negative and economically harmful image on the Oak Ridge region. The governors of Utah and Wyoming have both cited potential tourism losses as reasons for rejecting proposals for locating an MRS facility in their states.

The stigmatization of products has resulted in severe losses stemming from consumer perceptions that the products were inappropriately dangerous. A dramatic example is that of the pain reliever Tylenol. Mark Mitchell estimated that, despite swift action on the part of the manufacturer, Johnson and Johnson, the seven tampering-induced poisonings that occurred in 1982 cost the company more than $1.4 billion. Another well-known case of product stigmatization played out in the spring of 1989, when millions of consumers stopped buying apples and apple products because of their fears that the chemical Alar (used then as a growth regulator by apple growers) could cause cancer. Apple farmers saw wholesale prices drop by about one-third and annual revenues decline by more than $100 million.

Public Policy Implications

Technological stigma raises the specter of gridlock for many important private and public initiatives. The current practice of litigating stigma claims under the aegis of tort law does not seem to offer an efficient or satisfactory solution but rather points to a lack of policy options.

Project developers can, of course, simply pay whatever is asked for as compensation. However, for several reasons this option seems unwise. First, the pay-and-move-on option fails to distinguish between valid claims for compensation and strategic demands based on greed or politically motivated attempts to oppose a policy or program. Second, claims are often made for economic losses predicted to take place years or even decades into the future, despite the many difficulties inherent in forecasting future economic activities or social responses. Finally, this option fails to help us learn, as a society, to understand stigma and to manage it more effectively.

Another option is for proponents or developers to abandon projects threatened by stigma. This would be unfortunate in the many cases where the risks of proceeding are in fact reasonable. Our society's continued economic and social strength depends on its willingness to accept reasonable risks. The search for safer means to store hazardous wastes and to produce new goods for the marketplace requires us to face difficult trade-offs between new and old sources of risks, costs, and benefits.

One recent response has been to restrict communications that might produce or contribute to stigma. Apple growers in Washington State are suing Columbia Broadcasting System Inc. (CBS) for broadcasting what they claim were false messages about Alar on the network's *60 Minutes* program. The Alar scare has led the Florida legislature to pass a bill allowing Florida growers to sue anyone who says in public that fruits, vegetables, and other food products are unsafe for consumption but who cannot substantiate their claims with scientific evidence. Whatever their merits for specific cases, these anti-disparagement efforts are problematic for policy because they threaten the constitutional right of free speech and may inadvertently limit the ability of people and the media to discuss legitimate concerns about the safety of a product or technology.

Stigma effects might be lessened if public fears could be addressed effectively through risk-communication efforts. However, such a simple solution seems unlikely. In our view, risk communication efforts often have been unsuccessful because they failed to address the complex interplay of psychological, social, and political factors that creates a profound mistrust of government and industry and results in high levels of perceived risk.

We believe the best response is to recognize that stigma is the outcome of widespread fears and perceptions of risk, lack of trust in the management of technological hazards, and concerns about the equitable distribution of the benefits and costs of technology. Technological stigma should be seen as a rational social response to the multiple influences that produce it and therefore as subject to a variety of rational solutions. These solutions must involve thoughtful public policies and active public support. Adopting more open and participatory decision processes could provide valuable early information about potential sources of stigmatization and invest the larger community in understanding and managing technological hazards. This approach might even remove the basis for the blame and outrage that often occurs in the event of an accident or problem. Rather than being seen as unethical or evil, an industry might be viewed as unlucky or even shortsighted, avoiding the moral tainting that exacerbates a stigmatizing event. The active involvement of stakeholders, for example, might have enabled the nuclear industry to begin at a much earlier stage to plan for the disposal of nuclear wastes and to operate within a much less adversarial process.

In recent years, Rodney Fort and others have discussed the idea of offering insurance against the realization of potential stigma effects. We are not aware of any currently existing stigma-insurance markets. However, improvements over time in the ability of researchers to identify and define those factors contributing to stigmatization may make it possible to predict the magnitude or timing of expected economic losses. This could open the door to the creation of new insurance markets and to efforts for mitigating potentially harmful stigma effects.

In cases where fears are great and impacts are extraordinarily difficult to forecast, the developer may have to provide protection against potential losses from stigma. The creation of a stigma-protection option by the New Mexico state government, for example, could guarantee that the value of properties lying along the proposed nuclear waste transportation route would be maintained according to acceptable measures of the market. Such an indemnification strategy would provide an opportunity to monitor expected versus actual market behavior over time: Will the average sale prices of homes near nuclear waste transportation routes actually fall? Will buyers really go elsewhere? In addition, the existence of a program that guarantees market-based values for properties along a transportation route may in itself act to ease public fears and thereby diminish stigma.

Most importantly, the societal institutions responsible for risk management must meet public concerns and conflicts with new norms and new methods for addressing stigma issues. The goal should be to create arenas for resolving these conflicts based on values of equity and fairness. This undoubtedly will require a major reorientation for risk management, reducing the heavy reliance on technical expertise and creating new forms of partnership that allow the public expanded roles in decision making and effective oversight of the risk-management process.

Over time, these and other new policy initiatives may decrease the significance of technological stigma. At the moment, however, stigma effects loom large. In some cases, technological stigma may be so strong as to rule out the use of entire technologies, products, or places. The experience of the past decade has been difficult. The years ahead are likely to add to the growing list of technological casualties unless innovative public policy and risk-management processes are developed and implemented.

Acknowledgments

This paper has been funded by a grant from the Alfred P. Sloan Foundation and a shared Public/Private Sector Initiative award to Decision Research from the National Science Foundation (NSF; Grant No. SES-91-10592) and the Electric Power Research Institute (EPRI).

2 Stigma and the Social Amplification of Risk: Toward a Framework of Analysis

Roger E. Kasperson, Nayna Jhaveri, and Jeanne X. Kasperson

In March 1996, the British government announced the possibility of a link between a serious cattle disease, bovine spongiform encephalopathy (BSE), and a rare and fatal human neurodegenerative disease, Creutzfeldt-Jakob Disease (CJD). The announcement was prompted by the discovery of ten atypical cases of the disease, which usually afflicts people over 65, in patients under the age of 42. The government's announcement provided scant details of the relevant scientific data but noted that they were "cause for great concern" (O'Brien, 1996). The great concern did indeed quickly materialize in the form of an avalanche of press coverage, much of it highly dramatized and speculative. Following on the heels of a decade of ministerial denials, reassurances, and belittling of this potential hazard, the sudden about-face produced an instant crisis and a collapse of public confidence in the safety managers of the $3-billion British beef industry, symbolized by the March 21, 1996 headline in the *Daily Express*: "Can We Still Trust Them?" (O'Brien, 1996).

The results were speedy. Within days, the European Union imposed an export ban on all British beef and beef byproducts. Sales within Britain plummeted before rebounding partially. Meanwhile, the effects "rippled" to other countries, as consumption of beef from other source countries fell by fully 40% in France and Germany. The impacts also spread to other industry sectors, such as slaughterhouse workers, auctioneers, truckers, and beef export firms (Johnson & Vogt, 1996, p. 4). The long-term effects of the event remain unclear and it is uncertain whether the stigma affecting British beef will persist in the face of remediation efforts to reduce the hazard and to restore consumer confidence in Britain and on the continent.

The "mad cow disease" case typifies a special class of hazards, those that trigger intense media coverage and strong public concerns, high institutional attention, and large secondary or higher order consequences, what we have termed elsewhere "socially amplified" hazards (Kasperson et al., 1988). In such hazard cases, the connection of biophysical hazards to social processes can generate powerful signals to society either that a new hazard has appeared on the scene or that an existing hazard is more severe or difficult to manage than previously understood. The secondary or indirect effects may assume proportions that eclipse the direct, and more apparent, biophysical and health consequences, elude anticipatory assessments, and take by surprise those charged with managing the hazards as well as society at large. An important property of these socially amplified hazards is their potential for generating stigma-related effects for places, technologies, or products (Gregory, Flynn, & Slovic, 1995). In such cases, some critical hazard event, accident, or report sends a strong signal of an abnormal risk, and the ensuing negative imagery and attendant publicity become closely linked with the place, product, or technology, resulting in its stigmatization. The stigmatization, in turn, generates a series of adverse effects, greatly enlarging any negative consequences that would have occurred in the absence of such stigma.

Beginning with the seminal work of Goffman (1963), a substantial literature has emerged concerning the sources, types, and effects of social stigma. This work is primarily associated with the stigmatization of people arising from social interactions surrounding race and ethnicity, disease, mental illness, or handicaps (see the references, for example, in Jones et al., 1984). More recently, Paul Slovic and his colleagues at Decision Research (Slovic, Layman, Kraus, et al., 1991; Gregory, Flynn, & Slovic, 1995; Slovic, Flynn, & Gregory, 1994) as well as other researchers (Edelstein, 1987, 1988; Vyner, 1988) have focused attention on the relationships among toxic materials, pollution, potential contamination, and the stigmatization of places, products, and technologies. In particular, the Decision Research group, through its empirical studies of public perceptions surrounding the proposed siting of a high-level nuclear waste repository at Yucca Mountain, has explored important aspects or components of a more general theoretical framework or model of risk and stigmatization. Building explicitly upon this previous work, this discussion seeks further progress toward a general analytic framework.

Toward that end, we enlist the general conceptual base offered by the social-amplification-of-risk approach. The question addressed is: can that framework structure the stigmatization arising from risk in ways that clarify and integrate the factors and processes that affect the emergence, effects, and durability of such stigma? This discussion highlights the stigmatization of places arising from the presence of hazardous activities or facilities in these locations, but the approach is relevant to technologies, products, and industrial facilities. We begin by examining a broader and older literature—one that treats people's perceptions and images of places—and what it may suggest about the potential stigmatization of places.

Images of Places and Regions

A long tradition exists of geographic studies of environmental perception focused on people's experience and images of places (Burgess, 1978; Hewitt & Burton, 1971; Tuan, 1979). The meaning of "place" is often evoked through a discussion of "image," "environmental image," or "mental maps." Typically, this can involve the meaning of a place for individuals, the group, or outside people. When people communicate their experience of place, they often convey information about a locale and their feelings about it through images or verbal pictures. This is a combination of ideas and emotions evoked in the individual stemming from direct environmental experience of the place and from secondary and often media-based information garnered from diverse sources, transmitted images, and notable previous events.

Conceptualizations and terminology associated with analyses of the images of places cover a range of approaches. Even a frequently used term—"mental maps"—has quite varied usage. For Beck and Wood (1976), for example, mental maps are "personal views of the geographic structure of the world expressed in map form." "Image" is another commonly used term used to convey the totality of verbal descriptions of individual experiences, feelings, and attitudes towards a place or region (Burgess, 1978). Downs and Stea (1973) offer a more formal definition for the more universal term "cognitive mapping": which they view as a process composed of a series of psychological transformations by which individuals acquire, code, store, recall, and decode information about the relative locations and attributes of phenomena in their everyday spatial environments." They argue that cognitive maps are the base by which individuals develop a strategy of environmental behavior; they are cognitive representations that have the *functions* of familiar cartographic maps but not necessarily the physical properties of graphic models. Cognitive mapping expands upon the basic road map to capture a level of symbolism, including images, cognitions, and mental maps. Throughout, an "environmental image" or "image of place" whether held by the individual or a group, is viewed essentially as a process—a dynamic ongoing development. Measurement techniques largely convey the content of an image frozen at a moment in time (Stallings, 1975).

A "co-orientational approach" to analyzing the perception of different social groups (Uzzell, 1982) may bear upon place stigmatization. Here the environment becomes meaningful not only through physical confirmation but through social interpretation. Group communication and negotiation are essential to social processes. Thus, it is argued, one needs to study not only an individual's cognitive map of the environment but also how other members of the social group conceive of the situation and how agreement among different individuals and larger groups may be achieved.

The relationship between attitudes and environmental perception has also received attention. Downs and Stea (1973), for example, have distinguished among attitudes, preferences, and traits. Preferences, in their view, are less global and pertain to a specific object rather than a class of objects. They are also less enduring in time as compared with more stable and durable attitudes. Traits, on the other hand, develop when a given attitude has come to embrace a wide variety of objects over a considerable period of time. Craik (1970) regards an environmental trait as one that seeks to identify an individual's self-conception in reference to the natural and human-made physical environment. This idea of trait reverberates in the work of Golledge and Stimson (1987), who view cognitive maps as semantic long-term memory structures. Cognitive maps, they argue, should not be interpreted as a one-to-one "mapping" of an object or reality. Rather, as Garling, Book, and Ergezen (1982) have suggested, these semantic long-term memory structures occur as holistic images, semantic networks, or chunks and strings of information organized in a fashion akin to that assumed by various network models of memory. This idea resonates with information-processing theory in psychology and recent work in artificial intelligence.

Places, this research suggests, have connotative meanings that possess emotional, metaphorical, and symbolic value (Burgess, 1978). It is the "appraisal" aspects of image construction concerned with feelings, value, and meaning that give meaning to the informational aspect of perception (Downs & Stea, 1973; Pocock, 1974). Studies in general show that appraisal varies with the type and depth of personal involvement with a place. The sketchy maps of the tourist compared with the deeper, refined map of the resident suggest this difference. The well-known popular version of such maps, "A New Yorker's Idea of the United States of America" (Figure 2.1), exemplifies the former and highlights the types of distortion that can occur.

Popular culture, rather than formal education, often extensively shapes images of faraway places (Beyer & Hicks, 1968). Goodey (1973) has noted, for example, that misguided and inaccurate images of immigrant concentrations in the city emerge from interpersonal conversations and selective acceptance of media reports. Once popular maps and images of places emerge, they are slow to change in the face of new and contrary evidence (Thompson, 1969).

A last theme of this research tradition refers to *preferential perceptions*, preferences for movement towards or avoidance of particular places (Goodey, 1974). Two types of preferential perception studies (Goodey, 1973) exist—one pioneered by Peter Gould and concerned with mental maps of residential preferences for sets of locations, the other concerned with preferred areas rather than specific locations. One overall finding is that mental maps of local areas are often very detailed, whereas those for distant places are scanty, rely on sparse information, and are easily transformed into stereotypes (and presumably prone to stigmatization). In a study of student preferences for residential locations in the United States, Europe, and Africa stereotypes of places and personal experiences played a very influential role (Gould, 1966). Apparently, stereotyped traits of particular regions, locales, and cities may be ready fodder for the creation of place-based stigma.

One such example (Figure 2.2) came to our attention more than 25 years ago when one of us (REK) was preparing a collection of readings on political geography (Kasperson & Minghi, 1969). This interesting map of derogatory images of places in the Canton of Zurich, Switzerland provides an early example of empirical work documenting place-related stigma, one in which the importance of ethnic and religious groupings as well as distance relationships was amply apparent. And the stigma itself was quite compelling, for how many of us would want to live in a community regarded as a "stupid-bourgeois place"?

This previous body of work on environmental perception and mental maps has several implications for our current effort to build an analytic framework for risk-induced stigma. Unfamiliar or distant places may easily fall prey to distorted or stereotypical perceptions. Secondary accounts or media coverage will often be principal sources of image formation. Once perceptions of unfamiliar places are formed, they may become resistant to new or "corrective" information. The connotative meanings associated with places are not only cognitive but include emotional, metaphorical, and symbolic properties. Finally, such mental maps may resemble long-term memory structures involving holistic images, semantic networks, and information "strings."

With this as background, we now turn to the issue of stigma and its relationship to risk.

Map "A New Yorker's Idea of the United States of America"

Figure 2.1 Source: *The Saturday Review of Literature* (1936).

Map of Derogatory Attitudes

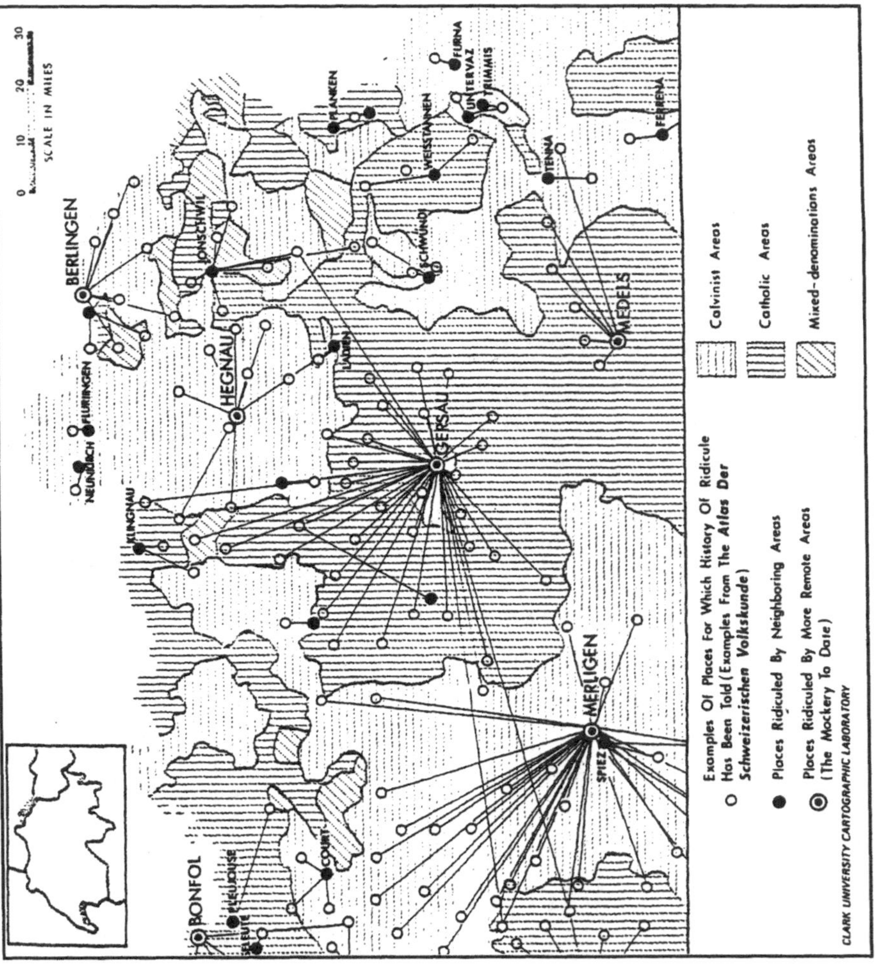

Figure 2.2 Derogatory images—Canton of Zurich, Switzerland. How extensive the reputation as a "stupid-bourgeois place"?

Risk and Stigma

The term *stigma,* as Goffman (1963) notes, originated with the ancient Greeks who used it to refer to marks placed on a person to denote infamy or disgrace. Current usage continues the original meaning but usually refers to an attribute of people, places, technologies, or products that is deeply discrediting or devaluing. Instead of the possessor's being viewed as normal or commonplace, the possessor is viewed as different, with this difference involving important qualities that set the possessor off as deviant, flawed, spoiled, or undesirable. Whereas stigma may be related to hazards and involve fear on the part of the beholder, stigma goes beyond the notion of hazard to refer to something that overturns or destroys a positive condition, and, accordingly blemishes or taints the possessor (Gregory et al., 1995, p. 220).

Marking of the possessor plays an essential role in stigma. The mark identifies and signifies the deviant status and typically has devastating effects on the person or place. The mark need not be physical but may be embedded in, and identifiable from, particular behavior, features, biography, ancestry, or location. It signifies some attribute of the possessor that is associated with the imperfect, or devalued, or dishonored status. Such marks come to arouse in outside observers strong feelings of repugnance, fear, and disdain. Also, the mark may become linked through attributional processes to responsibility, which is also seen as deviant and reprehensible. If these deviant disposition come to be viewed as central and intrinsic to the person or place, they can become an essential part of identity (Jones et al., 1984), as we argue below.

Following Goffman (1963), *stigmatization* in our usage refers to the process by which persons select an attribute of a person, place, technology, or product and denigrate the possessors of the attribute, discriminate against the possessor, and may even construct a stigma "theory" or "story" to explain the inferiority and its roots. Important to this process are (a) the selection of a negative attribute, (b) perceptions by others of the negative attribute, and (c) the resultant widespread devaluation of the possessor, frequently including labeling and communication of the labels.

Obviously, the context in which stigma arises and the dimensions of the marking process can greatly influence the vividness of the stigma and its effects. Jones et al. (1984, p. 24) identify six major dimensions that are particularly influential, as documented in a wide variety of empirical studies:

1. *Concealability.* Is the condition hidden or obvious? To what extent is its visibility controllable?
2. *Course.* What pattern of change over time is usually shown by the condition? What is its ultimate outcome?
3. *Disruptiveness.* Does it block or hamper interaction and communication?
4. *Aesthetic qualities.* To what extent does the mark make the possessor repellent, ugly, or upsetting?
5. *Origin.* Under what circumstances did the condition originate? Was anyone responsible for it and what was he or she trying to do?
6. *Peril.* What kind of danger is posed by the mark and how imminent and serious is it?

Edelstein (1988, p. 14), drawing on these dimensions and supplementing some of them by adding "level of fear," "responsibility," and "prognosis," analyzes what he terms "environmental stigma."

Since risk-induced stigma are at the center of this discussion, some comments on fear are appropriate. Fear, it has been argued, is a highly salient feature of stigma. Normally, studies of stigmatization of people have focused on those with mental illness or handicap, contagious diseases, or racial differences. Susan Sontag, for example, in examining personal experience of cancer, analyzes the ways by which Americans symbolically construct and respond to this disease: "Any disease that is treated as a mystery and acutely enough feared will be felt to be morally, if not literally, contagious.... Contact with someone afflicted with a disease regarded as a mysterious malevolency inevitably feels like a trespass; worse, like the violation of a taboo" (Sontag, 1978,

p. 6). AIDS is universally characterized as both mysterious and malevolent (Quam, 1990). For environmental problems, Mushkatel and Pijawka (1994) argue that the public stigmatizes environmental features that it views as repellent, upsetting, or disruptive. The source of the stigma may be a hazard with characteristics, such as dreaded consequences and involuntary exposure, that typically contribute to high perceptions of risk. Stigmatization can also occur due to a newly discovered or anticipated change in exposure to a toxic substance (Edelstein, 1987; Gregory et al., 1995). The sources of stigma can be direct, as with an increased incidence of cancer or declining market price of properties, or more indirect, as in an exodus of residents from a contaminated area or a decline in sales of a product manufactured in an "unhealthy" environmental area (Mushkatel & Pijawka, 1994).

The level of uncertainty associated with the hazard generally or events arising from one of the hazard stages may contribute to the stigmatization of a place or technology. Pollution and contamination events are often associated with high levels of uncertainty. The contamination is often indiscernible to direct sensory confirmation and carries the threat that it may exist but be "invisible." Many contamination events involve substances or synergistic interactions whose dose/response relationships are poorly understood. Some involve threats to particularly sensitive receptors, such as human reproductive systems, future generations, or hypersusceptible individuals or groups. Some involve ill-defined threats to the life-support systems of the local environment, as in contamination of groundwater. Finally, as Kai Erikson (1994) has noted, threats of toxic substances seemingly never end, as some level of residual effects remain, new outbreaks or dimensions of the hazard surface, and the debate and anxiety continue well after the cases are formally settled.

The stigmatization of places due to hazard events typically involves various characteristic properties, including the strong social amplification of the hazard or hazard event (Gregory et al., 1995, pp. 220–221). An initial event somewhere along the hazard chain emits a "signal" that a major threat has occurred or is imminent. Extensive media coverage interprets the meaning and projects risk signals, imputing blame, trustworthiness, vulnerability, and victimization. The events and media coverage mark the place and propagate its visibility to other places, so that the very identity of the place becomes tightly linked with the hazard and the associated negative and threatening imagery that arise. This change in identity is an essential property of risk-induced stigma and as we shall see below, is the stage of stigmatization to which indicators and measurement are most appropriately directed. The emerging stigma changes behavior related to the place, as in avoidance of the place by outsiders, and flight, self-deprecation, or anger by residents of the locale.

Because of the critical role that social amplification of the hazard plays in the stigmatization of place, we next turn to a brief description of this approach to hazard analysis and then probe more deeply into the application of this framework to stigma and its effects.

The Social Amplification of Risk

As conceived in this framework (Figure 2.3), the social amplification of risk begins with a risk event, such as an industrial accident or a chemical release. It also may emerge from the release of a government report that provides new

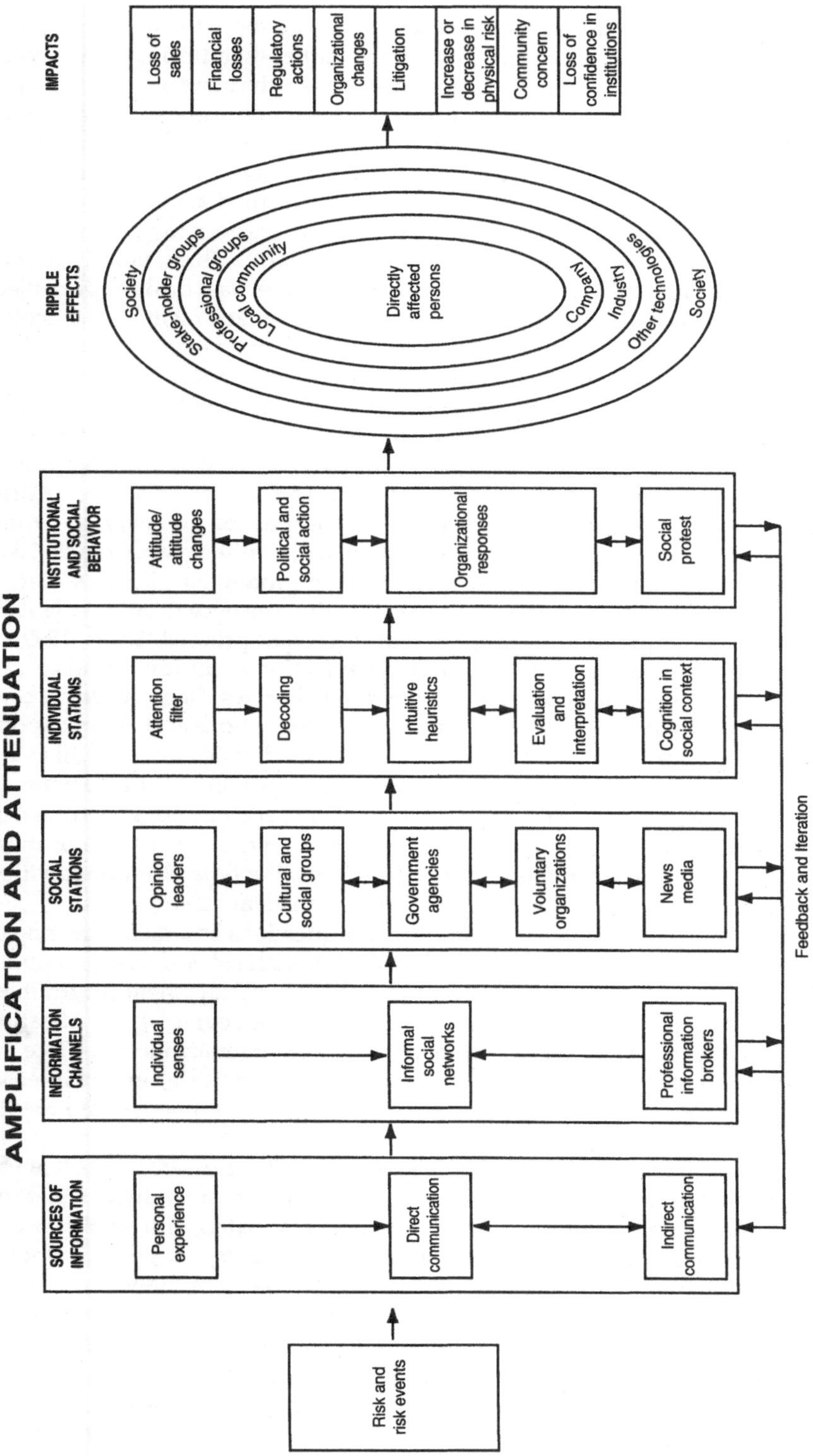

Figure 2.3 Amplification and attenuation.

information on the risk. Since most of society learns about the parade of risks and risk events through information systems rather than through direct personal experience, risk communicators, and especially the mass media, are major agents, or what we term "social stations," of risk amplification, and by inference, marking. Particularly important in shaping group and individual views of risk are the extent of media coverage, the information conveyed, the "framing" of the risk, the presence of risk "signals" in the media, and the symbols, metaphors, and discourse used in depicting and characterizing the risk.

The channels of communication are also important. Information about risk flows through multiple communication networks—the mass media represented by television, radio, and print media, the more specialized media of particular professions and interests, and, finally, the more informal personal networks of friends and neighbors. Of these, most is known about the mass media, and particularly their multiple and often conflicting roles as entertainers, watchdogs, gatekeepers, and agenda setters. In the context of stigma, they play an essential role in maintaining the visibility of the mark. It is also apparent that the mass media cover risks selectively, according those that are rare or dramatic—that is, that have "story value"—disproportionate coverage while downplaying, or attenuating, more commonplace but often more serious risks.

Social institutions and organizations take on prominent roles in society's handling of risk for it is in these contexts that the conceptualization, identification, measurement, and management of most risks proceed (Short, 1992, p. 4). In postindustrial democracies, large organizations—multinational corporations, business associations, and government agencies—largely set the contexts and terms of society's debate about risks. Institutions and organizations are major "nodes" of risk amplification and require detailed attention in gauging how society responds to risk. Risk issues are also important elements in the agenda of various social and political groups, such as nongovernmental organizations with environmental and health concerns. The nature of these groups figures in the definition of risk problems, the type of rationality that underlies interpretation, and the selection of management strategies.

The information system surrounding risk questions and the processing of risk by the various stations of amplification transmit *signals* to society about the seriousness of the risk, the performance of risk-management institutions, and, in this context, the hazardousness of place. The degree of amplification will affect the extent to which *risk ripples* and the stigmatization of place accompany a risk or risk event. Where social concern and debate are intense, secondary and tertiary impacts may extend beyond the people who are directly affected.

The consequences of risk and risk events, then, often exceed the direct physical harm to human beings and ecosystems to include more indirect effects on the economy, social institutions, and well-being associated with amplification-driven impacts. Risk ripples and secondary impacts carry the potential for the stigmatization of certain technologies or places. The mere thought of certain technologies, products, or places can conjure up enough negative imagery and emotion to render them tainted objects to be shunned and avoided. Nuclear energy and hazardous waste facilities are prime examples of stigmatized technologies or places now embroiled in controversy and public opposition. Biotechnology and chemicals also face some elements of such stigmatization. At the heart of such effects are the ingredients of the social amplification of risk—

public perceptions of great risk, intense media coverage of even the most minor incidents or failures, distrust of the managers involved, social-group mobilization and opposition, conflicts over value issues, and disappointments with failed promises. In the modern risk society, amplification-driven impacts, such as stigma-related effects, may mar, compromise, and diminish the potential benefits to society from economic growth and technological change.

With this overview of social amplification, we can now explore the application and focusing of the framework on stigmatization.

Risk Amplification and Stigmatization

Risk-induced stigma, particularly those that come to characterize particular places or areas, were not central to Goffman's (1963) original thinking. With the recent enhanced knowledge of risk, and the social amplification of such risks, it is opportune to explore alternative ways of structuring the processes by which risk generates social stigma, if only to conceptualize and clarify the contributing elements and how they interact. Here we draw upon our previous collaborative work with colleagues at Decision Research on the social amplification of risk to explore how a risk-stigmatization framework, might reflect current understanding. The intent, it cannot be overemphasized, is *not* to set forth *the* way of conceptualizing such complex relationships but to offer one view intended to contribute to the formulation of a robust framework of analysis. Our current conception appears in Figure 2.4 and the discussion that follows is geared to that schematic diagram.

DEFINITIONS AND MEASUREMENT

As suggested above, we define *stigma* as *a mark placed on a person, place, technology, or product, associated with a particular attribute that identifies it as different and deviant, flawed, or undesirable.* Here we are interested in stigma arising from risk-related attributes. In our conception, such stigmatization involves three stages:

1. The risk-related attributes receive high visibility, particularly through communication processes, leading to perception and imagery of high riskiness, a process that we refer to as the social amplification of risk;
2. Marks are placed upon the person, place, technology, or product to identify it as risky and therefore undesirable; and
3. The social amplification of risk and marking alter the identity of the person, place, technology or product, thereby producing behavioral changes in those encountering the imagery and marking as well as those to whom they are directed.

Accompanying this process will often be a story or narrative that interprets the evolution of the stigma and assigns responsibility or blame for its presence.

The attributes of risk, as we describe below, may emerge from any stage of the causal chain of hazard, whether it be risk events, exposures, or consequences. Since the emergence of stigma follows a similar process of attribute selection, visibility, marking, and changed identity, all these stages are relevant to identifying and measuring the stigma. Since the process is cumulative and registers its full impact when a change in identity occurs, as Goffman (1963)

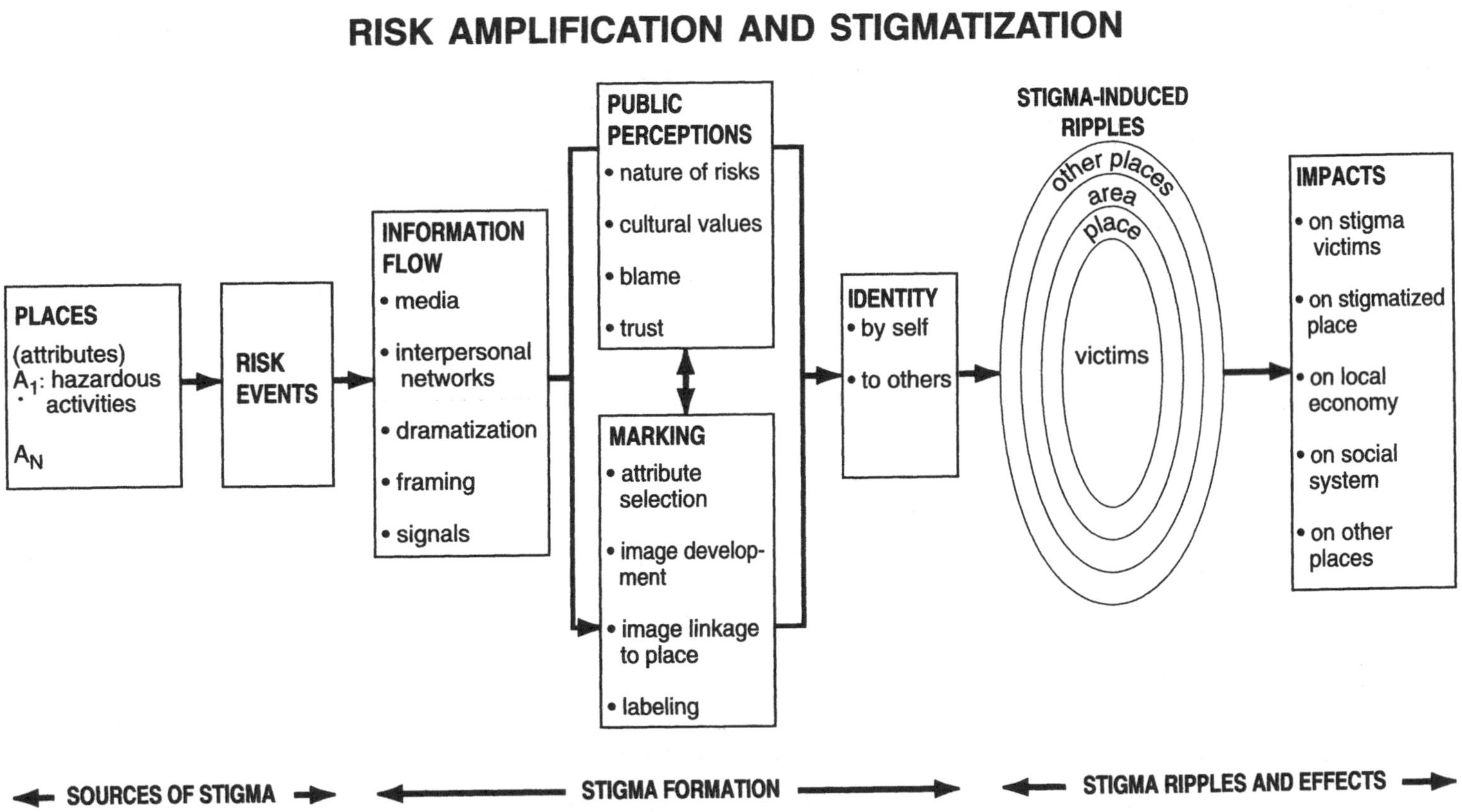

Figure 2.4 Risk amplification and stigmatization.

notes in his classic statement, identification and measurement are most reliable at that stage of stigma evolution. This parallels risk-analysis thinking where the risk is not actualized until risk consequences actually occur. But just as risk analysis has recently given much attention to risk characterization, which includes profiling the various stages in the evolution or development stage, so something like *stigma characterization* might chart the evolution of the stigma from risk events and experiences, the social amplification and marking, the change in identity, and the ensuing behavioral consequences. Each stage of the characterization would be subject to an assessment of intensity as well as an evaluation of potential contributors to the next stage of stigma formation. The development of indicators for each stage should also be an achievable goal.

SOURCES OF RISK-INDUCED STIGMA

Much of the literature on stigma addresses sources of stigma that lie in interpersonal or social relationships. Here we focus on places that are sites for risky facilities or activities that have the potential to produce risk events that command attention from the various "social stations" that process such threats. Risk events can occur anywhere across the "hazard chain":

They may be actual risk events or consequences but may also constitute new information, reports, or allegations concerning the hazard as a whole or a particular hazard stage. Some such risk events are socially rather than technologically induced. For example, a local environmental group that continually monitors experience in a community for "targets of opportunity" relevant to its political agenda may seize upon an event occurring elsewhere to sound an "alert" that a new risk has appeared locally.

The place involved will typically host a conglomerate of social and economic activities or *attributes* that define the nature of life at that site as well as the source of various beneficial and potentially risky activities. As noted above, most places accommodate rather inconspicuous mix of such activities or attributes, none of which comes to dominate the image of the place or its identity. In some cases, of course, an activity or attribute of the place, such as gambling in Las Vegas or automobile manufacture in Detroit, assumes such prominence as to shape extensively the identity of the place. In regard to hazards, a brew of environmental, technological, and social hazards typically defines a diffuse "hazardousness of place" (Hewitt & Burton, 1971) or "cartography of danger" (Monmonier, 1997) and no dominant image of a specific hazard emerges as linked to the place. Notable exceptions exist, however, as the very names "Three Mile Island," "Love Canal," "Times Beach," "Seveso," or "Bhopal" suggest, often associated with a catastrophic hazard or accident. Socially amplified hazards, such as the murder of foreign tourists in Miami, can provide a powerful source for stigmatization of a place. Beirut, Lebanon, once the "Paris of the Middle East," suggests how dramatically a positive image of a place can be transformed into a stigma through a succession of risk events. And New Jersey may struggle forever to live down its "cancer alley" image and pass itself off as a genuine "garden state."

INFORMATION FLOW

The emergence of broadly based stigmatization of places, technologies, and products depends upon extensive coverage of the risk and risk events at a place. As noted widely in the stigma literature, the *visibility* and *salience* of the offending or denigrated attribute is essential to the marking of the place, product, or technology. From past hazards research, it is known that the mass media exercise considerable selectivity in their coverage of hazards (Singer & Endreny, 1993). Risk and risk events compete for scarce space in media coverage, and the outcome of this competition is a major determinant of (a) whether a risk will undergo social amplification or attenuation in society's processing and disposition of the risk, and (b) whether the risk becomes central to the stigmatization of the place.

The mass media can also play a critical role in dramatizing and framing the risk problem or threat. Particularly important are both the extent of coverage and the particular "facts" selected and the language used to characterize the risk. Here the radioactive accident in Goiânia, Brazil, in which radioactive material obtained at a junkyard contaminated portions of the city in 1987 (International Atomic Energy Agency, IAEA, 1988; Petterson, 1988; Roberts, 1987), is particularly instructive. Initially, the accident received only minor attention in a casual report in a local newspaper, but on 1 October 1987, a highly sensational and lengthy São Paulo television broadcast initiated an intense period of dramatic and often exaggerated media coverage of the unfolding incidents and discoveries in the aftermath of the accident. Overnight, an army of reporters and camera crews descended on Goiânia to cover the tragedy. North American headlines spread the news of "deadly glitter," "a carnival of glittering poison," and "playing with radiation," thereby "marking" Goiânia. This media blitz generated extraordinary public concerns, with perceptions of enormous risk apparent even among people who had no contact with contaminated persons or materials. Similar observations obtain concerning the coverage of Love Canal, the Chernobyl and Bhopal accidents, and the mad-cow disease episode.

The mass media also play an important role in interpreting the meaning of the risk event. Such interpretation can provide the base for the subsequent stigmatization of a place or product. Elsewhere we have examined the flow of risk *signals* in the print news media in the area of the proposed Yucca Mountain nuclear waste disposal facility. Defining risk signals as "messages about risk or risk events that affect people's perceptions about the seriousness and/or manageability of the risk," we analyzed the appearance of such signals in five years of coverage in the *Las Vegas Review-Journal* (Kasperson, Perkins, Renn, & White, 1992). A few examples, chosen more or less at random, may suggest the potential of such signaling for altering risk perceptions, imagery, and place identity:

> Yet the drumbeat of opposition to the *radioactive grave* at Yucca Mountain rumbles on. (23 July 1986)
>
> In editorials where "pejorative" is not a no-no, we'll call it a *dump* or *a tomb* or a shaft full of *gunk* or anything else that comes to mind. (29 July 1987)
>
> If we are going down the road to nuclear *ruin, slopping the swill* from roadside to roadside, then let's get started. (12 February 1989)

> It is possible, he argues, that some poor country, down on its luck and saddled with debt, might gladly accept American *atomic garbage*, for a price. (14 July 1987)
>
> You don't even have to squint to see the handwriting on the wall—that *big R.I.P.* already scrawled on the crags of Yucca Mountain. (13 July 1986)

From such examples of risk signals, it is apparent that, in addition to providing ongoing flow of information about issues and events, the mass media also interpret the larger social meaning of what is occurring. So the media in a wide variety of subtle ways provide *messages* that shape the perceptions of readers and viewers about what has occurred. Such messages appeared explicitly in this case in the editorializing about the proposed nuclear waste facility in Nevada, the siting process under way, and the implications for people and their well-being. Such messages were embedded in the choice of headlines, the positioning of stories, and the selection and use of pictures, symbols, and metaphors to assist interpretation. The "flows" or "streams" of signals certainly affect, albeit in ways that are yet unknown, people's views of risks and the marking of places in which the risks occur.

PUBLIC PERCEPTIONS

Risk perception is a critical part of risk amplification and stigmatization. Much is known from past risk research about the characteristics of risks and risk events that are likely to generate strong public concerns and elicit high media coverage. Risk that are new, involuntary, potentially catastrophic, and that involve dread all tend to elicit strong concerns and reactions. If the risks connect to group agenda concerned with the technology, activity, or product, additional media scrutiny and social conflict may easily result. Some risks pose threats to deeply held values or social institutions; these, too, often evoke strong public reactions.

The responsibility of the managers of the facility or technology for the occurrence of the risk event, the extent to which they are viewed as blameworthy, and the character of their early response to the event are important compounding factors. It is sometime the case that a rapid and effective accident response combined with action to protect nearby residents and the community can minimize damage to public confidence in the managers. Much may also depend upon the history of relationships between managers of the facility and members of the host community. On the other hand, if the managers are clearly to blame, betray a history of failures, or seek to conceal their responsibility, then the risk event is likely to be strongly amplified through media revelations and intensified public concerns.

Lack of trust can be another compounding factor. If high levels of trust exist in those responsible for risk management, risk events may undergo only limited social amplification in media coverage and public perceptions. But if reservoirs of social trust have been drawn down, or worse yet if active distrust prevails, then even small events may generate high levels of public concern, particularly if the risk is one that publics fear. Certain hazards, such as hazardous wastes or radioactive materials, have broader contexts and histories of mismanagement that may contribute to social distrust.

Contending social groups and watchdog organizations, especially those based at the place in question, also can influence risk amplification and stigmatization. To the extent that risk becomes a volatile issue in a community or a source of contention among social groups, it may be vigorously brought to greater public attention and subjected to value-based interpretations. Polarization of community views and escalation of rhetoric often occur. New community members may be drawn into the conflict. These social conflicts outlive the particular risk event and can become anchors for subsequent risk debates, contributing to an image of the place and the emergence of risk-based stigma.

Finally, it is important to note that places do not exist in isolation. Rather, they are often linked in the mental maps of publics with the areas of which they are part or other places possessing similar attributes, including prominent risks. So in the amplification of risk and potential place stigmatization, geographical associations with societal experiences with risk may be a contributing element. Thus, the Yucca Mountain site's proximity to the Nevada Nuclear Test Site and linkages made with unsuccessful nuclear waste disposal facilities at other places, such as West Valley in New York State or the Hanford Reservation in Washington State, can fuel emerging perceptions, images, and interpretations. An interesting question is whether a stigma becomes associated with other nearby places or the broader geographic areas of which they are part. And does a "distance decay" attenuate stigma as one moves away from the stigmatized place or facility?

MARKING

Risk events, as major accidents and pollution/contamination cases demonstrate, wield a powerful potential to *mark* the places (and their inhabitants) in which they occur. Edelstein (1988, p. 14) argues, for example, that "stigma routinely accompanies the announcement of contamination and the identification of its boundaries." This marking occurs in many ways and does not require the Scarlet Letter A or a yellow star to identify the outcast. Mitchell, Payne, and Dunlap (1988, p. 98), for example, note that outsiders make wide use of an imaginary physical sign "glowing in the dark" to brand Richland (host community for the Hanford Reservation) residents. Richland people, in response, "celebrate" their imagery, or stigma, by adopting the term Richland High School "Bombers" and displaying a mushroom cloud insignia on their sports uniforms.

Marking involves the selection of a particular attribute of a facility or place and the deselection of other attributes. Accordingly, the risk event or series of events directs attention to the hazardous facility or activity at a place. Extensive media scrutiny, social debate and conflict, and public concerns—in short, the amplification process—raises to prominence a single attribute of place. Other attributes that otherwise characterize the place are pushed into the background. In the most intense situations, amplification dynamics reconstruct the image of a place around the risk attribute.

Telling evidence of such risk-based image formation is provided by the work of Slovic, Layman, Kraus, et al. (1991) on the proposed nuclear waste repository and the nuclear test site in Nevada. Using a method of "continued associations," images were elicited by use of telephone interviews. The results, as suggested by Table 2.1, indicate the extraordinarily negative imagery that such facilities or sites can come to have. The research also found that the nuclear weapons test site had led to a "modest amount of nuclear imagery becoming associated with the state of Nevada" (Slovic, Layman, Kraus, et al., 1991, p.693).

Table 2.1 Images Associated with an "Underground Nuclear Waste Storage Facility"*

Category	Frequency	Images included in category
1. Dangerous	179	Dangerous, danger, hazardous, toxic, unsafe, harmful, disaster
2. Death/sickness	107	Death, dying, sickness, cancer
3. Negative	99	Negative, wrong, bad, unpleasant, terrible, gross, undesirable, awful, dislike, ugly, horrible
4. Pollution	97	Pollution, contamination, leakage, spills, Love Canal
5. War	62	War, bombs, nuclear war, holocaust
6. Radiation	59	Radiation, nuclear, radioactive glowing
7. Scary	55	Scary, frightening, concern, worried, fear, horror
8. Somewhere else	49	Wouldn't want to live near one, not where I live, far away as possible
9. Unnecessary	44	Unnecessary, bad idea, waste of land
10. Problems	39	Problems, trouble
11. Desert	37	Desert, barren, desolate
12. Non-Nevada locations	35	Utah, Arizona, Denver
13. Nevada/Las Vegas	34	Nevada (25), Las Vegas (9)
14. Storage location	32	Caverns, underground salt mine
15. Government/industry	23	Government, politics, big business

*Basis: $N = 402$ respondents in Phoenix, Arizona
Source: Slovic, Layman, Kraus, et al., 1991, p. 689.

Further evidence of the marking of places from hazardous facilities can be found in focus group studies conducted by Mitchell et. al. (1988) in Richland, Washington. Testimony of the experience of Richland residents speaks to marking, imagery, and stigmatization:

> Female: Even my own people who live in Indiana won't come to visit me because they think they're going to be contaminated. I've lived here since 1944 and they're scared to come here. We can't even convince them, you know, that it's safe to come here. Now that's ridiculous. (p. 100)
>
> We had a gentleman call yesterday. He was considering a job here and he said: "I've got to convince my family it's safe. What can I tell them? They all think the plant's right here in town and the people glow in the dark." (p. 100)
>
> Female: Well, when our daughter graduated from college, she, of course, had a teaching certificate, had a job in Portland. And right away the people started to turn her against Richland including the principal of the high school. It was just terrible and they were brain-washing her about, you know, your father is crazy for working there, you know. They expected us all to have cancer and all these things. They really did a job on her to convince her that don't go there; don't go there. (p. 99)

Labeling is another means of marking places. Perhaps no example is clearer than the label "dump" to refer to hazardous-waste disposal facilities in the United States. Indeed, this is a case where opponents of such facilities have won the battle over language. A dump, in common parlance, is an open hole in the ground to which you back your car or pick-up truck and literally dump rubbish or refuse into the hole. It certainly has nothing in common with the highly engineered, multiple-barrier, encapsulation, and monitored facilities now being proposed for the disposal of radioactive and toxic chemical waste. Yet in our study of print media coverage of the proposed nuclear waste facility in Nevada, we found that the label "dump" dominated all other images and descriptors and had become the customary way to refer to the project. Indeed, nearly 1 of 2 news headlines during a 5-year period used the label "dump" as the project referent (Kasperson et al., 1992, pp. 45-48).

IDENTITY

The net effect of the amplification dynamics, imagery development, and labeling can be to alter fundamentally the identity of the place. As a place comes to be dominated by a single negative attribute while other attributes characterizing the place recede into the background or insignificance, identity is altered and stigmatized. And this altered identity is one that is highly tainted and discredited. The various dimensions of imagery blend into a holistic sense of the altered identity. Goffman (1963, pp. 2-3) refers to the emergence of a *virtual social identity* as opposed to an *actual social identity*; the former dominated by a single negative attribute, the latter composed of the various attributes that the place actually possesses. This altered identity is held not only by others who exist beyond the place but by the victims of stigmatization themselves, the residents of the now tainted and discredited place.

RIPPLES AND EFFECTS

As we described above in the general social amplification framework, the consequences of a risk or risk event can, if amplified, ripple to other places, facilities, technologies, and society as a whole, or even to future points in time. Risk-induced stigma greatly enlarge the potential for rippling. Such rippling has been apparent in a number of well publicized accidents. The Bhopal accident not only affected the victims at that city but produced significant changes in hazard management at chemical and other industrial facilities throughout India. Subsequently, widespread changes occurred in Europe and North America in industry management practices concerning inventories of hazardous materials, risk communication, and community preparedness. Similarly, the accident at Seveso, Italy, resulted in the Seveso Directive in the European community, with far-reaching changes in many countries concerning emergency preparedness, response, and information dissemination.

But certainly the accident at Goiânia, Brazil, offers a perhaps unsurpassed case of stigma-induced ripples and secondary consequences. The first weeks of the media coverage following the accident found more than 100,000 persons voluntarily standing in line for radiation monitoring. Within two weeks of the event, consumer concerns over possible contamination fueled a drop of some 50% in the wholesale value of agricultural production within Goiâs, the Brazilian state in which Goiânia is located, though no contamination was ever found in the products, a significant adverse impact was still apparent some eight months later. Meanwhile, the number and prices of homes sold or rented within the immediate vicinity of the accident plummeted; hotel occupancy in Goiânia, normally near capacity at this time of year, experienced vacancy levels averaging about 40% in the six weeks following the São Paulo television broadcast. One of the largest hotels lost an estimated 1000 reservations as a direct consequence of risk perceptions and stigmatization. And even a hot-springs tourist attraction, situated a full one-hour drive from Goiânia, experienced a 30-40% drop in occupancy rates immediately following the São Paulo television broadcast. Hotels in other parts of Brazil turned away Goiânia residents. Some airline pilots refused to fly airplanes carrying Goiânia residents. Elsewhere in Brazil cars bearing Goiâs license plates were stoned. Even nuclear energy as a whole in Brazil was affected, as several political parties used the accident to mobilize against "nuclear weapons, power, or waste" and to introduce legislation designed to split the National Nuclear Energy Commission into separate divisions.

Risk and Stigma in Modern Society

Remarkable as the Goiânia and "mad cow" cases are, where modern society has publics that are often risk averse to certain hazards (e.g., radiation, toxic chemicals) where suspicions about technology have mounted, and where information systems are highly developed, it may be expected that stigma-induced hazard disasters may assume greater prominence in risk management. The stigmatization of one of the world's major sources of renewable energy due to the accidents at Three Mile Island and Chernobyl and the subsequent paralysis in many countries in building new nuclear energy plants are telling evidence. The worldwide effects on the chemical industry of the Bhopal accident suggest the potential for rippling from one to many other countries. The "mad cow" case in England demonstrates not only how powerfully an entire industry can be affected by an amplified risk, followed by stigmatization, but how foreign producers in other countries can undeservedly be tainted as well. The lesson seems clear—some places and technologies may experience enormous indirect consequences that far exceed the direct consequences and define the societal experience of the risk or the event. We need to understand much better than we currently do this process of amplification, stigma-induced ripples, and the propagation of more indirect secondary and tertiary consequences. And we may anticipate that risk events followed by intense social amplification of the risks and associated stigmatization of technology, industry, or places may be recurring surprises that will exact high societal prices and will increasingly intrude upon the societal experience with risk.

Contamination Stigma

This section addresses the experience of a thing or place that is contaminated and thereby degraded and spoiled. Paul Rozin presents results from innovative experiments demonstrating the power of contamination through touch and mental association. Once an object is seen as contaminated by a loathsome source the adverse reactions have strong carrying power in time and despite explanations intended to remove the stigma effects. Michael Edelstein recounts a case of contamination of feed stocks and defines environmental stigma as a "discrediting [of] settings, places, objects, nonhuman lifeforms, and surroundings, as well as people associated with these environments." Theresa Satterfield describes how residents were affected by the contamination of their southern Georgia neighborhood by a pesticide plant over a number of years.

3 Technological Stigma: Some Perspectives from the Study of Contagion

Paul Rozin

A stigma is a negative feature that typically pervades and dominates an otherwise acceptable entity. The idea, described by Goffman (1963) as a spoiled identity, is usually applied in the interpersonal domain, but this basic "contaminating" feature can also be observed in the domains of eating (Rozin, Millman, & Nemeroff, 1986) and technology (Gregory, Slovic, & Flynn, 1996). Thus, exposure to high voltage lines seems to wipe out all of their great contributions to the convenient life of late 20th century America, the possibility of a leak of radioactivity seems to more than neutralize the great energy benefits of nuclear power, and a small amount of fat is felt to spoil the delight of a desirable food. I propose to explore the psychology of this type of stigmatization, based on the psychological concept of contagion or contamination. I believe we can learn something about technological stigma by studying how people think about food and food safety. There are reasons to think this might be a profitable line of thought.

First, many examples of technological stigma actually involve food and/or ingestion, via food technology. Thus, the list of possible situations of technological stigma by Gregory et al. (1996) includes many cases of ingestants. Second, technological stigmata usually focus on potential bodily harm; some technologically linked event or entity threatens bodily welfare by entering the body. The principal route through which the environment impacts on the body is oral, and hence involves ingestion. In general, people are particularly sensitive about traffic with strange, foreign, potentially dangerous entities, and the great majority of the most intimate traffic occurs through the mouth (Rozin, Nemeroff, Horowitz, Gordin, & Voet, 1995). Oral entry of toxins is a focus of fear; although some major feared items, like radiation, are promiscuous in their routes of bodily entry, even here, the thought of swallowing something radioactive is particularly upsetting.

With this in mind, I suggest a vicarious experience and analysis of an instance of food contamination. The contaminant (source of stigma) is the lowly cockroach (*Periplaneta americana*). While this eminently successful creature of nature is not in the technological domains, it is more successful than any technological innovation that I know of. Cockroaches are among the most successful and ubiquitous of animals, which certainly speaks to quality of design, albeit by mother nature. These little creatures, in interaction with the rest of the world, create many of the same effects as technological stigma. The touch of a cockroach can spoil a good meal. Indeed, viewed as a stigmatization, this meets many of the criteria for technological stigma proposed by Gregory et al. (1995).

Consider the following scenario (Rozin et al., 1986; Rozin, Fallon, & Augustoni-Ziskind, 1985). You are about to drink a glass of your favorite juice, when a playful friend drops an inch long cockroach into it. You display disgust, and in consideration of this, your friend deftly removes the roach with a spoon. He now informs you that you can drink your juice, but you refuse. "Why?" he asks. Your most likely response (if you are like our subjects) is, "It's dangerous. Cockroaches are dirty and disease vectors." Your patient friend is sympathetic, pours you some more juice in a new glass, and takes a dead cockroach out of a plastic box. He says, "OK, this is a dead cockroach which was completely sterilized by heat. It is safer than a spoon or glass." Before you can respond, he drops the dead sterilized cockroach in your juice and then removes it. "OK, surely you'll drink this safe juice?" You are disgusted, and refuse. (In fact your refusal is almost as strong as it was for the live/unsterile roach). Your friend points out that there is no safety issue here, and you have to acknowledge this. You squirm, with your prior reason invalidated. Ultimately, you confess that the juice has been spoiled just because it contacted a cockroach; it has been "cockroached"; what motivates rejection is psychological offensiveness, sometimes bordering on a perceived moral violation. Apparently, brief contact with a cockroach spoiled the juice. Is there anything we can now put in the juice (vitamins, antibiotics, your favorite form of alcohol?) that will neutralize the juice and make it desirable again? For most people, the answer is, "No!" Finally, we remind you that if a cockroach is really an extremely powerful contaminant, then you cannot rationally deal with the world at all, because cockroaches are ubiquitous, and surely virtually everything has touched a cockroach, or something that has touched a cockroach, or something that has touched something that has touched a cockroach! The only sane response to this is that you just don't think about such things.

There are five important things to learn (re technological stigma) from this often repeated scenario (in our laboratory, and often, in one form or another, in daily life). Each will be discussed in successive sections of this paper.

1. The response to the cockroach illustrates the principle of *contagion*.
2. Western/educated people (perhaps, all people) tend to prefer health/risk accounts to vaguer personal offensiveness accounts for their behavior and feelings (*medicalization*).
3. There is a strong human tendency to respond more powerfully to negative than positive events, such that many negative events produce essentially irreversible psychological consequences (*negativity dominance*).
4. Since the threats of contagion are ubiquitous, we manage life by ignoring them unless they are particularly salient, through a process like *framing*.

5. There is a strong human tendency to conflate moral and health risks (*moral-physical conflation*).

Contagion

The cockroach contamination sequence illustrates what has been called the "law of contagion." This law of sympathetic magic, originally conceived to characterize the thinking of "primitives," was described by the distinguished anthropologists Edwin Tylor (1871/1974), James Frazer (1890/1959), and Marcel Mauss (1902/1972). The basic idea is: Once in contact, always in contact. That is, when two objects touch, properties are passed permanently between them. My colleagues and I have shown that the law of contagion operates frequently and saliently among educated adults from Western-developed cultures (reviewed in Rozin & Nemeroff, 1990). Frequent sources of contagion are offensive potential foods (like worms and cockroaches), body products, and unsavory people.

Contagion seems to be a universal feature of adult human thinking; it is absent in very young children (Fallon, Rozin, & Pliner, 1984; Siegal, 1988). It is probably uniquely human. There are four principles of contagion that the cockroach example exemplifies (Rozin & Nemeroff, 1990):

1. Contagion requires or depends heavily upon physical contact. A cockroach *near* one's juice glass or mashed potatoes or a fly flying near one's mouth is much less potent than a contact experience. The act of sniffing/breathing constitutes a marginal sense of contact, since we actively take in air, and the real substance of the odorous object enters our body.

2. Contagion is permanent. Once contaminated, an object remains so indefinitely ("once in contact, always in contact"). The cockroached mashed potatoes or juice remains objectionable after one year in the freezer.

3. Contagion is dose insensitive. Very brief contact (as of a cockroach with juice) is very potent. Although there may be an increased negativity with increased contact, the effect is small, because the effect of brief contact is near ceiling (Rozin, Markwith, & Nemeroff, 1992).

4. The origin of contagion is unknown. Adaptively, it could be linked to either microorganism or toxin contamination of foods. On the positive side, contagion establishes the link that concretizes kinship, that is, shared blood. But, whatever its origin, this psychological "idea" has spread through the human psyche and its many interactions with the world. This process, taking place over cultural evolution, has been called preadaptation in biological evolution (Mayr, 1960). Something evolved for one purpose is used for another (see Rozin, Haidt, McCauley, & Imada, 1997, for more on preadaptation in relation to contagion, disgust, and cultural evolution). Contagion comes to be a component of thought whenever there is danger or offensiveness lurking; it is the partner of fear and disgust.

The cockroach example initially looks like a fear/danger situation but reduces, on analysis, into a disgust/offense situation. But the laws of contagion apply as well to fear and danger. Consider sodium cyanide, a poison that blocks transport of oxygen by the red corpuscles, and, in moderate dosages, causes a rapid death. It has been a favorite of executioners. Sodium cyanide is not a cumulative poison; it is disposed of by the body. Very low doses are harmless. In fact, low amounts are found in foods, such as apricot pits.

People do not want to drink anything out of a bottle that previously held sodium cyanide, even though it has been thoroughly cleaned, and traces of the poison would be harmless. (I have no organized, published data base supporting this claim, nor the claims that follow in this paragraph. However, I believe that (a) the claims are obviously true, (b) they are supported by pilot data we have collected, and (c) they are a weaker form of results we have published showing considerable reluctance to drink sugar taken from a bottle with a cyanide *label,* even though the subject saw sugar placed in the bottle and then put the cyanide label on, herself (Rozin et al., 1986; Rozin, Markwith, & Ross, 1990). They don't mind drinking from a glass/bottle that was near a bottle of sodium cyanide (contact principle), don't find that storing the glass for a long time reduces its negativity much (permanence), are almost as upset by a glass that had a little, sublethal amount of cyanide as one that was full of cyanide (dose insensitivity), and find that there is nothing one can put in a glass that produces the kind of positive psychological effects that would measure up against the negative effects of cyanide (negativity dominance). Here the motivation is clearly fear, and not disgust/offense. But fear of what? There is no risk. Furthermore, many of the same people put off by the washed cyanide bottle would probably be willing to suck an apricot pit, even after being informed that it has harmless trace levels of cyanide in it!

Contagious Essence

Since stigmata have contagious properties, and these account for much of the aversion and withdrawal that results from contact (or fear of contact), it is likely that understanding the nature of contagion would enlighten our understanding of stigma, whether interpersonal or technological. Carol Nemeroff and I have recently been exploring psychological models of contagion; that is, what (in the mind) is "passed" when two objects touch (Rozin & Nemeroff, 1990; Nemeroff & Rozin, 1994; Rozin and Weinberg, unpublished).

We have developed three types of models to explain contagion: association, material essence, and spiritual essence (Nemeroff & Rozin, 1994). Contagion can be conceived as a strong form of association. That is, Hitler's sweater is a reminder of Hitler, and since Hitler is an unpleasant thought, his sweater is rejected. There is no doubt that such associations occur; the question is whether they are sufficient to account for contagion. The answer is no. First of all, common sense tells us (as with cockroaches) that contact is worse than close association. Furthermore, we have determined, using AIDS as a stigma/contaminant, that a sweater, condo, or fork used by a person with AIDS is more negative than the same item that is owned (but not used) by the same person (Rozin, Markwith, & Nemeroff, 1992).

Material essence holds that some material substance is passed in contagion. The spiritual essence model holds that something is passed, but it does not have material properties (e.g., it cannot be washed out). It is extremely difficult to disentangle these three models empirically, because they usually make the same predictions. However, specific manipulations designed to negate or purify a contaminated object may differentially affect responses, according to one model or another. For example, material essence should be countered by washing or boiling, whereas neither spiritual essence nor association should be affected. On the other hand, repeated presentation of the contaminated object (extinction) or changing its visible appearance should reduce associations, but have minimal effects on either type of essence.

The most difficult to understand, and most resistant type of contagion, is spiritual essence. It behaves like material essence (transferred by contact) but does not have certain critical material properties. For example, it is resistant to washing and boiling. It isn't clear how one might reduce it; it is extremely hard to make Hitler's sweater acceptable. One possibility is to fight spirit with spirit, that is, put Hitler's sweater on Mother Teresa or Michael Jordan. This is the only manipulation that we believe, theoretically, would reduce spiritual essence, and we have evidence to support this belief.

Our principal study on models of contagion (Nemeroff & Rozin, 1994) discriminates between material essence and other models. In-depth interviews with 36 American adults were carried out, in which (imagined) sweaters and other objects were contaminated in a variety of ways were subject to "purifications" of various sorts (e.g., washing, boiling, color change, and reknitting to change the appearance, subsequently worn by the subject's favorite person, etc.). On the basis of the pattern of results, we classified the operative model in the subject as material essence or other (spiritual/associative). A variety of contagious sources were used, including an evil person and a person with hepatitis. To simplify, comparing the hepatitis and evil sources, we found that two thirds of the subjects adopted a material essence model for hepatitis contamination, and a different (spiritual/associative) model for evil contamination. This is what might have been expected. However, one sixth of the subjects had a material essence model for both hepatitis and evil, and another one sixth had an associative/spiritual essence model for both sources. The results suggest that more than one "model" may exist in one head, but that some people rely principally on one model. For those with a spiritual/associative model for hepatitis, it is notable that washing, boiling and other material manipulations did not substantially reduce the negativity.

The implication of this for technological stigma is that it is important to understand the nature of the underlying fear or offense with respect to designing ameliorations.

Medicalization

Virtually all of our educated American subjects refuse a "cockroached" beverage. Almost all of these subjects explain their refusal (see example in the introduction) in terms of health risks. The sterilization procedure shows them that this account is false. We do not know whether the same resort to health accounts would be present in less educated Western or non-Western subjects. However, the subjects' "preference" for medical-scientific, as opposed to more value-laden accounts, seems quite widespread among educated people, at least in Western culture. The same holds true for accounts of why there is reluctance to touch people with AIDS, or even to live in houses previously owned by persons with AIDS. In scholarship, there is a tendency to prefer such accounts, as for example, the apparently baseless claim that pork avoidance in the ancient Hebrews had to do with the avoidance of trichinosis (the trichina is killed by cooking). It is quite possible that explanations of technology fears in terms of health or environmental risks may also mislead and mask what are perceived to be less "rational" accounts.

Negativity Dominance

It is extremely difficult to reverse or neutralize the cockroach contamination we described at the beginning of this paper. The reader might chose to think about what operations on the cockroached juice would render it neutral or desirable. Contagion is negatively biased. Although there is surely positive contagion (grandma's ring, Michael Jordan's sneakers), negative contagion is more common and more potent. As a Nebraska car mechanic put it, "A teaspoon of sewage spoils a barrel of wine, but a teaspoon of wine does nothing for a barrel of sewage." Although there are entities that are positively potent in trace amounts (e.g., vitamins, antibiotics), they do not have the psychological potency of their negative brethren.

Negativity dominance in contagion is present in two of the cultures with the most salient, structured, and powerful contagion systems. Among the Hua of Papua New Guinea, the act of eating is thoroughly imbued with the passing of negative and contagious essences. Although there is much positive contagion, as, for example, benefiting from eating food prepared by or partially consumed by certain close relatives, negative contagion is much more powerful (Meigs, 1984). In Hindu India, the food contagion rules play a major role in expressing and maintaining the caste system. While consumption of food prepared by those in a lower caste has a strong social "lowering" effect, consumption of food prepared by those in higher castes has no substantial elevating effect (Appadurai, 1981; Marriott, 1968).

These types of findings on contagion, plus the phenomenon of loss aversion (Kahneman & Tversky, 1979, 1984), and the powerful effect that single moral lapses can have on the reputations of generally admired people in public life, all argue for a general principle of negativity dominance in human life (Rozin & Fallon, 1987; Rozin & Nemeroff, 1990; Rozin & Royzman, 2000). This principle, applied to technological stigma, may help to account for the net negative response to innovations that have many positive effects, but with public suspicion about negative effects (e.g., nuclear power, electric power lines, pesticides, genetically engineered organisms).

Framing

Cockroaches are ubiquitous. There is little that has been around for periods of months or more that is totally innocent of cockroach contact. Cockroach contagion is potentially crippling, as is the fact that we are constantly breathing in the air breathed out by others, sitting where they sat, using their silverware at restaurants, etc. In order to function in the world, some limits must be placed on contagion. Various "defense mechanisms" may be at work, including the familiar denial and rationalization. A major way of coping with contagion falls under the term framing, as used by Kahneman and Tversky (1979, 1984). One basically ignores it, relegates it to the background. This works well so long as the source of contamination, or the act of contamination, is not especially salient or the contact is minimal. Thus, most people use their hand to turn the knob on the door leading out of a public lavotory, yet surely even a little thought of what has recently touched that knob would be unsettling. If not, imagine having to turn the knob with your mouth instead of your hand!

Attention is called to sources of contamination by the media or everyday conversations or observations. However, these events typically recede in salience, allowing life to continue. Of course, there is a major individual difference variable. Individuals, in the United States, vary extensively in their sensitivity to contagion (Rozin, Fallon, & Mandell, 1984; Haidt, McCauley, & Rozin, 1994). Contamination sensitive people with obsessive-compulsive disorder represent a particular extreme. Whether people high in disgust and contagion sensitivity are more likely to respond to technological stigma remains to be seen.

Another means of coping with the omnipresence of contagion is to develop rituals to cancel it and/or rules to set limits upon it. In different cultures, washing or a variety of incantations can be employed to contain or reverse contagion. Limits can be expounded which relegate minimal contagious contacts as insignificant. For example, the Jewish kashrut tradition (the laws governing eating, sometimes referred to as "kosher") is vulnerable to crippling contagion effects. Molecules from a neighbor's pork barbecue may fall into one's kosher stew, or molecules of one's neighbor's milk may fall into one's kosher meat dish. There is clearly a level of microcontamination that cannot be defended against. The kashrut tradition "solves" this problem by legislating a contamination threshold: Essentially, if contamination occurs by accident, and if the contaminant is less than 1/60 of the volume of the contaminated entity, then the latter remains kosher. Unfortunately, the majority of kosher Jews that we tested, although they knew this "law," continued to feel offended by technically legitimate microcontamination and claimed they would refuse such food (Nemeroff & Rozin, 1992). Arbitrarily stated boundaries do not reliably map onto feelings. Kosher Jews' personal sensibilites are generally of lower threshold than the law stipulates.

Moral/Physical Conflation

Both physical and mental (soul) danger (via offense) have been implicated as consequences of contagion. It is but a short step to the addition of moral taint. There is widespread reluctance, among Americans, to wear a sweater that had previously been worn by Adolph Hitler. This suggests a negative moral transfer. We have already shown that if one scratches away a presumed health risk (the cockroach in the juice), one may find a psychological (offense) fear behind it. Fear of moral taint also often masquerades behind a presumed physical threat. In an inquiry on this subject, we asked undergraduate students to rate how they would feel about wearing a brand new unisex sweater of a style they like for one day, using a scale of 0 (dislike extremely) to 100 (like extremely), with 50 as neutral. Subsequently, they rated how much they would like to wear an identical sweater, after it had been worn by a target person, and was then thoroughly laundered. A healthy man was rated a mean drop of 20 points, a person who lost his leg in an auto accident which was not his fault was rated down 33 points, a man with tuberculosis was down 64 points, and a healthy, convicted murderer sweater dropped 62 points from the new sweater baseline. The murderer is assumed to be a pure moral taint (Rozin, Markwith & McCauley, 1994).

AIDS (one of the principal stigmata of our times) produces a strong negative effect (mean, –66). The response to AIDS may be decomposed into four sub aversions to: a stranger (man), a misfortune (accident victim), an illness exposure (tuberculosis), and a moral taint (murderer) (Rozin et al., 1994). The results from this study indicate that the negativities associated with misfortune, illness, and moral taint, share a common core. Thus, the negative judgments of subjects for these three types of contamination are positively correlated (in the .25 to .46 range). There are a number of possible accounts of this correlation; one is that one cannot easily disentangle moral taint, illness threat, and misfortune stigma/contamination.

The important point here is that if there are moral components to responses to stigma, then health assurances alone will not solve the problem. The history of health movements in America consistently involves a linkage to moral issues (Brandt & Rozin, 1997). Health issues often become moralized (Rozin, 1997); that is, they take on moral characteristics. Individual attitudes or choices move from a domain of preference to a domain of values. This has clearly happened in the last few decades in the United States, with respect to cigarette smoking (Rozin & Singh, 1998). Our analysis of moralization suggests that two important contributors, in the United States are the Protestant ethic and the invocation of harm to children (Rozin, 1997).

In a way, moralization is the opposite of medicalization. The former converts physical risks to moral ones, while the latter describes moral or psychological events as physical risks. In either case, there are problems with accepting the first account given by individuals, a caution that extends to reactions to technological stigma.

Benign Nature and Malevolent Humans

My final example from the food domain does not derive from contagion, but does have direct parallels in the study of technological stigma. Human intervention seems to be an amplifier in judgments on food riskiness and contagion. That is, Americans seem to believe that nature is benign or benevolent, and humans are malevolent. Hence, the popularity of the word "natural" and the natural foods movement. This stance is quite bizarre. Surely, more lives are lost to natural than to man-made disasters in the world. By all evidence, humans are kinder than nature. More often than not (unlike nature), they pity and often aid the unfortunate.

Ames, Magaw, and Gold (1987) show lay overestimation of carcinogenic risk from pesticides, and under-estimation of the risks from natural carcinogens. According to their evaluation, the risk of cancer caused by natural carcinogens is higher. Analyses of questionnaire data by Marc Spranca (1998) also indicate a reverence for nature and a sense that it implies safety. Thus, natural water is rated preferable to processed water even if it is stipulated that are chemically identical. The natural bias is also illustrated by my prior suggestion that former cyanide bottles are strongly shunned, whereas sucking on an apricot pit is not. It is also present with technological stigma (Slovic, 1987).

Some Examples of Stigma/ Contagion in the Food Domain

I want to close with two examples of unreasonable fears in the food domain. Both have to do, in different ways, with technological stigma.

FAT AS TOXIN

Fat is becoming a frightening, and even immoral substance in American society. It is associated with both ill health and animals. Fat, of course, is natural, and it is the technology that removes it, generating more favored (if less tasty) foods. Two related patterns of thought seem to enhance the fear of fat. One is the public's inability to think of non-monotonic functions. The idea that a little fat might be good, and a lot might be bad, seems foreign and unecessarily complicated to many people (Rozin, Ashmore, & Markwith, 1996; see also Gregory et al., 1995, for similar results in the domain of environmental toxins or "chemicals"). A second conception relates to contagion and its property of dose insensitivity. Since fat is considered bad, any contact with it seems to transfer the badness. We have found that a substantial minority of Americans treat fat as a contaminant and believe that a diet completely free of fat is healthier than a diet with a very small amount of fat (Rozin et al., 1996).

Recently, Stein & Nemeroff (1995) have shown, in an Asch impressions study, that Arizona college students believe that someone who consumes a high fat/sugar diet is a morally less satisfactory person than someone whose diet focuses on fruits and vegetables!

MAD COW DISEASE

My second example is mad cow disease. I want to raise it because it seems to involve virtually all of the features that I see as relevant to enhancing fear, and all those that Slovic (1987) and Gregory et al. (1995) identify as contributing to technological stigma.

By a human intervention, a disgusting food, raw meat, is given to cows. These cows may thus become infected with an insidious, poorly understood, new, microscopic disease vehicle (prion), which hides for years to decades in the body, and then attacks the mind. We have here involvement of human intervention (unnatural), a disgusting food, interaction with another animal food (animal foods are much more likely to generate strong affect), a prospect of eating (oral incorporation of this food), an invisible cause, a delayed effect, a catastrophic outcome, an effect on the brain, and something that is out of an individual's control, in the sense that the presence of the disease agent cannot be detected. What is surprising is not that there is a mad cow fear in Europe, but that the Americans, champions of such fears, have yet to get seriously engaged. Oddly, this hasn't been of as much interest to Americans as Alar in apples or Chilean grapes. Perhaps Mad Cow Disease is seen as a European problem.

Conclusion

I hope to have raised some issues of relevance to technological stigma. I note that just as there is no opposite to disgust, there is no opposite to stigma. There is no potent force that can be applied to erase disgust or stigma. I think we have to understand contagion, disgust, fear, stigma, and risk perception better before we can make intelligent suggestions about how to control or modulate technological stigma. It is odd that a very small risk attached to a very positive technological process causes such a fuss, just like a small indiscretion by a generally honest and effective politician (assuming such exist) seems to have ruinous consequences. I think, along with Gregory et al. (1995), that many first-order responses by well meaning public agencies to reduce technological stigma may not work too well. We have to understand this as deeply embedded in human nature. Before dealing with the stigma, we have to know what its nature is.

Acknowledgments

Preparation of this paper was assisted by NIDA grant R21-DA10858-0 to Paul Rozin.

4 Crying Over Spoiled Milk: Contamination, Visibility, and Expectation in Environmental Stigma

Michael R. Edelstein

Introduction—Environmental Stigma

This chapter introduces the construct "environmental stigma" as a tool for social analysis. A case study, written in 1984 but previously unpublished, is used to show a radiating chain of stigma beginning with the accidental contamination of silage with oils containing polychlorinated biphenyls (PCBs), the consumption of the feed by dairy cows, the discovery of PCBs in cows milk bottled as a raw milk product, efforts to reduce the contamination in the herd and milk, the decision of health officials that the herd was sufficiently purified, the publicity of the event, and the resulting discrediting of the milk product, the dairy, the farmer, and the cows.

Environmental stigma is analogous to social stigma. Social stigma is a process of changing in a negative way or discrediting the perception of a human victim "marked" by some observable characteristic such as a noticeable deviance, a flaw or limitation, a sign of being spoiled or some other undesirable characteristic (Jones et. al., 1984, p. 6; Goffman, 1963). The term stigma is commonly used in the social sciences to explain the perception of members of such visibly deviant groups as convicts, the physically handicapped, and minority ethnic or racial populations. Such groups suffer from a "spoiled identity" or deep discrediting due to their socially undesirable variation from expectations (Goffman, 1963, p.5).

Environmental stigma involves a parallel process for discrediting settings, places, objects, nonhuman lifeforms, and surroundings, as well as people associated with these environments (Edelstein, 1991, 1988, 1987, 1984). Environmental stigma is evident, for example, in situations where feared environmental conditions pertain, as with "contaminated" places, media (soil, air, water), or substances and products (e.g., milk or apples). An anticipatory stigma may be created when a future hazard threatens a place or community, as with the proposed siting of a hazardous facility.

Environmental stigma is closely intertwined with social stigma. Environmental stigma has its social effects, as when victims of toxic contamination are ridiculed for their misfortune, much as all victims are looked down upon. Social stigma also contains environmental components, as in the "other side of the tracks" phenomenon. The interplay between environmental and social stigma is evident in the practice of zoning, where decisions are made about what people and activities are considered to be peripheral or feared—evidenced in public concern about low income housing, hazardous industries, and community facilities for stigmatized groups (see Perin, 1977). Carried to its extreme, in making decisions about dangerous facilities, stigma may embody deep social prejudices, as in the phenomenon of "environmental racism," where exposure to disproportionate hazards is suffered by socially devalued groups or aggregates as the result of deliberate social decisions (see, for example, Bullard, 1994a).

There is also a general recognition that environmental stigma can smear one's identity merely by association, or what Goffman (1963) termed a "courtesy stigma." Because places reflect identity (Heider, 1963), information about environments may influence judgments about people associated with the environments (see Ichheiser, 1970, Goffman, 1963). This importance of environments to the attributional process suggests that, beyond a direct fear of a stigmatizing condition in its own right, there is a concern that any association with the marked setting may serve to mark oneself as well (Edelstein, 1987). Victims are not only stigmatized for being victims, but may literally be viewed as themselves contaminated, carriers of the affliction as if pollution were contagious. Thus, Chernobyl victims were apparently actively shunned in the former Soviet Union and atomic bomb victims were shunned in Japan after World War II. Such stigma reactions are not uncommon after communities face groundwater, air or soil pollution due to illegal dumping, industrial or commercial facilities or landfills. As a result, I have argued that environmental stigma is an inherent property of toxic contamination (Edelstein, 1988).

The theory of Environmental Stigma had its origins in the 1981 study of residents of a contaminated community in south New Jersey, where the neighborhood and its residents were rendered social outsiders by the drawing of the contamination boundaries (Edelstein, 1982, 1988). It was further developed in the study presented here, written in 1984 as an expert report for litigation brought by a victim of contamination. In examining our case study, we pay particular attention to two key factors in the process of environmental stigma: the fact that a mark must become "visible" to be stigmatizing and the expectations of the "audience" to whom the mark is made visible. The importance of these components of the environmental stigma process is evident in the case of Homestead Dairy and its owner, Dave Cito, presented here in its 1984 form with only minor editing.

The Homestead Dairy Case

In 1984 the law firm Martin and Snyder asked the author to explore the social and psychological issues relating to the Cito/Homestead Dairy pollution incident. Of particular interest was the extent to which "stigma" related to the incident could account for the loss of business suffered by the dairy after the incident. The analysis focused upon a small set of articles and transcripts that represented a file of media coverage of the event.[1] Field data contemplated as a second phase of research was never collected. The resulting report served as the basis for a deposition in litigation brought by the dairy owner against Monsanto Corporation, the dominant manufacturer of PCBs. The case was dismissed at trial. The findings were never presented to the court.

In brief, the Cito/Homestead Diary PCBs case involved the following chain of events. The Homestead dairy in Niwat, Colorado was a third generation operation run by Dave Cito [Alvarez, 1983]. About 60% of the milk from the dairy was bottled under the name Homestead dairy and sold as "old fashioned unhomogenized milk" to health food stores [Alvarez, 1983, n.d.-a], with the remainder sold in bulk. The dairy served a region stretching from Wyoming to Colorado Springs [Alvarez, 1983].

On March 4,1981 PCBs were discovered in the milk by the Colorado State Health Department. The probable origin was 1,500 tons of corn silage bought from Tonaka Farms, where the PCBs allegedly were an additive in hydraulic fluid which leaked from a conveyor belt [Alvarez, 1983; Noriyuki, n.d.-a]. Levels of PCBs never were sufficient to force Cito's milk off of the market, and health officials considered the milk to be safely under standards for PCBs once the most contaminated animals were culled from the herd. However, once word of the incident was leaked to the press in late April, publicity about the pollution caused retailers to stop buying the product [Alvarez, n.d.-b].

Subsequent to the discovery of pollution, Cito's herd was reduced from 300 head to 105, as contaminated cows were removed from milking and killed, and bottling operations were ceased [Alvarez, 1983]. With a drop of sales from 800 gallons a day to five gallons over a period of just a few months, the Homestead dairy was closed [Noriyuki, n.d.-a; Associated Press (AP), 1981b]. It later reopened to sell milk wholesale out-of-state for cheese production [Crary, n.d.]. An auction in December 1983 failed to attract bids for the Cito cows high enough to offset $200,000 in debt [Alvarez, 1983]. Cito was forced back into business [Alvarez, 1983].

Stigma

This report attempts to develop a social and psychological perspective for understanding why people stopped buying Cito's milk. In order to do this, it is argued that the contamination by PCBs resulted in stigma. As used in psychology and sociology, the word stigma normally refers to an individual or group perceived as "deeply discredited" because of certain characteristics. These characteristics often involve "an undesired differentness from what we had anticipated" (Goffman, 1963).

Although we most commonly use the term in reference to people's perception of groupings such as the handicapped, ethnic or racial groups or convicts, the concept of stigma can be applied broadly. Stigma always involves a victim who can be identified as "marked" (deviant, flawed, limited, spoiled or generally undesirable) by an observer. When the mark is noticed, it changes in a negative and discrediting way how the observer sees the victim (Jones et al., 1984).

[1]For citations in brackets only, see "References to Media File" at the end of this chapter.

In the context of the Cito/Homestead Dairy case, stigma implies the following. When Cito's cows became contaminated with PCBs, the potential for stigma existed. But since neither the public nor Cito knew about the contamination initially, business continued as usual. Even when the health department informed Cito of the PCBs and steps were taken to address the problem, the public still did not know of the contamination. The mark was identified, but it was concealed. It thus was not a factor in how people perceived Mr. Cito, his dairy (and cows) or his milk product.

However, once the pollution incident became known, people saw Homestead milk differently. This difference, as compared to their former perception, to their expectations and to the perceived quality of competing milk, was negative and discrediting. It accounts for the discontinued use of Homestead milk, and presumably a different view of the cows, the farm and Cito himself. All were victims of stigma.

To fully understand the nature of this stigma, we will now examine each of the elements discussed above. We first have the raw materials of the pollution case: a contaminant (PCBs) and a product that was contaminated (milk). But in order to understand how this contamination was perceived, we must also examine the way that the pollution became visible (i.e., how it was publicized) and the characteristics of the dairy's customers that night influence how they interpreted the publicity. We must give some thought to the dynamics of threat perception which might further account for the customers' interpretation of the publicity as indicating a threat. Finally, we must return to the nature of this perception: its being a stigma and the targets and thus victims of the stigma.

The outline for the ensuing analysis is built around these points. Specifically:

Raw materials. Any reader of the media coverage of this event would know certain basic "facts" about the case. Thus, contaminated grain was fed to the cows, resulting in the detection of PCBs in milk during routine testing by the state health department. Two elements of this scenario are particularly important to our analysis. One is the property of the product, the other the property (as recognized) of the contaminant. In simple terms, what is the significance, on the one hand, of the fact that the product is milk and, on the other hand, of the fact that the contaminant is PCBs?

Customer characteristics. The perceptual biases of the customers may influence how they perceive this message. This is particularly important in an incident where we are concerned less with the real danger than with the perceived threats of exposure. Thus the characteristics of the Homestead customers, their expectations and concerns become relevant to the analysis. Of special interest are health food customers, accounting for about 60% of Cito's business.

Visibility. Since the pollution of the milk, per se, was not noticeable to consumers, consumers never directly detected a change in the quality of this product. Instead they learned of this change from the media. The "visibility" of the pollution thus stems from the amount of publicity, its content and how the message is presented. How is the pollution incident understood by the public?

Perceiving threat. The elements which contribute to the perception and judgment of threat and the decision to stop using Homestead milk are examined.

Stigma and its victims. The overall focus of this analysis will be how the customers perceived the milk sold by Homestead dairy, the dairy itself and the owner(s) of the dairy as a result of the pollution incident. The principal argument made is that stigma resulting from knowledge of contamination by PCBs of the milk represents a plausible and even likely explanation for the loss of Homestead customers.

Analysis

RAW MATERIALS

PCBs

It is not necessary to provide a scientific overview of PCBs for this report. We are dealing less with scientific facts about PCBs exposure than about consumer perceptions about such exposure. Therefore, basic to the argument advanced here is the notion that exposure to PCBs is a potentially stigmatizing condition. To establish this argument, it is necessary to suggest that what people know about PCBs is likely to establish a negative "set" toward any product (or person or company) associated with exposure. In a latter section, we look at the visibility of the Cito contamination and the kinds of information carried by media reports. Even someone who had never before heard about PCBs would have learned a great deal from these reports.

The Homestead case was not the first occasion in which the public might have learned about PCBs in food. A Monsanto letter sent to its product dealers on February 9, 1970, acknowledges publicity about the spread of PCBs throughout the environment (D. Olsen, personal communication, 1970). A government document from 1976 reports six PCBs cases in more than six states beginning in 1971 which involved farm animals. Another 17 cases listed involved other types of chemicals in a similar context (see Reich, 1983). Indeed, PCBs along with other compounds like polybrominated biphenyl (PPB) and dichlorodiphenyltrichloroethane (DDT) have been recognized as ubiquitous contaminants of humans as well (Barr, 1981).

Among incidents in which PCB contamination received wide-spread attention was the 1968 "Yusho" contamination. More than 1,000 Japanese developed severe illnesses because bran oil they ingested had been contaminated with PCBs (Wickizer, Brilliant, Copeland, & Tilden,1981). PCB contamination of Lake Michigan (Wickizer et al., 1981) and the Hudson River have been widely reported, as has attempts to regulate consumption of contaminated fish.

An incident in the western United States in 1979 involved the spread of PCB contaminated food products across the United States and as far away as Canada and Japan. A punctured transformer leaked into a hog rendering room in Billings, Montana. PCB-laden waste products were made into meat-meal used in animal feeds. More than 1,000 ppm of PCBs were found in feed fed to chickens in Franklin, Idaho, from where eggs were marketed locally and chickens sold to distant soup makers. Epidemiological research revealed a wide distribution of PCBs in the food chain from this one source. Breast milk samples correlated significantly with consumption of the eggs, but were not abnormally high given national studies of PCBs in breast milk (Drotman, 1983).

Possibly the most famous contamination case involved a related and similar-sounding chemical, PPB. In 1973 and 1974, this chemical contaminated thousands of Michigan dairy farms (Wickizer et al., 1981). First identified on a single dairy farm, the incident involved a lone farmer fighting "the system" until a mix-up in a chemical and feed plant was discovered. Because of this mix-up, a

PCB-based fire retardant had been added to feed instead of a similarly packaged ingredient. As the extent of the contamination became known, some 800 farms were quarantined (Hatcher, 1992) and thousands of cattle and pigs, millions of chickens, tons of animal feed, millions of eggs, and thousands of pounds of cheese, butter, and dry milk products were destroyed. Protests by farmers involved attempts to capture media attention through such practices as dumping dead cows on the state capital steps (Reich, 1983; see also Coyer & Schwerin, 1981).

Among the various concerns over human health impacts, possibly most frightening was the August 1976 report by Michigan health officials that PPB had been detected in the breast milk of 96% of nursing mothers (Hatcher, 1992). As part of the same testing program, nursing mothers had the option of being tested for PCBs. Results revealed that all breast milk tested had at least trace amounts and half had contamination nearly equal to or above FDA tolerance limits for PCBs in cow's milk (Wickizer et al., 1981).

This protracted case, along with others that have occurred, has received attention in scientific circles, as well as in the media. For example, a television movie called "Bitter Harvest" was based upon the experiences of Rick Halbert, the farmer first involved in the PBB case. An advertisement for the movie shows actor Ron Howard holding a calf. The headline reads: "True-Life Tale About Danger to Millions." The description of the movie refers to a chemical… accidentally introduced into the environment" ["True- life," n.d.]. Such media reports and presentations of contamination events, along with reports of laboratory tests on animals, have raised a series of complex issues for the public as well as the scientific community. These include questions about the safety of breast milk, levels of risk associated with exposure, health effects, and the ability of various agencies to respond effectively to such incidents (Barr, 1981; Reich, 1983; Hatcher, 1992; see also Levine, 1982; Edelstein, 1982, 1983; Gricar & Baretta, 1983, note also the Hawaii milk contamination cover up case, see Smith, 1982).

Milk

The product contaminated is likely to be a factor in the perception of those judging the significance of the event. For example, studies of well water pollution suggest that this event is disruptive for reasons beyond fear of health effects. Water is a life requisite and is assumed to be readily available in support of the American lifestyle; its disruption results in a major interference with this lifestyle (see Edelstein, 1982). Unlike water, milk is most often purchased in a packaged form; it is a product. Normally, there is no dependence upon one fixed source for one's milk.

Milk is often advertised for its wholesomeness and nutritional value. Milk is heavily consumed by children as well as by expecting and breast-feeding mothers, raising the possibility of enhanced concern among parents over any possible contamination of this food. Hatcher (1982) reviews the benefits of breast-feeding, examining a range of issues associated with contamination of breast milk specifically in the case of the PBB exposure discussed above. It would appear that some nursing mothers in the Boulder area were concerned about similar dangers resulting from their consumption of Homestead milk [KOA-TV, 1981a, 1981b].

Homestead Milk had an image of being purer than most commercial milk. In a sense, it offered a promise that was broken, suggesting that it might no longer be trusted to retain and deliver its formerly expected degree of purity. If contamination could happen once, it inherently becomes a property of the dairy's image; it might happen again!

If we mix our raw materials, how might PCBs and milk interact in consumers' perceptions? While it has long been recognized that perceived risk can be a factor in consumer decisions (see Cheron & Ritchie, 1982), no specific studies were located of perceived risk due to food contamination by chemicals.[2] Nevertheless, one might draw some rough generalizations from work by Slovic and his colleagues on perception of 90 hazards. Two of the factors underlying these perceptions were (a) severity of threat (from "no dread" to "high dread") and (b) knowledge of the threat (from "familiar" to "unfamiliar"). Graphs summarizing two studies of these factors reveal that threats such as water fluoridation, food coloring, saccharin, sodium nitrate, food preservatives and food irradiation registered as fairly unknown risks of less than moderate severity (food irradiation was nearly at the midpoint of severity with the other threats decreasing in severity in reverse order listed). In contrast, threats such as chemical disinfectants, asbestos, chemical fertilizers, herbicides, pesticides and DDT are seen as more than moderately serious (with DDT the most serious, decreasing in reverse order listed) and fairly unknown as threats. While PCB contamination of milk was not included among the 90 hazards, one might deduce that it would at the very least involve a fairly unknown risk of moderate severity (The Royal Society, 1983, pp.132–136).

CUSTOMER CHARACTERISTICS

The biases of the customers and consumers of the milk are important in understanding how the messages discussed above are perceived or decoded. It is likely that some consumers are more sensitive and sensitized to impurities in their milk. While no specific studies were located detailing the attributes of health food users, some general characteristics can be surmised from the author's own experience, from attendance at a major conference on "Food and Energy'" and from a few relevant sources.

Values appear to be closely related to the criteria by which consumers choose products (Pitts & Woodside, 1983). One study of environmental values suggests that health food users probably fall within the overall group of people ascribing to a more natural lifestyle. People favoring such a lifestyle responded positively to such items as these (Kameron, 1975, p. 89): "You can't replace the natural goodness in artificial coloring, flavoring, and vitamins"; "It's better not to eat too much artificial processed food; stick to the basics," and "The best hope for increasing your life span is to learn to live in a natural, less artificial way." Adherents to this natural lifestyle tended to be younger people, concerned with environmental issues yet unlikely to act on this concern. Instead, they were more likely to seek escape in the outdoor environment (Kameron, 1975).

The extent and philosophy of such a group are further elaborated in the concept of "voluntary simplicity" (Gregg, 1977; Elgin & Mitchell, 1977; Elgin,

[2]This case study predated the fascinating case of Alar, which created widespread consumer antipathy toward apples tainted with a chemical ripening agent.

1981). A marketing report by the Stanford Research Institute in the 1970s found a growing number of Americans concerned with simplifying their lives. Practitioners of voluntary simplicity see their attitudes toward the foods they consume as part of their personal growth. Accordingly, they "tend to shift their diet away from highly processed foods, meat, and sugar and toward foods that are more natural, healthy, simple, and appropriate for sustaining the inhabitants of a small planet" (Elgin, 1981, p. 347).

This group is also very concerned with holistic health and preventative medicine (Elgin, 1981). Indeed, health is generally seen as central to the well-being of Americans. However, the norm is to not recognize the importance of health until illness calls attention to it (Campbell, 1981). Users of health food can thus be seen as varying from the norm. They are concerned about health in an active, ongoing, and preventative way. Rather than relying upon the commonly accepted "disease model" of health (e.g. "Conquer cancer through research"), they ascribe to an environmental model of health (e.g. risks are common in the environment; only persistent vigilance in how one lives will minimize danger). For health food users as a group, therefore, health is not merely consumed, it is actively managed. Nutrition is thus seen as a key ingredient to overall health. "Inner environment" is of great concern, as may be the outer environment. Careful control of how the latter affects the former is exercised. Health is a key "end value" and control of nutrition is viewed as an important means to this end. Thus, unlike the general public (see Kassarijan, 1982), health food users appear to use nutritional information in making purchase decisions.

It is also likely that different consumers vary in their personal risk standards. Thus, we see health food users rejecting supermarket food consumed by the majority because it is either insufficiently pure in ingredients and/or because it is felt to be unhealthy. That this group of consumers feels at risk from minute quantities of food preservatives, artificial coloring, and other chemical additives in food is indicative of their risk standards. Furthermore, foods such as white rice, white flour, and white bread are rejected for being neither pure nor comparatively healthful. Beyond avoiding impurities, the health food preventative-nutrition strategy seeks the maximum benefit from the natural qualities of food.

Dickson (1981) posits a difference between technocratic and democratic risk assessment which is useful in understanding the variance between how the health department viewed the PCB contamination as opposed to how the health food consumers may have viewed it. The technocratic approach makes a best estimate of risk as a basis for regulations which minimize economic cost. It seeks to avoid overestimating risks in a way that would damage the private sector. In contrast, the democratic approach uses the victim's perspective to judge the probability of technological hazard. Caution involves placing safety before profit by not demanding conclusive proof of a hazard before preventative action is taken. Those most likely to bear consequences are given the right to weigh the risks and select a course of action (Dickson, 1981; see also Edelstein, 1983).

The point here is that people in general are likely to weigh risks differently then are regulators. But health food users in particular are likely to maximize this difference such that they allow a minimum amount of threat. Rather than requiring conclusive proof that something is dangerous, the suggested possibility leads to cautious avoidance of the source of danger. While normative and technologically-based practices may be rejected (i.e. that milk be homogenized), different and possibly more stringent expectations for purity and healthfulness are applied.

The nature of the vigilance used by such consumers is also of interest. Thus, they are likely to read about and discuss purity and health issues related to food. A perusal of *Prevention* magazine demonstrates an active search for information on nutritional strategies for achieving health. It is likely that social networks also play a key role in providing information in the Boulder area. The health food store may be more than just a place to buy things. Often it is a place to learn about new products and health strategies. It may be a key node in an overall communication network.

Such motivated consumers may also have a greater ability to follow through on their concerns than do normal consumers. They may have background knowledge about possible threats. They demonstrate the financial and logistical ability to act in accordance with their risk perceptions, purchasing often more expensive goods from stores that may not be as convenient as conventional food stores. Importantly, their values, beliefs, attitudes, and intentions appear to match their actual behavior. As suggested by Elgin (1981), such behavior possibly invokes a means of coping in a personal manner with environmental pollution. If toxics are truly ubiquitous in our environment, then only the most vigilant practice of selective avoidance might protect one from this threat.

It is interesting to note that not all natural foods are free of toxins. Beyond general poisons such as botulism and aflatoxin, raw milk is viewed by some as unsafe because of its potential for carrying a dangerous form of salmonella (see "New Evidence", 1984). Yet, such "natural" hazards do not appear to be as much feared as "human-caused" or "technological" hazards (for a more general discussion of these distinctions between hazards, see Edelstein, 1982; Baum, Fleming, & Singer, 1983).

It might be argued that this pattern of food usage is based upon "chemophobia," "cancerphobia," and other types of stereotype and prejudice. Thus, the rejection of Homestead milk despite its being legally "safe" may be just an extension of the rejection by health food users of a whole range of legally safe products that they view as tainted. Just as this milk product was part of one stereotyped category initially (pure products), it now belongs in another (contaminated products). The dangers of using Homestead milk may not have warranted its disuse. However, once "labeled" as contaminated it was lumped into a category of foods to be avoided (see Allport, 1954 and Ichheiser, 1970 for discussion pertinent to such categorizations).

The fact that some 60% of Cito's business involved health food stores suggests that he was dealing with a highly sensitized set of consumers who acted on their concerns about purity in milk initially by using his product and, after its attributes changed, by not using his milk. Other contextual factors may be important here as well. For example, the scale of the operation was regional. The persistence of the stigma may thus be greater than for a national product. The particular susceptibility of Homestead Dairy to stigma is discussed later.

VISIBILITY

As a broadest statement, one can conclude that PCBs in the environment is stigmatizing under the conditions that its presence is made visible to people who know or learn about its attributes. Visibility thus is a prerequisite for stigma to occur. This inference assumes that the term "PCB" has meaning in its own right, and that this meaning will be affected by the amount and type of publicity.

It would appear that the media gave this case sufficient coverage to allow for most consumers to become familiar with the contamination (see reference list). For the purposes of this analysis, we confine ourselves to this media "visibility." Thus, we will examine available samples of television and print media about the Cito case in order to approximate what consumers heard and knew about the case. Additionally, it would appear that Boulder has a local health culture which may provide for informal communication channels by which people learned and formed opinions about the incident.

While visibility per se is important, one must further examine the nature of the "message" to be perceived. The subject of the message must be considered, including the two "raw materials" of the situation discussed above. Thus, the content message "PCB pollution" is likely to have a certain meaning for readers and listeners who know the term separate from how it is presented. And, presumably, "polluted milk" might have a meaning different than that accorded to other contaminated products. The message's content additionally goes beyond the product and the contaminant, containing information about the cause and victims of the pollution as well as the response by various parties.

The presentation of the message, involving such factors as the headline, pictures, and the use of emotionally charged words, further influences the meaning accorded to the visible message. For example, research on risk perception demonstrates that subtle nuances of presentation can directly affect estimates of threat (see Slovic, Fischhoff, & Lichtenstein, 1982). Other factors influence the believability of the warning. Williams (1964) notes that perception of warnings is affected by such factors as ambiguity, generality, and incompleteness of message as well as the perceived credibility of the source (see also Edelstein, 1982).

An analysis of eight articles and one television report [AP, 1981a; Noriyuki, n.d.-b; Ross, n.d.; *Times-Call*, n.d.; UPI, n.d.] from the time of the first publicity of the event suggests a number of insights (see Table 4.1).

Headlines contain two elements, the name of the contaminant and the product affected. Thus, any reader familiar with PCBs would quickly link them as a contaminant with the food, milk. Very early in the story some description of PCBs is given. This generally involves some linkage of PCB exposure to cancer. Of the nine sources, three mention cancer in the first paragraph, two in the second paragraph, two in the third paragraph, and one each in the fourth and fifth paragraphs.

That this can reasonably be considered as a threat warning is evidenced by the meaning of cancer. Expanding upon Sontag's analysis of cancer as serving metaphorically to transfer horror to other things, Ablon (1981, p. 7) notes that: "Cancer carries the mystique of death. It is as Sontag notes, 'The disease that doesn't knock before it enters...a ruthless, secret invasion.'" Fear of cancer has been identified in interviews with toxic waste victims, as have a range of ad-

Table 4.1 Summary of Nine Media Reports from about the Time that the Cito Incident Became Public

Item 1: AP, "PCB discovered in Longmont milk." *Colorado Gazette*, April 29, 1981.

Summary by paragraph:

1. Health investigators find "high levels of the chemical PCB in the milk" at Homestead Dairy. Sixty five cows ordered removed from production.
2. State health department says level of PCBs does "not exceed safe levels."
3. "No danger to consumers" according to health official.
4. Cito stopped milking and even destroyed most contaminated cows.
5. PCBs "found to cause cancer in laboratory animals." Uses of PCBs are listed "before being banned as a health hazard."
6. Cito's cows ate contaminated feed.
7. Question raised by official as to whether feed accounts for the level of PCBs found.
8. The chemical was detected in a February test.

Item 2: Michael Ross, "PCBs Discovered in Feed for Dairy." *Camera*, undated.

Summary by paragraph:

1. "Polychlorinated biphenyls, suspected cancer-causing agents" were discovered at Homestead dairy.
2. First found in February.
3. "99% of the herd could be contaminated."
4. Health official says PCBs were found during routine testing.
5. Officials will follow up with further tests.
6. Cito will destroy or not milk the worst cows.
7. The cows are at half of the legal safe limit.

8, 9,10. The source of the PCBs was feed.

Item 3: Picture caption, *Daily Times-Call*, undated.

Caption: "Chemical PCB, a proven carcinogen in laboratory animals, was detected in milk." A number of cows may have to be destroyed.

Note: Picture of milk bottles.

Item 4: AP, "PCB-contaminated milk closes Longmont dairy." *Greeley Tribune*, May 2, 1981.

Summary by paragraph:

1. "Low-levels of PCB in milk" forced dairy to close.
2. Drop from 800 gallons/day previously to 5 gallons.
3. Cito says "it goes fast."
4. Health test in January detected PCB "a suspected carcinogen"
5. Contamination is spreading.
6. Cito says health food customers "panicked."
7. PCBs traced to feed.
8. Tanaka cleaned up the feed plant, but Cito is still left with contaminated cows.
9. Cito will try for a loan to buy new cows.

Item 5: UPI, "PCB reported found in milk marketed by Boulder dairy." Source and date unknown.

Summary by paragraph:

1. Quantities of PCBs were found in a dairy's milk, "although not in excess of government standards."
2. PCBs, "a cancer-causing compound," appeared in testing.

(table continues)

Table 4.1 *(continued)*

3. A health officer reported that levels did not exceed "acceptable permissible limits."
4. The dairy is identified.
5. PCBs in the milk is indicated by PCB levels in the cows' body fat. Some cows have been removed.
6. Ironically, the milk had been sold to health food stores.

7-9. The source of the PCBs was silage.

10-11. The health officer reports that the agency is monitoring so that "at no time has milk gone for public sales that exceeded (state) standards."

12. Standards take into account the accumulation of PCBs in body over time.
13. PCBs are everywhere in the environment. Homestead milk is over the background level.
14. Consumer official says that there is little chance that milk over the limit was sold.
15. The milk was below the standard.
16. Testing for PCBs occurs every six months.

Item 6: Duane Noriyuki, "PCBs found in Homestead Dairy milk." *Times-Call*, undated.

Summary by paragraph:

1. Chemical PCBs, "proved to cause cancer in laboratory animals," detected at dairy.
2. While a number of cows were over the standard, a health official says the milk was safe because of dilution.

3-4. The official says that cows over the standard were removed. The state level is 1.5 ppm while Cito's herd ranged up to 9.6 ppm.

5. Cito discusses his problems.
6. The official says that cows removed from milking will be tested to see if they can be slaughtered.

7-12. Discussion of PCBs at Tanaka farms.

13. PCB level is said to be high for Colorado, but not nationally.
14. Produce grown at the Tonaka farm is okay.
15. Homestead had bacteria problems the prior year. Note: Pictures of Cito's cows eating and of Cito. Caption notes that 13 head produced milk over state levels of PCBs.

Item 7: Duane Noriyuki, "PCB-plagued Homestead dairy to close." *Times-Call*, undated.

Summary by paragraph:

1. Dairy to close after "traces of the chemical PCB" found in the milk. EPA withholding information on the source of the PCBs.
2. Health officials say the amount of PCBs in the milk is not a health hazard.
3. PCBs "have been proved to cause cancer in laboratory animals." Uses of PCBs are mentioned. Homestead milk never exceeded the limits, but individual cows did.
4. Cows above the limit were removed and are being tested for possible butchering.
5. Two cows have been killed of those tested.
6. Cito says people stopped buying his milk when "news broke" about PCBs.
7. Tanaka has been cleaned up. Their produce was not affected.

(table continues)

Table 4.1 *(continued)*

8-11. Discussion of Tanaka.
12. This is the first time that PCBs were traced in Colorado milk.
13-15. A health official discusses the delay of the announcement about PCBs at Homestead, noting there is "no health hazard."

Note: Pictures of dairy farming at Homestead before incident.

Item 8: Linda Cornett, "PCB-Tainted Fodder Spoils Dairy Farm's Future." *Denver Post*, May 1, 1981.

Summary by paragraph:

1-2. Cito's misfortune.
3. Test found "the presence of Polychlorinated biphenyl (PCB) a suspected carcinogen."
4. Cito had never heard of it before.
5. Discussion of source.
6. PCB was used as a lubricant until it was banned; "researchers have established that PCB causes cancer in animals."
7. Discussion of the Tanaka cleanup.
8. Homestead Dairy's background.
12. Cito hopes to milk the PCBs out of his herd.
13. He is attached to the cows like they were pets.
14. He needs money; he is not getting help from the government.

Note: Picture of Cito and of cows. Says he was forced to close the dairy "in wake of bad publicity over milk contamination by the suspected carcinogen, PCB."

Item 9: Newscenter 4 script, KOA-TV Channel 4, Denver, April 28, 1981

Summary by paragraph:

1. Dairy cattle "contaminated by the chemical PCB." Health agencies have been investigating "potentially contaminated milk for months."
2. 65 cows have PCBs in their milk. "PCB is a potentially dangerous compound known to cause cancer in laboratory animals."20% of the herd have levels that demand that they be taken out of production.
3. Corn silage was contaminated with the PCB.
4. "Milk containing PCB has been sold to the public, but health officials emphasize there's no reason to believe that any of that market milk has contained dangerous levels of PCB, levels above those limits set by the federal government."

5-6. Discussion of whether meat from the cows can be used.

Note: Other television coverage appears to follow a similar outline over the course of this and the following day.

ditional health fears. Children are the focus of particular concern. Much as cancer, other feared illnesses may be manifested with a long latency period. This contributes to an altered view of the future, focused upon sickness and death (Edelstein, 1982).

A variety of qualifiers are also provided in the analyzed media (see Table 4.1) which might mitigate against the perception of threat. There is a consistent reference to the fact that PCB levels in milk did not exceed government regulations, often provided in the context of a quote from a health official. In two cases, this is found in the first paragraph, in three others in the second paragraph, and in one in the fourth paragraph. One story does not provide the qualifier until the seventh paragraph, one omits it altogether and several return to the question of safety for additional discussion later in the article. In four cases this qualifier was presented before the description of PCBs as carcinogenic. Additional qualifiers also are included. In describing PCBs as carcinogenic, an adjective such as "suspected" is commonly used. Several of the references to cancer come in the context of discussing the effects of PCBs on laboratory animals, possibly diminishing implications for human cancer for some readers.

In nearly all the stories, readers or listeners were also made aware of the health officers' attempts to remove the most contaminated cows from milking. They were variously informed that the contamination was diluted enough to pose no danger, that accumulation in humans is taken into account in the standards and that Cito hoped to milk the PCBs out of his herd. At the same time, they were informed that some cows could not even be slaughtered for meat because of excess PCB levels and had to be killed and buried. Several stories noted that the health investigation had been going on for some time beginning with a routine test and that government had not sought to publicize it because no danger was involved. This latter point may have suggested an opposite conclusion to some, mainly that government had attempted to cover up the incident. Another consistent element in the stories is the discussion of the source of the PCBs. This ties into a further point, the portrayal of Dave Cito as an innocent victim. We return to this point in a later discussion of victimization in stigma.

PERCEIVING THREAT

Once made public, the contamination appears to have dominated the way that people thought about Homestead milk, Cito and his business. Rather than exhibiting the "shades of gray" analysis suggested by the health regulators, health food consumers appeared to use an "all or none" evaluation. In an attempt to understand this judgment, we will review the process by which threat is perceived and acted upon.

The Royal Society (1983) notes two approaches for studying the assessment of risk. One involves the direct measurement of perceptions, an approach not possible here. However, risk may also be inferred from people's behavior. Therefore, in reconstructing the following process of threat perception leading to action, we begin with a concrete behavior—the cessation of purchases for Homestead milk that followed the announcement of the problem.

For purposes of discussion, we assume that the decision to stop using Homestead milk was made by the consumer. In reality, health food stores may have stopped stocking it, removing consumer choice. In this case, it is probably safe to assume that store owners reflected their customers' concerns and expectations. Thus, analyzing the perception of threat and the apparent ignoring of qualifiers to the threat is instructive regardless of the actual role of the consumer. In order to accomplish this analysis, we will distinguish between the actual perception of threat based upon the media accounts, the judgment that the threat was serious and the decision to act upon the threat in order to remove it.

Perception of Threat

Threat perception is more than a cognitive exercise involving logical reasoning. Emotion plays a major role in how we respond to the "hot cognitions" which come into play as we think about "vital, affect-laden issues" (Janis & Mann, 1977, p. 457). Affect is central to Janis' theory of threat perception (see Beck & Frankel, 1981). This theory asserts that the motivation to protect oneself from health threats depends upon how much fear is generated. Too much fear may be more than the person can handle, leading them to avoid coping with the threat. Similarly, threat that is too low would not motivate protective actions. Fear levels in between these points, however, would be expected to generate actions to deal with the threat.

An interesting and relevant study of the PBB contamination in Michigan appears to offer support for the importance of fear in responding to toxic exposure (Hatcher, 1992). Thus, a questionnaire given to a sample of breast-feeding mothers who had their breast milk tested for PBB was examined in light of the amount of the chemical they had been told they evidenced. Hatcher found that those scoring higher in their level of contamination were more likely to deny the threat and its possible consequences (exhibiting what Janis called "defensive avoidance"). In contrast, those with lower levels of exposure were more likely to attempt to master the problem.

Beck and Frankel (1981) argue that, while there has not been experimental support overall for a fear drive, high threat is more persuasive than low threat. However, reducing fear is not seen as the primary reason that people undertake protective actions after receiving health threat warnings. Instead, a competing theory developed by Leventhal (Beck & Frankel, 1981) asserts that protective actions result from the desire to control the source of danger rather than the resulting fear. Rogers (1975, see also Beck & Frankel, 1981) developed a "protection motivation theory" which suggests that three factors are considered in deciding whether to heed or ignore warnings (see also Janis & Mann, 1977; also Radelfinger's 1965 summary of Rosenstock's health model). These factors can be illustrated in light of the Cito case:

1. How serious is the threat? The threat of cancer is probably viewed as extremely serious by Homestead customers.
2. What is the probability of this outcome? Health officials maintained that the probability is so low as to not violate their definition of safety. Consumers may have perceived themselves to be more susceptible.
3. How effective will protective action be? Discontinuance of using Homestead milk would effectively eliminate any further expected exposure. The actions of the health officers and Cito may not have been as convincingly protective.

Weinberger, Greene, Mamlin, and Jerin (1981) studied the first two of these factors in comparing heavy, moderate and ex-smokers. They found that heavy smokers willing to reduce their habit to a moderate level tended to focus upon the awareness of potential harm while smokers willing to quit had the perception that they were personally susceptible to ill effects. For them, the harm was not merely abstract, but it was a personal threat.

In applying these findings, we must keep in mind that, unlike smoking cigarettes, drinking Homestead milk is not addicting. Thus, giving up milk use may not have required a sense of personal susceptibility to the dangers. However, for health food users who may have totally quit their consumption of perceived unhealthy and impure foods, this may have been more than an abstract danger. Just as they avoid other foods that they perceive as contributing to a health risk, they avoided Homestead milk.

Fishbein and colleagues (see Beck & Frankel, 1981) argue that acceptance or rejection of a health threat message depends on the message's impact on primary beliefs about the consequences of an event. We have already advanced the argument that health food users believe that consumption of impurities can be unhealthy. They already believed or were prepared to believe that PCBs cause cancer. Thus, it is neither surprising that the threat message was taken as a warning even though it was not intended as such by health officials nor that reassurances by these officials were ignored.

Beck and Frankel (1981) further extend the protection motivation model to take account of the effectiveness of protection. Thus, fear can he controlled if a person perceives that they have available an effective response to the threat ("response efficacy") and that they are able to successfully carry out this response ("personal efficacy"). Control of fear is thus a function of the control of danger (see also Bandura, 1983; Peterson & Seligman, 1983).

A final factor in responding to perceived threat, thought by Fishbein and others to be particularly predictive of protective behavior (see Beck & Frankel, 1981), is intention to seek protection. For example, intention was found to be a good predictor of whether people sought tuberculosis testing (Wurtele, Roberts, & Leeper, 1982). Intention is an important factor in looking at health food user behavior as well. Thus, health food users implicitly state their intentions in their practice of consumption (i.e. choosing the safest and purest products). That they would subsequently seek protective action based upon a threat is not at all surprising—it is in full keeping with their intention to selectively choose products. In fact, given their public and private commitments to this selectivity, it might have been particularly "dissonant" for these consumers to continue using a product which violated their standards. Rather than rationalizing the continued use, it was easier to be consistent to their standards by avoiding the product (see, for example, Heberlein & Black, 1981).

Judgment of Threat

Having perceived a threat, the consumers of Homestead milk had to next judge the severity of the risk. Given their vigilant concern with health food, they may have done this in a concrete, factual and strategic way. Chaiken (as cited by Wood & Eagley, 1981) terms such an approach a "systematic information-processing strategy," in which the message's arguments are understood and evaluated. However, lay people generally lack statistical evidence regarding risks and must make judgments based upon what they remember hearing or observing (Slovic et al., 1982). In doing this, they may employ a"heuristic information-processing strategy" in which cues outside of the message are interpreted through simple decision rules (Chaiken as cited by Wood & Eagley, 1981). Such "heuristics" serve to reduce the complexity of the issue to a point where it can he managed (Slovic et al., 1982, see Kahneman, Slovic, & Tversky, 1982). It may be instructive to examine some of the heuristic biases which might have affected the consumer's judgment of threat.

We estimate the likelihood of a threat in light of what most readily comes to mind (e.g. is most "available" or imaginable). In the same way that the movie *China Syndrome* may have prepared people to increase their risk estimates of the subsequent Three Mile Island incident, consumer's judgments about risk from Homestead milk may have been affected by various influences. Many health food users are attentive in an ongoing way to reports of risks from food. It is understandable then that they interpret this incident within the framework suggested by their previous knowledge of contamination. Judging the severity of the risk in light of previous incidents which were attention getting (dramatic and salient) or frequent, the consumers may have applied to this incident the expectations suggested by the other incidents. From a heuristic viewpoint, error may have occurred due to unfair generalization. Also the level of risk from events receiving media attention or attention within a "health culture" may have distorted its importance beyond the statistical estimates of risk (Tversky & Kahneman, 1982; Taylor, 1982; Johnson & Tversky, 1983). Discussion of low probability hazards may make them more imaginable and memorable; the perceived risk may increase despite what other evidence indicates (Slovic et al., 1982). Even knowledge of one past event (such as the PBB contamination or the Love Canal incident) could serve as the basis for a judgment (Read, 1983).

Taylor (1982) distinguishes between several means by which an availability bias might occur. We may recall information which is unrepresentative of the situation at hand. Or we may adopt stereotypes which distort our evaluation of the situation. Finally, some information might be highly salient, such as colorful, dynamic or distinctive information which disproportionately engages our attention and affects our judgment. This is invited by media coverage which reflects a bias toward violent catastrophic events rather than more common frequent ones (Slovic et al., 1982). Personal experience with the problem also can contribute to an availability bias, as when the consumer or someone they know or know of experiences a certain outcome (Christensen-Szalanski, Beck, Christensen-Szalanski, & Koepsell, 1983). Given a strong local health culture, it is likely that Homestead consumers had access to dramatic examples which suggest unhealthy outcomes from contaminated food. Affect also influences risk judgments. Thus, a negative mood induced by one story might carry over to an entirely different topic and enhance perceptions of risk (Johnson & Tversky, 1983). Certainly, all of these heuristics may have contributed to perceptions of risk from Homestead milk.

We have already discussed the question of why the qualifiers were ignored. "Heuristics" suggests several additional points. First, concrete emotionally interesting information has a greater power to generate inferences than does abstract information (Nisbett, Borgida, Crandall, & Reed, 1982; Anderson, 1983). Thus, the information that PCBs cause cancer is comparatively concrete as opposed to the numerically based discussion of safe risk levels; one can more easily develop associations based upon this information; it has more meaning. Finally, in this regard, a difference exists between the "pseudocertainty" of avoiding PCBs by avoiding known tainted milk and the "probabilistic protection" offered by health officials regarding the safety of continuing to drink it. The prospect of harm in this comparison favors the seemingly more solid alternative (Slovic et al., 1982, p. 480).

Furthermore, we can assume that the information that PCBs cause cancer confirms the past beliefs of health food users whereas the information that milk containing PCBs is safe does not fit with such beliefs. People inherently select confirmatory evidence and are very critical of disconfirmatory information (Ross & Anderson, 1982). The prior belief "perseveres" because evidence that would refute it is ignored (Anderson, 1983). In other words, once we see an uncertain situation in a particular way, it becomes harder to reinterpret it in competing ways. This is in part due to the "anchoring" of risk estimates to one's original values (Tversky & Kahneman, 1982). Also, many of the consumers in question were people who had committed themselves to "health food." Citing Festinger's discussion of dissonance in decision making, Janis and Mann (1977, p.15) note, "Once a person has committed himself, there is 'less emphasis on objectivity and there is more partiality and bias in the way in which the person views and evaluates the alternatives'."

Also affecting the influence of the threat qualifiers were consumers' perceptions of the source. Thus, the competence, honesty, clarity and the appropriateness of action of health officials would have been noted (see Edelstein, 1983). When a communication appears to have been biased, this bias is weighed carefully in considering the message (Wood & Eagely, 1981).

Were consumers rational in rejecting Homestead milk when health officials considered it to be safe? Certainly, the health officers seemed to view "consumer attitude" as irrational ["True-Life," n.d.]. Yet, conflict between "technical" and "acceptable" risk is general to all technological controversy (Nelkin cited by Regens, Dietz, & Rycroft, 1983). Doubts about the rationality of health officials have also been raised (see Levine, 1982; Edelstein, 1983). And besides, whether the consumers used strategic or heuristic thinking, either path had the same result: the loss of the market for Homestead milk.

A fairly simple explanation of the consumers' decision to act involves the consistency of their thinking (see Heberlein & Black, 1981). Thus, spending more for milk that fits one's values results in a partial conflict for the consumer; they are doing the "right thing" but at a high cost. Upon learning that Homestead milk isn't so pure after all, there is no justification for purchasing it.

A more complex framework for decision making has been advanced by Janis and Mann (1977). The model has been applied to situations such as predicting response to threats from Mt. St. Helens (Perry, Lindell, & Greene, 1982). In the first stage of the model, some challenge or opportunity which tests a current decision is appraised. In our case, the information that Homestead milk contained PCBs represents such a challenge, particularly to consumers whose prior decision making had led them to seek out particularly pure and healthful food. An appraisal is made of the risk. If the risks are not seen as demanding any change in behavior, then the person continues their prior practices without any psychological sense of conflict. This would have been the case if the qualifiers to the threat message had been accepted and consumers had continued to use the milk believing that it was safe. However, we can assume that Homestead customers did perceive the threat to be serious enough to demand at least a consideration of changing their consumption pattern.

A second stage of decision making involves surveying various alternatives. Here a number of possibilities exist. First, no better alternative may be found, causing people to continue to use the milk out of resignation and what Janis and Mann term "defensive avoidance." Second, no better alternative may be found but perception of threat may increase because of this failure, particularly if same deadline exists. This situation may result in panic on the part of the consumers. Panic offers one possible explanation of the Homestead case. Cito himself indicated this view [AP, 1981b].That a dairy is particularly vulnerable to this type of decision making is evident from the importance to a dairy to keep selling its milk (demanded by both the need for cash flow and the ever present milk flow). Health food store owners not wishing to have Homestead milk on their shelves might understandably have canceled their orders leaving Cito with no market on extremely short notice. Finally, when an adequate alternative exists that entails no major disadvantages, "unconflicted change" can result. Particularly if another source for unhomogenized milk was available, there was no reason not to change brands.

Janis and Mann (1977) argue that none of the above approaches to decision making is likely to lead to a full comparison of alternatives. All may lead to a final outcome which is open to challenge. Consistent with the discussion of health food users, another strategy of decision raking, vigilance, may have been in use. Vigilant decision makers take the time to survey and evaluate, searching for the best alternatives. It is presumed that consumers of health foods are generally vigilant, seeking the mast pure and healthful products. Either through their own research or by relying on the vigilance of health food store owners, these consumers tend to make a very selective decision regarding nutrition.

The result of a strategy of continuing vigilance is the ongoing retesting of decisions. In a third stage of decision making, alternatives are weighed, such that one looks to see if other alternatives are better and also whether requirements can be relaxed. Through this process, the best alternative is identified. Janis and Mann (1977) make reference to Simon's distinction between "optimizing" and "satisficing" strategies of decision making. Thus, while the average milk consumer presumably feels that regular store-bought milk is "good enough," users of Homestead milk (and health food users in general) seek to optimize their nutrition. Accordingly, selection of foods is given extensive thought as part of an overall strategy for maintaining health. Such optimizers would be less likely to relax their requirements, unless they had no choice. Through vigilant evaluation, they apparently rejected use of Homestead as unacceptable. Either another product was seen as being as good as Homestead had formerly been, allowing for continued optimization, or a less desirable alternative was selected which was still better than the new perception of Homestead. In either case, the decision may well have been part of an overall strategy of vigilance. In this sense, it cannot be seen as a defective decision. Rather, as noted earlier, more stringent risk standards embodied by health food users may have led to an evaluation of the risk which varied from that utilized by the health officials who pronounced the Homestead milk safe.

STIGMA AND ITS VICTIMS

The above analysis has shown that milk is a product about which we have certain expectations regarding purity. Homestead milk was marketed so as to further heighten those perceptions, and it was purchased largely by customers with expectations for quality and purity which exceeded the average. These customers' extreme vigilance regarding the purity of their food was evidenced by their behavior—their willingness to pay more for food in which they had particular confidence.

One of the concerns likely to be noted by those vigilant over the purity of their foods is the possibility of contamination by toxic substances. The chemical PCB has received adequate historic attention in the media to be so recognized. Additionally, even one unfamiliar with this particular compound might react negatively to the idea that an inappropriate "chemical" was found in the milk. This is particularly true for those concerned with purity. Those concerned as well with "health" would react to the association of PCBs with risk of contracting cancer. Parents would be expected to be particularly vigilant (see Edelstein, 1982). We have shown that media presentation of the incident contained both a questioning of purity and of healthfulness of the Homestead product. Thus, even one unfamiliar with PCBs per se would attain a sufficient familiarity from the media to plausibly raise concerns about both purity and health. We have further noted that those highly vigilant regarding issues of purity and health in foods might be likely to discount the qualifiers present in most of the media accounts. Thus, the assurances of no risk are overlooked in favor of attention to possible dangers.

We next examined the process by which this threat was likely to have been considered. We suggested that the customers were likely to have viewed the media coverage as suggesting a real threat that could not be ignored. Because a fairly straightforward alternative action was available to these customers, they chose this alternative behavior. This involved stopping the use of Homestead milk. Any costs involved in this decision were presumably small compared to the simplicity of this step in addressing the threat. While some concern regarding milk already consumed may not have been addressed through this action, there was little to be done about that. Future contamination, in contrast, could easily be controlled by avoiding the product.

As a means of capturing these points in one concept, we will return to our discussion of stigma. Earlier we defined stigma as a negative change in the way one perceives another because he or she notices that the other is "marked" (deviant, flawed, limited, spoiled or generally undesirable).

The Victims of Stigma

One key characteristic of stigma is that the mark tends to be seen as a fundamental characteristic of the victim, so that the victim's identity is "spoiled" (Jones et al., 1984; also Goffman, 1963). This would be true of all of our victims in this case—the milk, now seen as fundamentally spoiled; the dairy, no longer trusted to deliver pure milk; the cows, seen as having PCBs in their fat and thus undesired; and even Cito, viewed as a victim of circumstance, a man who has lost control of his fate and thus likely to be blamed by at least some observers for his downfall.

In using a social concept such as stigma to generalize to objects such as the farm, the cows, and the milk, it is useful to recognize two normal differences between the social and the object (Ichheiser, 1970). Objects are neither expressive (they lack feelings) nor do they act in a way that can be morally judged as right or wrong. Despite these differences, there is some utility to viewing these objects as stigmatized. At the least, their stigma feeds Cito's victimization.

Janoff-Bulman and Frieze (1983) present an overview of victimization due to particularly negative life events. They note of people exposed to such events that their coping resources are taxed both because of life threatening occurrences but also because these events are unusual. The events thus shatter the victim's basic assumptions about the world, giving way to new perceptions marked by threat, danger, insecurity, and self questioning. The authors specifically discuss three such assumptions often shattered by victimizing events: the belief in personal invulnerability, the belief in a meaningful, understandable, just, and controllable world, and the view that one has the power to guide one's life (see also Edelstein, 1982).

The above analysis meshes with the stigma literature in its analysis of blaming the victim. Thus, it is noted that victims are often blamed because of a view that "in a just world people generally get what they deserve and deserve what they get" (Lerner as paraphrased by Jones et al., 1984, p. 59). Janoff-Bulman and Frieze (1983) note that as a means of preserving their own sense of invulnerability, people see others as responsible for their own fate. Related discussions of blaming victims can be found in the fledgling literature on toxic exposure (see Levine, 1982 and Edelstein, 1982) and technological disaster (Baum et al., 1983). Levine and Edelstein discuss isolation and pressure upon two communities which experienced toxic waste incidents, Love Canal and the Legler section of Jackson, New Jersey, respectively. What about the victims of the Cito case?

CITO AS A VICTIM

The perception of Cito is likely to be complex. On one hand, he is a victim, consistently portrayed as innocent and helpless by the press. Several of the news articles on the incident show Cito's picture, often with some expression of confusion or defeat [Alvarez, 1983; Cornett, 1981; Noriyuki, n.d.-b]. His misfortune is described as being "through no fault of the dairy owner" [KOA-TV, 1981a]. We are told that Cito had never heard of PCBs "until we had it" [Cornett, 1981]. We are told of his family history and his personal success as an "outstanding young farmer and rancher" who bought out his father to build a dairy business so successful that he had hoped to be soon out of debt and to retire early [Cornett, 1981], but "now, through bad luck, Cito faces starting over" [Cornett, 1981].

But Cito is also responsible for allowing his customers to be exposed to, at the minimum, a violation of promises of purity, and, at the maximum, a physical danger. It is likely that at some level he was blamed for not being vigilant enough in protecting their interests. He also is subject to a resulting loss of trust, in that his failure to protect people in the past suggests a similar inability to protect them in the future. Thus, his plans to work the PCBs out of the herd through several years of steady milking [Cornett, 1981] might be greeted with some skepticism by customers evaluating whether to continue with him. How would they know when the herd had been purified? And what does this say of his milk in the interim?

It should be noted that Cito's sister, apparently the co-owner of the farm [Cornett, 1981] received little media attention. Symbolizing the accidental agent of harm, Cito is potentially subjected to the revulsion and rejection we feel toward carriers of contagious illness (see Jones et al., 1984). He faces the inherent emotional reactions to symbolic dangers posed by people that we see as deviants. Jones et al., cite Douglas' (1966) observation of how such deviants are viewed: "A polluting person is always in the wrong." (1984, p. 230). While the statement was intended to refer in a more general manner to those polluting a culture symbolically, the phrase is unfortunately appropriate in our more specific context.

THE DAIRY AS A VICTIM

As the site of pollution, the producer of the tainted milk and the home of the tainted cows, Homestead Dairy was sufficiently associated with the contamination to become the subject of stigma. Given the definition of corporations as individuals in our legal system as well as the close identification of Homestead as the historic family business of the Cito's, it is reasonable to speak of stigma applied to this business. Media accounts make frequent reference to the closing of the dairy. In fact, some references to loss of business and the dairy's closing [e.g., Alvarez, 1983; Cornett, 1981; KOA-TV, 1981a, 1981b; Noriyuki, n.d.-a;] might have played the role of creating a "self-fulfilling prophesy" in that reportings of such loss might have influenced additional customers to withdraw their business.

MILK AS A VICTIM

The events surrounding Homestead dairy did not taint milk in general, although it is possible that some observers became concerned with the commonly acceptable levels of PCBs in milk because of this incident. The stigma was basically confined to one subclass of milk, that identified as Homestead "old fashioned unhomogenized milk" [Alvarez, 1983]. There was a boundary around the extent of the contamination, and thus stigma. This milk was identified as "different" and "discredited" in the press (despite an attempt to portray the milk as no worse than other milk—a damning by faint praise). A clear behavioral attempt to avoid the milk is seen on the part of customers. In these ways, as will be elaborated shortly, the milk's "identity" was subject to stigma.

THE COWS AS VICTIMS

If one draws an analogy between the contamination and a plague, the cows are viewed as infested. The "infection/contamination" is carried in their milk. It may appear to be strange to include the cows in a listing of stigma victims, but in all ways they were perhaps the most victimized of all. A review of media coverage of the incident reveals that the cows received extensive attention. Typical of the television scripts was the following, apparently voiced-over a view of Cito's herd:

> These cows don't know it, but they are still the focal points of an intense investigation. Because the of PCB contamination in their milk, much of the herd at the Homestead Dairy has already been destroyed. [KBTV, 1981]

One consequence of this stigma is seen in a story from the *Times-Call* reporting on Cito's attempt to sell his remaining herd. The article, entitled "Homestead Dairy:`Stigma' forces auction, but no one is buying"[Alvarez, 1983], describes Cito's attempt to cease dairy operations. Because bidding on his herd was too slow, he decided to go back into bulk milk production under a different dairy name.

ADDITIONAL VICTIMS

It is noted later that the health food stores that carried Homestead milk were in danger of acquiring what Goffman terms "a courtesy stigma" (Goffman, 1963, p.30). Thus, their association with the Homestead product left them in danger of "sharing the discredit" brought to the product. Presumably the health food dealers' fast action to remove the milk from their stores helped to break this association and preserve their own reputations for protecting their customers' health. Otherwise, they were in a roughly analogous situation to Cito's at a higher level of the food chain: they had purchased tainted food which was fed to others who were thus at risk of being contaminated. It would be interesting to know if they were at all blamed by their customers for a lack of vigilance because they carried Homestead milk.

The Stigmatizing Process

Jones et al. (1984) suggest six criteria for examining stigma. We will attempt to use these criteria to examine the basic question, "Why did people stop using Homestead milk?"

DISRUPTIVENESS

The realization that one may have been put in peril by an accidental exposure to something toxic is inherently disruptive. Thus, contamination is a challenge to the routine assumptions we make about life, leading to a marginal situation where things are no longer routine (see Jones et al., 1984 for a parallel discussion of other stigmatizing events). The disruption here, then, is more than just the unavailability of a formerly used product, it is the sense that an event on a rural farm has somehow reached out and touched us. This point is amplified later.

CONCEALABILITY

In the Cito case, the presence of PCBs in the milk was not noticeable to customers. It was only detectable to chemists. Realizing, apparently, that news of the PCB contamination would lead to stigma, the health officials delayed announcement of the contamination [Noriyuki, n.d.-a]. Win Franklin, a local health official, is quoted in one article discussing the need to disclose details of Cito's case to confirm a news leak:

> But it raises an interesting philosophical question about whether we have the right to release information that could be damaging... if we made a press release every time we found a rodent hair in a can of tomato sauce, we'd fill a magazine. [Crary, n.d.]

Once it became public through the media, contamination was no longer concealable. For those vigilant against personal pollution, we might expect that the PCB contamination became the sole salient attribute of the milk. This process of "obtruding" upon attention (Goffman, 1963, p. 5) would make other attributes of the milk comparatively irrelevant. In other words, the characteristics of Homestead milk that caused people to be pleased with it previously were now overshadowed by this one new factor. Effectively, the milk now tasted differently, even if in fact there was no measurable change in its actual taste.

The revealing of a formerly concealed attribute further complicates this stigma. Jones et al. (1984) note that the discovery of abnormal conditions where appearances are normal may contribute to a perception of deceit and danger. Thus, it might be guessed that the fact that consumers drank Homestead milk without being able to detect any change in its quality only to find that it had been contaminated adds another dimension to their reaction. They had been deceived in believing in the milk's purity and healthfulness.

Interestingly, when the incident was made public, health officials claimed that the milk sold by Homestead was safe. One official even noted that PCBs occur "everywhere in the environment" [UPI, n.d.]. Discussions of safe limits implied that low levels are regularly found in milk. The difference was that Homestead milk could no longer conceal its being tainted. While, in fact, it may have been clinically equivalent to the milk of other dairies, it was "spoiled." As health official Franklin put it [Crary, n.d.], "The problem is consumer attitude. He (Cito) is well within the established limits (for PCBs), but once word leaked out, he was out of business over night."

The vulnerability of Homestead milk to this type of stigma pressure stems in large part from its absence of anonymity. Bulk milk has no identity. The dairy which produces it likewise has no identity. Thus, some new cows and a cleanup of the pollution would adequately have mitigated the physical problem for a bulk dairy with no further consequences. But Homestead milk had an identity. It was sold in labeled bottles. Consequently, it received a new defacto "label"—POISONED.

AESTHETIC QUALITIES

As the above suggests, the PCB contamination is visible both as a physical "deformity" of the milk and as a "character blemish" (see Goffman, 1963, p. 43). Thus, the milk could be "safe" from the standpoint of a health official measuring parts per trillion of PCBs and at the same time continue to be spoiled to consumers responding to the character blemish. At the same time, an invisible deformity is perceived, possibly accompanied by the same kind of "gut feeling" reaction felt when confronting other types of deformity (see Jones et al., 1984).

Could Homestead overcome such a stigma? The complexity of this question is suggested by the invisibility of the contamination. Just as one could not see when the pollution began, they cannot see that it has ended. The result of the stigma is that the milk cannot be made to look normal again. Because one's senses are useless as a guide, the perception of quality is truly subjective. The pollution may be "out of sight," but it is not "out of mind."

Furthermore, Homestead was not "normally" pure milk. The "cream-like milk" [Crary, n.d.] was purported to be purer and more unspoiled than most milk. It certainly was more expensive. To overcome its stigma, it not only had to attain the same quality standards met by all milk, it had to again be seen as different and better. But this claim was discredited by the contamination, effectively leaving Homestead open to the perception that its milk had been sold under false pretense, albeit unknowingly. The apparent loss of trust in the product resulting from this incident would be difficult to reverse.

ORIGIN

Also to be considered in overcoming the stigma would be the question of origin (Jones et al., 1984). Was the victim responsible for the mark? While Cito was consistently portrayed as an innocent, such that the incident was not his fault, he had not been able to prevent it, despite his higher claims of purity for his product. And, if Cito failed once to protect consumers from possible harm, why have confidence that the problem would be fixed and prevented from reoccurring in the future? The point here was that the events in question were hardly natural. Rather, they were human-caused, with all the potential for blame, anger, and concern about unfairness and inequity that accompanies such events (see Edelstein, 1982; Baum et al., 1983).

THE COURSE OF THE MARK

It would appear that consumers believed that the pollution was not merely temporary. Despite their being informed that the condition was transient and controllable, the stigma associated with the milk continued, possibly fed by the continuing publicity. Perhaps because the consumer can not themselves detect the presence of PCBs in milk, their beliefs about the course of the contamination would be key to their response (Jones et al., 1984). One might conjecture about these beliefs. For example, it would appear that the pollution was now a property of Cito, his farm, his cows, and his milk, rather than the situation. Thus, one might well believe that once one has sold polluted milk, they are "a polluter"; that once an incident such as this occurs on a farm, that farm is "polluted"; that once Homestead milk is labeled as contaminated milk, that this becomes a continuing property of the milk—it is "off."

Similarly, beliefs about the effectiveness of the remedial actions would bias perceptions about the course of the pollution. First of all, consumers using a more sensitive risk level than do regulators may have interpreted the regulators' statements about contamination below the legal standards as suggesting a condition of danger rather than safety. And, if the regulators and Cito had been unable to avoid such an incident, why should they be seen as able to remediate it or prevent its recurrence? Skepticism about government assurances that pollution has been adequately mitigated has been noted elsewhere (Levine, 1982; Edelstein, 1982). And, if people had followed the story of Love Canal, they might have been influenced by some of the questioning of remediation raised by citizens from that location (see Gibbs, 1982; Levine, 1982).

PERIL

Jones et al. (1984) cite a source as commenting that "the essence of stigma is fear" (p. 65). Thus, one would expect that the greater the fear generated due to a stigma, the greater the rejection that might occur. For customers expecting a pure product, the perception that Homestead could only provide unpredictable and erratic quality was presumably seen as a threat. As long as they continued to use this product, they were dependent upon Cito's unproved ability to keep his milk pure. The only way to remove this power from Cito was simply to avoid his product.

The peril in this case is apparently cancer as spread through the medium of Homestead milk. One night draw the analogy of a plague to capture the revulsion directed at the source of danger. Cito's cows had a new type of plague which was contagious after a fashion if one ingested PCBs from the cow's milk. Noting the hostility and social rejection that accompany plagues, Jones et al. (1984) note the existence of irrational contagion fears, such as the fear that cancer is contagious. They further note that the "threat of being contaminated...may exist in a more virulent fashion than the real peril posed would justify" (p. 70). Thus, as we have already seen, customers appeared to interpret media coverage as suggesting a greater threat than the highly qualified messages actually appear to convey.

In part this threat stems from possible actual exposure to the contaminated milk. After all, the media might depict mournful looking cows awaiting tests to determine whether they might be slaughtered for food or merely shot and dumped into a hole in the ground. But just as these innocent cows had eaten tainted silage, Cito's customers had been drinking contaminated milk. What were the risks for them? Media reports indicate that inquiries about safety came from such disparate consumers as mothers of nursing babies and users of the cow manure for gardening [KBTV, 1981].

Part of the perceived threat in stigma situations stems from the anxiety associated with the recognition of our own vulnerability (Jones et al., 1984). This may have acted in two related ways in the Cito case. On one hand, customers may have focused upon the victimization of Cito and his dairy. Thus, if a supplier of a pure product could inadvertently be confronted with contamination, so might they. Avoiding Cito and his product thus helped avoid reminders of this vulnerability. For health food retailers, a more pragmatic concern may relate to this fear of vulnerability. Because a dealer is trusted to look out for the purity interests of their clients, they may have feared not only selling Homestead milk, but also being in any way associated with it. They did not want to also appear vulnerable, coming to share Cito's stigma. This view is supported by a report of the health food owners' apparent anger toward Cito, who indicated, "They were ready to kill me" [Crary, n.d.].

On the other hand, customers may have focused upon the victimization of the Cito products. Thus, if Cito's pure product had been contaminated, what was the significance of this misfortune for the likelihood that other "pure" products were also at times tainted, often without their knowing it? It would be interesting to know whether Cito's customers became suspicious of other foods due to this incident. Thus, there is a painful symbolism in this contamination for people extremely vigilant about purity and health. The message in the Cito incident is that it may be illusory for them to think that they can exercise their expected degree of control over environmental factors which might adversely affect their health. Thus, precisely the greatest fear for this group of customers, a fear that they actively spend time, money, and effort avoiding, comes true in the Cito case.

Given this vulnerability, one might expect there to be a greater tendency to believe the reassurances of the health officers. An interesting light is cast on this question by Jones et al. (1984) when they discuss how people weigh the costs of not changing their behavior around the stigmatized. "As a general principle it can be stated that the greater the perceived risk associated with not modifying one's behavior, the greater will be the influence that the belief has on the belief holder's and, in turn, the target's behavior"(p. 187). This formulation helps to explain the discounting of the health qualifiers which indicated that there was little danger from drinking the milk. Accordingly, if the belief that exposure to PCBs can lead to cancer, a greatly feared outcome, is shared generally, then it should make little difference whether a person believes that the probability that they will get cancer from the milk is high or low. The cost of erring in the direction of exposure is so high that it is not worth incurring regardless of the probability. Cancer incurred by an improbable exposure event is just as undesirable as that incurred from a probable exposure event.

Thus, it can be assumed that Cito customers were very likely to take the association of PCBs and cancer seriously. They believe that this association exists. Given this, they are not concerned with statistical nuances of risk (see Kahneman et al., 1982). If drinking the Homestead milk increased their likelihood of getting cancer by 1/1,000 or 1/1,000,000, it would make little difference. Given the chance of avoiding the danger, most people would be expected to make this choice unless they became entangled in what Janis and Mann (1977) term defensive avoidance. But for actively vigilant health food customers, as has been argued. it is completely understandable that a changed behavior to a fairly easy alternative (neither selling not consuming Homestead milk) represented a rational and effective means of minimizing danger.

Conclusion

We see in the Homestead dairy case a clear example of stigma associated with an environmental contaminant and the places and life forms it was identified as affecting. In this case, the contaminant's importance was enhanced by the medium of exposure, milk. A key factor was the nature of the milk product, its being marketed under a brand name as having higher quality and purity. That product connected the contamination to users of this product, consumers with unusually high expectations for purity and healthfulness. Despite these ingredients, however, the stigma would never have occurred were it not for the publicity received after word of the contamination was "leaked" to the press. Otherwise, government officials had acted to detect the contamination and notify the farmer. Cito had acted to cull his herd and reduce the levels of PCBs. The incident would have been over were it not for the visibility given it by the media.

In subsequent work, the author has examined numerous other facets of environmental stigma (see Edelstein 1987, 1991, 1993; Edelstein & Kleese, 1995; and Edelstein & Makofske, 1998). But the issues raised in the Homestead dairy case study represent universals relevant to any examination of the environmental stigma phenomenon, from closely parallel instances, such as the Alar case, to divergent instances involving the siting of hazardous chemical and radioactive wastes.

Acknowledgment

I would like to thank Dave Cito for his willingness to review the manuscript for accuracy.

References to Media File

Note: Access to this Media File is available through the author.

Alvarez, R. (1983, December). Homestead Dairy: "Stigma" forces auction, but no one is buying. *Times-Call*.

Alvarez, R. (n.d.-a). Dairyman's lawsuit blames Monsanto for PCBs in milk. *Times-Call*.

Alvarez, R. (n.d.-b). Monsanto corporation denies responsibility for contaminated milk. *Times-Call*.

Associated Press (AP). (1981a, April 29). PCB discovered in Longmont milk. *Colorado Gazette*.

Associated Press (AP). (1981b, May 2). PCB-contaminated milk closes Longmont dairy. *Greeley Tribune*.

Cornett, L. (1981, May 1). PCB-tainted fodder spoils family dairy farm's future. *Denver Post*.

Crary, D. (n.d.) "Consumer attitude" now local dairy's main worry. No source.

KBTV. (1981, May 6). Denver, CO.

KOA-TV. (1981a, April 28). Denver, CO.

KOA-TV. (1981b, April 29). Denver, CO.

Noriyuki, D. (n.d.-a). PCB-plagued Homestead dairy to close. *Times-Call*.

Noriyuki, D. (n.d.-b). PCBs found in Homestead Dairy milk. *Times-Call*.

Ross, M. (n.d.). PCBs discovered in feed for dairy. *Camera*.

Times-Call (n.d.). [Picture caption].

True-life tale about danger to millions. (n.d.). [Picture caption].

United Press International (UPI). (n.d.). PCB reported found in milk marketed by Boulder dairy.

5 Risk Lived, Stigma Experienced[1]

Terre Satterfield, Paul Slovic,
Robin Gregory, James Flynn, and C. K. Mertz

Introduction

Marshall, Georgia is situated in Pecan County in southern Georgia.[1] It hosts an historically Black college, a population of 5,000, and a very limited stock of inexpensive housing. Railroad tracks and a major thoroughfare separate the Alouette Chemical Works plant and an adjacent African-American neighborhood from the town's more prosperous residential and commercial center. The Alouette Company began operating in 1910 as a lime-sulphur plant, later becoming a supplier of arsenic-based pesticides for agricultural, lawn and garden markets (M. Hillsman & M. Krafter, personal communication, July 29, 1996). Locals refer to the plant as "the dust house," a designation that invokes the particulate matter that once permeated neighborhood air and life. A ditch carrying untreated waste from the plant traveled through the adjacent neighborhood until it was covered in the late 1970s. Adult residents of the neighborhood recall playing in the ditch as children while their parents were said to have waded across the ditch to avoid the longer walk to the plank bridges at the ends of each block.

For most of its history, the plant was owned and operated by a prominent local White family; it was sold to a corporate chemical manufacturer in 1985. In 1986 the state Department of Environmental Quality (DEQ) requested that the company clean contaminated areas within the commercial facility where arsenic had adhered to the soil on plant property. Nothing was said to the predominantly African-American residents living nearest the plant at that time. In 1990 the site was recommended to the U.S. Environmental Protection Agency (EPA) for listing on the National Priority or "Superfund" list. Three years passed before the EPA notified affected citizens and issued cleanup orders to the plant. Beginning in 1993, residents of the plant neighborhood learned that several

[1]Portions of this study are excerpted from a paper published in *Human Ecology Review*, vol. 7, no. 1, pp. 1–11. Copyright 2000 by the Society for Human Ecology. Reprinted with permission.

[2]All person, place, and company names cited herein have been altered to respect the privacy of those involved.

probable carcinogens, in particular arsenical compounds, had permeated the soil in neighborhood yards and the dust inside local homes. Testing in 1994 through 1997 on the plant property and throughout the adjacent neighborhood indicated dust- and soil-based arsenic levels of 15 to 800 parts per million (ppm) despite the cessation of arsenic production during the mid-1980s. The plant grounds include hot spots of up to 30,000 ppm. The background level for arsenic in comparable geographic regimes was judged to be about 7 ppm. Chronic arsenic exposure has been associated with skin, lung, liver, bladder, kidney, and colon cancers (Agency for Toxic Substances and Disease Registry, ATSDR, 1990); arsenic is also believed to be a cancer "progressor" as is benzene and asbestos (Steingraber, 1997, p. 244). A 1996 study conducted in Marshall, Georgia by the ATSDR concluded that significant dangerous exposures had occurred in the past, but that current post-remediation levels of exposure were not dangerous to residents (ATSDR, 1996).

Risk scholars recognize that physical harm results from exposure to chemicals, heavy metals, and/or radioactive isotopes, and that the social and psychological experience of that harm is both fully rational and central to the risk experience (Slovic, 1987, 1992; Edelstein, 1987; Erikson, 1994; Kasperson, 1992). A prominent extension of risk work to which this volume is devoted, studies of technological stigma, commonly defines risk consequences in fiscal or market terms. Stigma occurs when certain products, places, or technologies are identified by the public as dangerous and subject to avoidance given their affiliation with health risks (Gregory et al., 1995). The primary experimental evidence for technological stigmas is the correlation of negative cognitions about a place product, or technology—word associations, imagery, affective descriptors, and perceived risks—with detrimental changes in consumer behavior (Flynn, Kasperson, Kunreuther, & Slovic, 1997; Flynn, Peters, Mertz, & Slovic, 1998). Risks prone to this designation are those the public views as dreaded, potentially fatal, involuntarily imposed, or regarded as beyond individual control (Slovic, 1987). Ultimately the stigmatized object becomes an epicenter from which severe economic impacts emanate. The millions in lost revenues incurred by Johnson and Johnson in the wake of fear about further Tylenol poisonings, the decline in land values near nuclear facilities, and the devaluation of real property alongside electromagnetic fields are classic cases of (respectively) product, place and technological stigma (Mitchell, 1989; MacGregor, Slovic, & Morgan, 1994; Slovic, Layman, & Flynn, 1990a).

Defining stigma in terms of market impacts is logical to the extent that economic viability and public acceptance are necessary conditions for the commercial development of modern technology, such as nuclear power. Nonetheless, a focus on pecuniary impacts sustains a model of stigma that implicitly narrows the definition of impact to altered purchasing habits or fluctuating market values. If human stigma responses are reduced to those where consumer spending drops to *avoid* Alar-suspect apples or buyers whose worries prompt them to *think negatively about* housing purchases, something of the "complex interplay of psychological, social and political forces" (Gregory et al. 1995, p. 222) that produces stigmas is lost.

In contrast, a model that recognizes the full social expression of stigma has the potential to accommodate the important association between the stigmatizing of a technology or place by external society and adverse effects on the people most immediately affected.

The relationship becomes more pertinent in light of recent findings about the disproportionate presence of technological hazards in socially stigmatized (especially minority) communities (Bullard, 1990a; Szasz, 1994; Johnston, 1994, 1997; United Church of Christ, 1987; see also Zimmerman, 1993). Those historically subject to social stigmas—defamation due to race, class, or economic status—are often those contemporarily subject to technological hazards, and thus in some circumstances, stigma.

This chapter demonstrates the effects of stigma on one community subjected to the experience of contamination. The research in Marshall shows that the experience of living in a contaminated and stigmatized place includes both physical and psychological invasions. Neighborhoods are structurally altered; domestic routines are profoundly disrupted and long-time residents come to be haunted by the inversion of home as a safe haven, an inversion that insinuates itself into thoughts about health and leads to the nagging fear that one's body has been infected by toxic substances. Residents notably invoke their sociopolitical experiences of racism, of being socially marginalized, to interpret how it is that they are viewed by the outside world, to explain why some citizens are protected from contaminants while others are not, why their concerns go unheard, or how it is that they are blamed for the economic woes of the larger community. This study suggests that these opinions may be tied to the defeating social climate that can accompany the experience of contamination and thus warrant study as symptoms of the link between technological and social stigmas.

Methods

In the spring and summer of 1996, 206 questionnaire-based interviews employing open and closed-ended questions were administered to 66 past and 140 current residents of the contaminated neighborhood. Interviewees were selected from over 600 past and current residents listed as plaintiffs in litigation pending against the Alouette plant. Plaintiffs included all but a few past and present residents of the plant neighborhood who were (a) traceable, (b) had lived in the neighborhood for at least 5 years, and (c) were said, by a medical doctor, to have clinical signs of arsenic exposure. Interviewees (all 206) were selected not at random but because they lived or had lived in the houses closest to the plant and/or because their house or yard had already been tested for the presence of arsenic. Only one of the 206 interviewees currently works at the plant and fewer than 10 have ever worked at the plant for more than three months. All but 3 of those interviewed were African-American, although a larger proportion of the 600 litigants (approximately 5%) are White.[3]

[3]Relying on a litigant sample is admittedly problematic. On the one hand, the legal team did not exclude anyone who fit the above criteria and reported to me that only a very few (less than 10) of all traceable past and present residents declined participation. At the same time, current residents refer to an earlier period (see page 80) where more Whites resided on the periphery of the plant neighborhood. This seems to suggest that more Whites should have been included in the litigant list. Nevertheless, the sample for this chapter was drawn from the areas closest to the plant and included those whose properties were regarded by EPA and litigant experts as appropriate for contaminant testing. These areas represent neighborhoods that are currently, and were historically, primarily African-American.

Twenty-six of the 140 people referred to here as current residents moved or were moved in response to the news about contamination. The other 114 (of 140) still live in the neighborhood. The second group of people referred to here as past residents (66) include only those people who left the neighborhood well before (often many years before) the news of contamination broke. Most in this latter subset of interviewees live in comparable though not contaminated communities elsewhere in rural Georgia. They do not otherwise differ from current residents with regard to age, gender, or race: The mean age of past residents is 46.3 years; present residents' mean age is 46.9 years.[4] Thirty-nine percent of all present residents are male, 61% are female. Thirty-five percent of past residents are male, while 64% are female.

Questionnaire items were developed with reference to the literature on social responses to technological hazards, and on the basis of background ethnographic interviews conducted by the first author. Questionnaire items were pretested and when necessary re-written for simplicity and ease of administration. The instrument included word-association tasks, affective ratings, reported behaviors, and opinions about remediation procedures. The questionnaire was read aloud to each interviewee and answers were recorded by the interviewer. Questionnaires were administered by nine African-American school teachers all of whom were trained as interviewers. Many of the teachers had taught in the neighborhood but none of them lived there. After the questionnaires had been completed, approximately 15 follow-up interviews were conducted by the first author. This last group of interviews was, again, open-ended.

The Stigmatization of Place

RECONFIGURING HOME AND ENVIRONMENT

Community studies have documented the physical deterioration of contaminated places including the potential for infrastructural, social, and psychological upheaval that follows a disclosure of contamination (Edelstein, 1988; Fitchen, 1989; Erikson, 1994). In Marshall, Georgia, multiple houses on each of the blocks closest to the plant were purchased by the company, torn down and/or encircled with chain-link fences. The *hazardous–keep out* signs that hang on the fencing inform residents that the fractured landscape they occupy is no longer, and perhaps never has been, safe. The soil on the plant-purchased lots remains too con-taminated for habitation (the plant is not obligated to clean its purchased properties) which negates the potential for rebuilding the neighborhood's residential infrastructure. Neighborhood gardens, fruit trees, and farm animals (e.g. chickens and some goats) were removed from properties registering 30 ppm of arsenic or greater. Remaining residents see the fences and signs appearing where neighbors once lived and conclude that perhaps their properties are also unsafe; consequently, they cease to garden or trade locally produced fruits and vegetables. The overall inability for neighbors to maintain the quotidian behaviors that typify a comfortable

[4]Thirty-four percent of the resident group are between 18 and 39 years of age, 43.6% are between 40 and 59, and 20.7% are 60 or older (remaining unknown). Thirty-three percent of nonresidents are between 18 and 39 years of age, 39.4% are between 40 and 59, 19.7% are over 60 (remaining unknown).

domestic routine—to garden, permit children to play outside, complete yard work, visit neighbors, etcetera—represents a "collective trauma...a blow to the basic tissues of social life" that "impairs [any] prevailing sense of communality" (Erikson, 1994, p. 233).

Residents also portray their immediate neighborhood as a "ghost town" of vacant lots and the aesthetic quality of the neighborhood as "concentration-camp like."[5] Houses are uneasily occupied, devoid of the intrinsic merits of home as a safe haven from the predicaments of public life. Betty Fields thus prefers to stay late at her job rather than face "going home to my **arsenic house** [where] I can't breathe." Her neighbor, Helene Johnson, finds only that her home "feels like a trap...like there's something hiding in the shadows waiting to jump." Many feel there is little they can do to protect themselves, a defenselessness articulated by Leroy Roberts as the feeling of: "living in a place I'm afraid of, like it's [the contamination] coming in the cracks." Long-term neighbors regard these insults as historically rooted, a continuation of decades of plant encroachment into residential territory given the meteoric rise of the plant's productive capacity after the second World War.[6]

Individual expressions of "feeling trapped" or feeling "unable to breathe" should not be mistaken as idiosyncratic, indicative only of exemplars of severe impact. Word-association tasks, credited for revealing the content and thought pattern of the respondents' minds without the complication or burden of discursive language (Szalay & Deese, 1978), confirmed that both past and current residents define their environs in extremely negative terms. Respondents were asked to provide image or word associations for context specific prompts (fences, soil, dust, etc.), and subsequently rated their responses using a five-point affective rating scale: very bad (–2); bad (–1); neutral (0); good (+1); or very good (+2). The rating scores for each stimuli and a sampling of the consistently immoderate image content are displayed in Figure 5.1.

Seventy-eight percent of respondents rated their associations with the fenced-in areas in the neighborhood as highly negative ("very bad" or "–2" on the affect scale), whereas 81.6% and 84.3% of respondents, respectively, rated images associated with "soil" and "dust" as highly negative. Across all three stimuli, no single item generated a combined very positive, positive, and neutral response in excess of 14.0%. The apparent absence of neutral responses, which usually include synonyms and visual or sensory descriptors (e.g., dimension, color, sound, etc.), is distinctly revealing in that responses of this kind would be expected in circumstances perceived as benign or generally less threatening. The logical coherence to these affective scores is that the stimuli closest to home and thus closest to one's physical body (dust inside a house and soil immediately outside a house) are rated more negatively than are more distant stimuli (such as fenced-in lots).

---

[5]All quoted, unreferenced speech is derived directly from word-association tasks and interview notes.

[6]In the United States, the post-1945 production of synthetic organic chemicals accelerated exponentially and by 1955 had captured 90% of the agricultural pesticide market. By the early 1990s there were 860 active pesticidal ingredients registered with the federal government (as compared to 32 ingredients in 1939). They are disbursed into more than 20,000 products (Steingraber, 1997, p. 95).

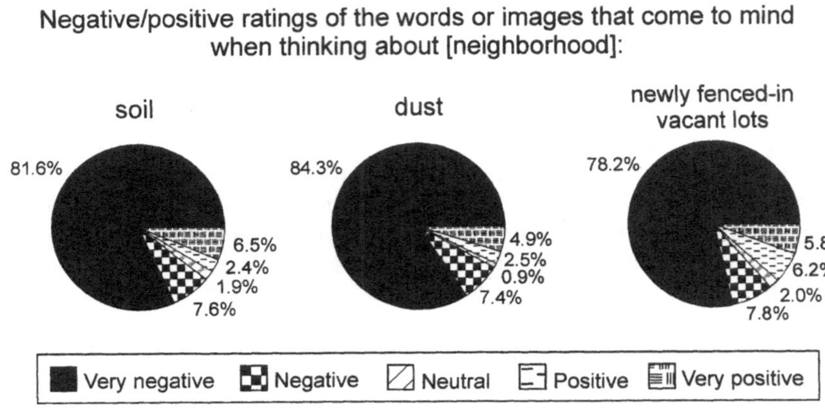

In their own words . . .

I'm living in a time bomb • suffocating in prison • a silent killer • Death Valley • living in death • everyone that lived in that house died of cancer • imprisonment • feel like an unwilling subject in an experiment • confusion • I hate it • invasion • children in danger • hazardous • bodily harm • poison plants • infected • someone dying of cancer • we have poison in our system

Figure 5.1 Image/word associations and affect ratings (N = 206).

AVOIDANCE BEHAVIORS

The decayed sense of safety within and around the homes is confirmed, equally, by parallel efforts of residents to avoid activities that normally comprise the acts of everyday life (Edelstein, 1988). Current residents were asked whether they found themselves unable to do some activities given concern about the plant. If the response was affirmative, respondents were then asked if the avoided behaviors were missed a great deal, missed slightly, or not at all missed ("I don't miss it," "I miss it slightly," or "I miss it a great deal"). The majority of residents reported changes in their domestic routines. Responses were nuanced and residents distinguished restrictions that were extremely bothersome from those that were less so. Table 5.1 demonstrates activity avoidance attributed to the plant, and reports frequency distributions for those who missed the avoided activity "a great deal."

The response frequencies reflect clear distinctions between restrictions. Residents are much more likely to avoid ordinary activities like opening a window on a breezy day (79.8%) or sitting in the yard on a nice day (74.6%) than less frequent or necessary activities such as going under the house to repair something (44.7%), going up into the attic (47.4%), or allowing children to play in exposed ditches at the edge of the neighborhood (43.0%). When asked which activities respondents "miss a great deal," a similar pattern emerges. Commonplace activities generally associated with a pleasant sense of domestic environment are those most heartily missed. These include opening windows on a breezy day (84.6%), sitting in the yard on a nice day (74.6%), and allowing children to play in the yard (72.6%). Alternately, activities such as walking near the remaining, though distant, open ditches (29.0%), or

Table 5.1 Activity Restrictions: Residents

Activity	"I do it less often because of the plant"	"I miss it a great deal"	Percent of total sample[a]
Opening the windows in your house on a breezy day	79.8%	84.6%	67.5%
Sitting in your yard on a nice day	74.6%	84.7%	63.2%
Yard work	66.7%	64.5%	43.0%
Flower gardening	65.8%	70.7%	46.5%
Allowing children in your care to play in your yard	64.0%	72.6%	46.5%
Investing money or time to improve the quality of your house or fix something that is broken	63.2%	66.7%	42.1%
Allowing children in your care to play in a friend's or relative's yard that is near the plant	62.3%	71.8%	44.7%
Walking near the open ditch	54.4%	29.0%	15.8%
Visiting someone whose house or yard is said to have high arsenic levels	50.9%	51.7%	26.3%
Going up in the attic of your house	47.4%	53.7%	25.4%
Going under the house to fix something	44.7%	47.1%	21.1%
Allowing children in your care to play in uncovered ditches	43.0%	34.7%	14.9%

Note. Percentage who do an activity "less often because of the plant," who miss the activity "a great deal," and the percent of total respondents who agreed to both ($n = 114$).
[a]Percentage of total sample who do the activity less often because of the plant and reported that they "miss it a great deal."

allowing children to play in those ditches (34.7%) were "missed a lot" by a minority of respondents.

Embodied Stigma

Alterations in household routines signify the inclination of individuals to protect their physical bodies. Worry about bodily harm is often regarded as the defining feature of toxic emergencies: the fear is that contaminants have been absorbed into one's tissues and perhaps the genetic material of survivors (Erikson, 1990, p. 121; see also Edelstein, 1988; Oliver-Smith, 1996; Kroll-Smith & Floyd, 1997). In Marshall, Georgia residents were forced to interpret these fears while haunted by the image of remediation workers protected from exposure to contaminants, an invading army of cleanup contractors and soil-testing technicians, each of whom benefited from the prophylactic suits used in industrial hygiene. This other-worldly attire seals face, head, body, feet, and hands from external contaminants. Workers also were protected and physically distanced from soil and dust through the use of immense backhoes and hep-o-vacs (backhoes assist the removal of contaminated topsoil, while hep-o-vacs function as powerful dust-extracting vacuum cleaners). Such acts of caution are understandable under the circumstances, yet the symbolic weight of these protected workers lingered in neighborhood residents' discourse, and helped articulate poignant misgivings. Visually compelling recollections of heavy machinery and "suited knights" seemed to say that the residents ought to have been safeguarded these many years, that the residents' bodies were already "poisoned" rendering protection futile, or, more cynically, that the residents were a socially disposable population, unworthy of protection in the first place.

Congruent with this symbolically charged backdrop of protected workers versus vulnerable residents, the interview notes reveal the markings of residents' physical selves. Residents learned to regard the long-familiar patches of atypical skin color and density on different parts of their bodies as evidence that contaminants were systemically present. Hyperpigmentation, hypopigmentation, and hyperkeratoses manifest as epidermal discolorations and lesions, constitute the primary clinical sign of chronic inorganic arsenic exposure (ATSDR, 1990). ATSDR physicians and clinicians examined the health records of 274 current and past residents for signs of exposure. A subset (n = 75) of this group showed evidence of simultaneous occurrence of hyperkeratosis, hyperpigmentation, and hypopigmentation. Though clinically associated with exposure, these signs are not expertly defined as health risks unless they progress to cancer (ATSDR, 1996, pp. 3-6). Those diagnosed with skin cancers as well as those merely suspicious about the implications of their symptoms treated their skin discolorations as constant reminders that their physical well-being was potentially amiss. During interviews, individuals would draw attention to their "spots," point them out, or absentmindedly press upon them as though they were a kind of worry bead, a point of reference that redirected thoughts to the consequences of contamination.

Toxicologists speak of "body burdens," the sum total or physical history of exposures through all routes of entry (inhalation, ingestion, skin absorption) and through all sources (food, air, water, office building, etc.; Steingraber, 1997, p. 236). Denizens of the plant neighborhood refer instead to the burden of worry, worry about health, childhood exposures, and especially the heightened expectation of pending disease. Eighty-eight percent and 83% of all respondents define themselves, respectively, as "worrying a lot" about "birth defects in children" and "the impact of the plant on my health." Every child with asthma and every virus is thought to be symptomatic of something larger, more foreboding: "Am I going to come down with something in my throat and die?" Individual bodies have become physically inscribed (i.e., marked) in the eyes of the owners; atypical pigmentation, perceived risks, and socially mediated fears about health have, together, gotten under the collective skin of neighborhood residents (Erikson, 1994). Residents thus come to regard their lives as "one long lethal injection" or "feel that they are something that will slowly kill" them.

These observations are corroborated by the vast majority of respondents reporting a deep sense of dread—a quality well-documented as central to lay characterizations of toxins (Slovic, 1987)—as well as persistent thoughts about the inhalation and ingestion of contaminants. A full 94.2% of past and current neighborhood residents agreed that thinking about the contaminants left them with "a creepy, frightened feeling," while 90% of current residents

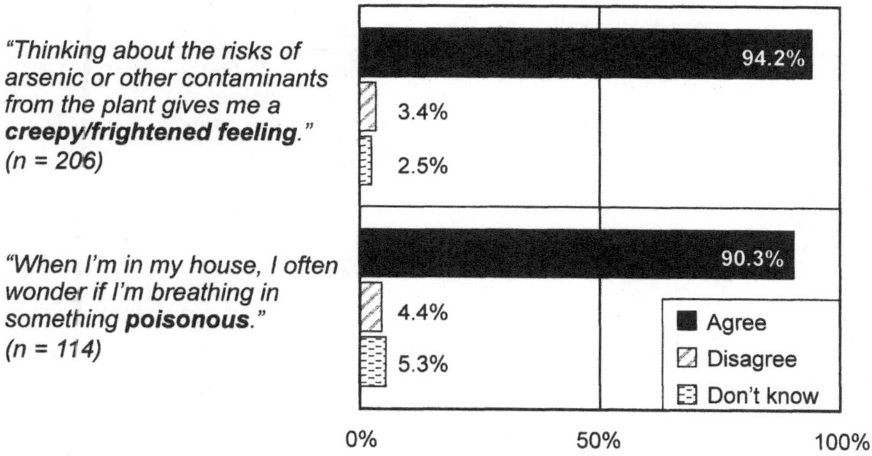

Figure 5.2 Psychometric dimensions: fear/dread. Note: "Agree" category is "strongly agree" and "agree" combined. "Disagree" category is "disagree" and "strongly disagree" combined.

agreed with the statement: "When I'm in my house, I often wonder if I'm breathing in something poisonous." Figure 5.2 demonstrates these findings graphically.

Older residents carry the additional burden of prior wounds and the unexplained deaths of loved ones. Further, the opportunity to reconsider old griefs in light of recent knowledge about contamination is, for many, unavoidable. Mary Aimes is in her late 60s. Her first child, a daughter, lived only 20 days—the result of a heart defect. Her disabled adult son died of asphyxiation in 1982, the result of a severe allergic reaction to "something" in the air. Mary's "bad nerves" began after the release of information about contamination and the concurrent threat that she might be moved from her home.

> You don't worry about it if you don't know, but once you know it makes you remember everything that happened before. . . . All these things I remember. I have nightmares about them now. Like when [as a child and teenager in the late 1940s and 1950s] men from the plant would knock on doors in the middle of the night and tell me and my family to leave the house immediately. There was a leak at the plant. They had giant gas masks, like creatures from outer space. They would tell us we had to run, and my mother would try to get all of us up; I was the youngest. When they told me I had to move [due to remediation], I woke up one night in the middle of the night, like as if my mother was trying to get me out of the house. I don't know [Mary stops herself] it's almost more than a body can stand after a certain age.

Mary's psychological and bodily peace is greatly disturbed by this recurrent nightmare and anxious ruminations about the premature deaths of both her children. Her fixation on the "middle of the night" memory has a particular capacity to crystallize and recreate a pivotal moment of horror for her, and is indicative of the "intrusive" states that characterize trauma (Herman, 1997, p. 38).

Table 5.2 Stress-Related Symptoms ($N = 206$)

Symptom	Symptom "bothers me a lot"[a]	Believe plant is the cause[b]	Percent of total sample[c]
Low energy	85.4%	59.1%	50.5%
Lower back pain	68.4%	41.8%	28.6%
Headaches	68.4%	60.3%	41.3%
Body weakness	65.5%	65.2%	42.7%
Memory trouble	64.1%	50.0%	32.0%
Nervous/shaky feeling	63.6%	62.6%	39.8%
Sore muscles	61.7%	44.9%	27.7%
Trouble getting breath	60.2%	73.4%	44.2%
Tense/keyed up	59.7%	60.2%	35.0%
Heart/chest pains	59.7%	58.5%	35.9%
Heaviness in arms/legs	57.8%	54.6%	31.6%
Depression	53.4%	62.7%	33.5%
Easily annoyed/irritated	52.4%	52.8%	27.7%
Nausea/upset stomach	51.9%	70.1%	36.4%
Trouble concentrating	51.5%	49.1%	25.2%
Heart pounding/racing	51.5%	62.3%	32.0%
Hopelessness	51.0%	74.3%	37.9%
Feeling trapped	49.0%	77.2%	37.9%
Confusion	48.5%	51.0%	24.8%
Faintness/dizziness	48.5%	58.0%	28.2%
Fear	44.2%	64.8%	28.6%
Others do not understand you	43.7%	35.6%	15.5%
Easily hurt feelings	42.7%	38.6%	16.5%
Feeling lonely/alone	41.7%	44.2%	18.4%
Avoidance due to fear	40.8%	67.9%	27.7%
Blaming yourself	37.4%	40.3%	15.0%
Crying easily	33.5%	40.6%	13.6%
Temper outbursts	26.2%	46.3%	12.1%
Critical of others	25.7%	47.2%	12.1%
Poor appetite	22.8%	55.3%	12.6%

[a]Percentage who answered "yes" to being bothered a lot by the symptom or problem.
[b]Of those who are bothered "a lot," percentage who believe the plant is the cause.
[c]Percent of total sample who are bothered "a lot" and believe the plant is the cause.

Extreme distress of this kind is unusual though most residents speak at length about their diseased life histories, and typically enumerate kinship ties and deaths-by-cancers in the same breath ("He was my uncle, he died of bladder cancer, and my sister died last year from breast cancer," and so on). The reporting of physiological expressions of stress was equally common. A majority of current and past residents reported suffering from "nausea," "feelings of hopelessness," the "feeling of being trapped," "nervous/shaky feelings," and the feeling of being "tense or keyed up." Over 60% of the subset of respondents who reported being "bothered a lot" by these symptoms attributed their symptoms to the plant. This did not, however, preclude a credible tendency to attribute other symptoms to noncontaminant causes. Only a minority of respondents reporting symptoms of lower back pain, crying easily, or temper outbursts subsequently attributed their sufferings to the plant (Table

5.2, column 2). Similarly, only one symptom, low energy, was reported by a slim majority of all respondents (50.5%) both as "bothering them a lot" *and* as "caused by the plant" (Table 5.2, column 3). Table 5.2 depicts both the distribution of symptoms and the subset of respondents who thereafter attributed their symptoms to the plant.

Sociopolitical Stigma

Stigma is a discrediting judgment that in turn evokes a response from those stigmatized (Goffman, 1963; Jones et al., 1984; Gregory et al., 1995). In contaminated communities the complex interplay between technological and social stigmas constructs a tangled mass of attributional actions and re-actions. That is, we can speak of those "constructing" the stigma versus those managing it, we can speak of the racial stigmatization that is likely at play in minority communities versus the technologically derived stigma that residents simultaneously project and suffer because of the plant. Some of this complexity is clarified by acknowledging two basic points. The first is that the occupant of a stigmatized environment can suffer damage simply because of association with that place. This "suggests that beyond a direct fear of a stigmatizing condition in its own right, there is a concern that any association with the marked setting may serve to mark oneself" (Edelstein, 1987). To this end, residents consciously worry that they are viewed by the outside world as socially contaminated, contagious and therefore unfit as members of the larger human community. Consider by way of example Marvia Lou Smith's characterization of herself as chaffing under media's occasionally ghoulish eye.

> People come through here now and you see them outside with TV cameras taking pictures and all that. I reckon they said: well what kind of neighborhood is this that has fences and barbed wire. That must be a bad neighborhood. They bad folks that got fences up around here.

Marvia faults both the physical consequences of remediation (fencing, barbed wire) and the media's amplification of those effects (see Kasperson, 1992) for the negative light they cast upon herself and her community.

Troubling reflections of this kind co-exist with a second basic point—that contamination events often involve the stigmatization of the already stigmatized. Exposure to environmental hazards is not random but rather selective of social and economically vulnerable populations. Risks are not distributed equally across social groups, there is a greater-than-average likelihood that the victims of hazardous technologies will be people of color and/or those occupying the economic margins of society (Bullard, 1990a; Johnston, 1997). At the same time, those living in environmentally degraded contexts are often subject to psychosocial debasement and dehumanizing innuendo (lazy, ignorant, backward) that destroys self-esteem and the motivation of individuals to control their destiny (Appell quoted in Johnston, 1994, p. 10).

In Marshall, this fusion of social stigma and environmental risk engulfs local disputes about the consequences of exposure. To this end all talk about "the plant" is somehow also talk about race. Arguments about the nature of legitimate evidence for injury, the appropriateness of different compensatory actions, or the logic of soil testing were invariably framed as "concerns that would have been addressed" or events that "never would have happened in a

White neighborhood." In particular, most residents believed the plant and the EPA ignored pertinent local input that might have ensured a mutually agreeable plan for the testing of soils and thus cleanup. EPA engineers posited a linear model of contaminant dissemination; properties immediately adjacent to the facility were tested as were those radiating outward from the source. When a safe property was encountered, testing would extend one or two houses further and then cease. It was assumed that all further properties were safe.

Locals opposed this model by insisting that wind patterns, the ditch's history of flooding into some properties and not others, the plant's trucking routes through the neighborhood, and the historical tendency for employees to carry contaminants into their homes via soiled work clothing had each contributed to an erratic dispersal of contaminants. Widespread discontent of this kind was expressed by survey respondents: 71.8% disagreed with the contention that "EPA experts considered all the important ways in which chemicals traveled from the plant into the neighborhood" while 74.8% believed that the EPA did a poor job of "testing for contaminants in the neighborhood." The dismissal of local concerns was eventually tempered by the hiring (on behalf of residents) of outside experts who confirmed a more extensive pattern of contaminant dissemination; the EPA subsequently verified these findings with further testing by their own technical staff.

Racist motives were also attributed to the EPA's procrastination regarding the distribution of knowledge about contaminants. The time lag between the 1990 Superfund listing and the 1993 official proclamation of exposure (a fact noted in this chapter's opening paragraph) was widely interpreted as an act dismissing Marshall's Black community as peripheral and thereby unworthy of urgent attention. Further, Black residents cite a late 1980s exodus of White residents from the plant neighborhood as evidence that knowledge of contamination was divulged well in advance to White residents. The suspicion is that White residents knew about the contamination early on and thus sold damaged residential properties at "good prices" to unsuspecting Blacks.

Representatives of Marshall's White community deny the persistent accusations of racism, and instead accuse (Black) plant-neighborhood residents of acting against the plant for "easy" economic gain via the several pending litigation efforts. Residents of the plant neighborhood are also censured by more affluent locals (White and some Black) for denigrating the town's reputation and its commercial prospects through exaggerated and false claims of plant-derived health impacts. Other White residents are not critical per se, but fear the repercussion of voicing support for those in the plant neighborhood. They fear being socially isolated because of perceived disloyalty toward their White peers (including the plant's founding family) or for being "too close" to the town's poorest and racially stigmatized residents.

Local African-Americans' pointed critiques of testing procedures and the racist undertones of interactions between local citizens and responsible parties can be read as healthy, pro-active signs of resistance to economic and racial stigmatization (Schwab, 1994; Szasz, 1994). Yet the impressions from field observations confirm something different. Neighborhood residents often appeared to be overwhelmed by a pervasive mood of hopelessness, a few resilient activist voices aside. The neighborhood's emotional landscape was marred by despair and a resignation not unlike the psychological numbing described in Lifton's (1967) work on radiation poisoning. Similarly, Jones et al. (1984)

defined the "essence of the stigmatizing process" as producing "devastating consequences for emotions, thought and behavior" (p. 4). The argument is that marked individuals are often unsuccessful at maintaining positive self-regard when the "evaluations elicited from other people [are] disproportionately negative" (Jones et al., 1984, p. 111). Other scholars of power and subordination have defined this defeated disposition as a "quiescence" of political participation despite a relatively open political system (Scott, 1990, p. 71).

In order to obtain some indication of the injuries of racism as they apply to political will, Srole's (1965) political alienation questions were modified to fit the Georgia context. The responses produced suggestive results. Compare, especially, responses of current residents with those offered by prior residents. These demographically similar groups differ from one another to the extent that current residents have lived through the full range of consequences of exposure—the parade of suited hygiene experts, exacerbated racial tensions, battles for voice in decisions about remediation, and, most dramatically, the resonating presence of a denuded landscape signified as hazardous—while prior residents have faced these events from a more removed and thus arguably protected position.

Both current residents and prior residents demonstrate an impaired sense of political efficacy. This impaired political efficacy is more prominent among current residents than prior residents on each of four questions, though only one of these differences is statistically significant at less than .05. Figure 5.3 demonstrates that current residents are more likely (by 10.0%) than prior residents to disagree that *local officials really do care about what I think*; less likely (by 12.9%) to believe that *people like me have a say about what will be done about*

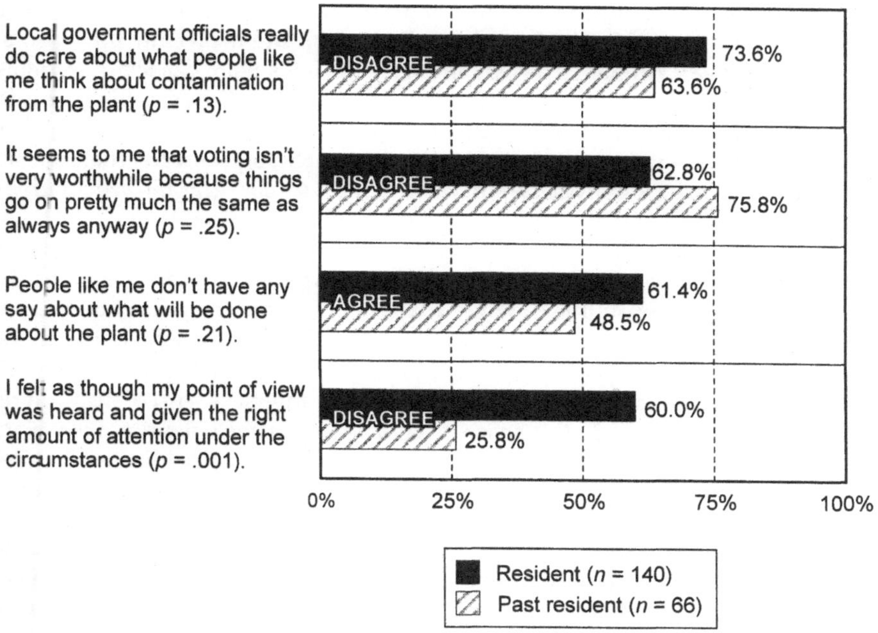

Figure 5.3 Expressions of political efficacy.

the plant; and much more likely to disagree with the suggestion that their *point of view was heard and attended to* (by 34.2%). Both respondent groups disagreed with the contention that voting was no longer "worthwhile," though prior residents were more supportive of voting (by a margin of 13%) than were current residents. The combined findings capture something of the flat affect about political efficacy expressed by both groups. The between-group differences suggest, however, that current residents share a greater sense of defeat with regard to political processes than do prior residents. Given that the two groups are demographically similar, save for current residents' greater exposure to plant and clean-up specific events, it is plausible that remediation procedures have had some effect on the loss of democratic control expressed by current residents.

Discussion

This chapter began with the contention that the personal trauma of toxic exposure merits a central position in theorizing about technological, product or geographic stigma. An expanded theory of stigma requires an understanding that extends well beyond the measure of market losses or adverse behavior by consumers. Accordingly, we considered the ravaging of home, neighborhood, and individual well-being that characterize Marshall's contamination events. An overwhelming majority of residents adjacent to the chemical plant think only negatively about soil, home, and neighborhood. Individuals change their daily routines, close windows, rest uneasily both inside and outside their homes, and abhor the "concentration camp" aesthetic that has taken over their lives. Implicit and explicit definitions of home as a place that promises safety for self and family, as an affective anchor in an otherwise chaotic world (Fitchen, 1989), are supplanted by the fear of dust in the attic and the feeling that "something will slowly kill me." The fear among Marshall's plant-adjacent residents is a state of mind that "gathers force slowly and insidiously, creeping around one's defenses rather than smashing through them" (Erikson, 1994, p. 21). This insidious "creeping" quality is evident in the psychological recoil that follows the sight of workers in hygiene suits and in individuals' graphic articulations of invasion (e.g., "My life feels like one long lethal injection").

Both body and place assist the reflective processes fundamental to human thought. The body is the means by which we experience and apprehend the world (Merleau-Ponty, 1962), while place (as in home, neighborhood, environment, etc.) is a basis for direction and self-reflection, for who one is in the larger social world (Basso, 1996). In Marshall, Georgia, the physical experience of a contaminated neighborhood and body intersect with disturbing reflections about the self. In this sense, the hazard signs, the emergence of vacant lots, and browning of the neighborhood can be understood as discrete injuries and as vehicles that repeatedly summon, indeed trap residents, in a vacuum of negative reflections. Dramatic changes in the landscape become insistent reminders of the presence of contaminants, forcing those who live there to cognitively register and re-register the possibility of "poison in [their] systems."

These reflections interact with larger sociopolitical realities. In the contaminated neighborhood studied here, worry about one's health or the safety of one's home merged with racial discrimination from some sectors of the town's White community, with anguished musings about denigrating the portrait of one's neighborhood and its residents on television, with implications about the "worthiness" of protecting remediation workers but not residents, and with experts' rejection of local complaints about remediation or the testing of soil. This combination of affronts encourages resignation among residents who define themselves as *not* cared for, listened to, or able to have a say in what will be done about the plant.

Ultimately, the Marshall, Georgia experience can enhance our understanding of the contamination experience and of stigmas. Experienced risk refers here to the contamination experience, that is the physical, psychological, and social consequences of exposure. These are direct reactions to hazardous environmental stimuli. Stigmatizing influences consist instead of signals that exacerbate the experience of contamination. The origin of stigmatizing impacts is in part media-fueled, as suggested by Kasperson et al. (1988), and as evidenced by one woman's response to the presence of camera crews in her neighborhood. More importantly, the Marshall, Georgia context demonstrates unremittingly that public agency (EPA, ATSDR) efforts to remedy hazards often contribute to the experience of stigmas locally. "Remedies" for protecting exposed communities (e.g., the stripping of vegetation, the removal of contaminated properties, the invasion of "suited knights," and/or the relabelling of pigmentation patterns as exposure symptoms) can foster the very fears they ought ideally to alleviate.

Finally, in this context one must come to some understanding of the combination of racial and technological stigmatization because persons of color are often those most immediately impacted by such hazards. We know from Goffman's (1963) early work that visible minorities already need to "manage" their "spoiled" identities. In minority communities faced with the ramifications of extant hazards, pre-existing experiences of racial stigmatization can constitute a dominant lens through which the new experience of contamination and technological stigma passes. Technological and social stigmas can thus form an ugly loop, where each follows and so intensifies the impact of the other. A more comprehensive, interactive, and socially astute model of technological stigma would acknowledge this interplay and thereby seek to define the links and causal relationships between social stigmas, technological stigmas and the local experience of contamination.

Nuclear Stigma

The section on "Nuclear Stigma" examines the potential for stigmatization of places associated with existing and potential radioactive contamination. Slovic and his colleagues report on their development of methods and measurements to gauge the potential for stigma effects. They find that nuclear images contain a record of perception of risk and stigmatization in reference to associated places. Jenkins-Smith reviews the work of Slovic et al. and proposes a revised model of imagery and stigma. The revisions focus on predispositions to evaluating information with stigma potential, such as ideological guidance, and the existing set of images of places that can modify the impact of new images. Easterling reviews the stigma studies undertaken to understand visitor behaviors to Las Vegas, Nevada. He finds the potential for adverse, stigma-related visitor impacts but does not find confirmation that these effects will occur. Hunsperger addresses the practical problems of measuring and placing a cost on properties contaminated by plutonium from a nuclear weapons production facility. He presents the approach he used in a litigation case.

6 Perceived Risk, Stigma, and Potential Economic Impacts of a High-Level Nuclear Waste Repository in Nevada[1]

Paul Slovic, Mark Layman, Nancy Kraus, James Flynn, James Chalmers, and Gail Gesell

New Orleans, Louisiana—This is New Orleans! Air conditioning ... Al Hirt ... Andrew Jackson ... antebellum plantations ... antiques ... Antoine's ... Arnaud's ... Audubon Park ... bananas Foster ... Basin Street ... Battle of New Orleans ... bayous ... Bourbon Street ... breakfast at Brennan's ... Cafe du Monde ... cafe au fait and beignets ... Cajun ... Canal Street ... chicory coffee ... "cities of the dead" ... Commander's Palace ... courtyards ... Creole cuisine ... Dixieland ... Duelling Oaks ... French Market ... French Quarter or "Vieux Carre" ... Galatoire's... Garden District ... Lafitte ... lace balconies ... Lake Pontchartrain ... levees ... Longue Vue Gardens ... the Longs of Louisiana ... Mardi Gras ... Old Absinthe House ... oysters Rockefeller ... Neville Brothers ... pecan pralines ... Pete Fountain ... Preservation Hall ... Ramos gin fez ... riverboats ... shrimps ... St. Charles streetcar ... Storyville ... Streetcar Named Desire ... Sugar Bowl ... Superdome ... Uptown ... vood-doo![2]

Introduction

The National Environmental Policy Act passed by Congress in 1969 established the requirement for an environmental impact statement for any "major federal actions significantly affecting the quality of the human environment." Section 101 of this act states that its major purpose is not only to maintain environmental quality but also to "fulfill the social, economic, and other requirements" of U.S. citizens. This requirement for a social and behavioral

---

[1]Reprinted from *Risk Analysis, 11*, 1991, pp. 683-696. Copyright 1991 Society for Risk Analysis. Reprinted with permission.

[2]Images characterize places, as is illustrated by this introduction to the announcement for the 1989 Annual Meeting of the American Psychological Association (APA) in New Orleans, Louisiana (APA, 1989).

science component with every environmental impact statement led to the birth of a field known as social impact assessment (Finsterbusch, 1980; Freudenburg, 1986; Wolf, 1977). The present study attempts to demonstrate the importance of perceived risk in social impact assessment, within the context of a specific project—the proposed national repository for disposal of high-level nuclear waste.

In December 1987, the U.S. Congress amended the Nuclear Waste Policy Act (NWPA) and authorized the U.S. Department of Energy (DOE) to determine whether Yucca Mountain, Nevada, is a geologically sound and technically feasible site for disposal of high-level nuclear waste. If the site passes a set of prescribed technical criteria, a repository will be constructed there to dispose of nuclear waste from the nation's commercial power plants.

Much effort has been, and will continue to be, devoted to characterizing the physical and biological risks associated with construction and operation of this unique facility, which must safely contain a large volume of highly radioactive material for a time period that is twice as long as recorded human history. Socioeconomic risks, though less studied, are also important. This paper addresses the following question pertaining to social impacts: What is the potential for a high-level nuclear waste repository at Yucca Mountain to have adverse economic effects on the city of Las Vegas and the State of Nevada during the period of constructing and filling the repository (approximately 40-60 years)?

The economic impacts of concern to us here include reduction in short-term visits to the city and state by vacationers or conventioneers, effects on long-term residents (moving out of the region, reduced in-migration of retirees), and reduced ability to attract new businesses. Assessment of these impacts is obviously important to citizens and officials of Nevada, who need to know what economic consequences to expect if Yucca Mountain is developed as the repository. Information about possible economic impacts may also be relevant to the final decision itself, regarding the acceptability of the Yucca Mountain site.

Empirical research on this topic faces some major obstacles, however. Changes in scientific knowledge and changes in public opinion are inherently difficult to forecast. For example, both scientific and public views about the risks of nuclear energy have changed dramatically since the "Atoms for Peace" program began in the 1950s. An obstacle to survey research is the fact that people may not really know how the repository will affect their future preferences and decisions. For example, asking people to project the repository's impacts on vacation decisions to be made many years hence may, in effect, be asking them to "tell more than they can know" (Nisbett & Wilson, 1977). Studies by Baker, Moss, West, and Weyant (1977) and West and Baker (1983) indicate that answers to questions about the impact of nuclear facilities on future behavior may not be trustworthy.

Despite these difficulties, there are theoretical reasons to expect that the repository may produce adverse economic impacts. In this study, we develop a method for assessing impacts that is not dependent on direct questioning of people who are unfamiliar with the decisions of concern here. We then use this method to assess the potential impacts from a repository at Yucca Mountain.

Background and Theory

Adverse impacts from the proposed Yucca Mountain repository may be expected to result from two related social processes. One has to do with perceptions of risk and socially amplified reactions to "unfortunate events" associated with the repository (major and minor accidents, discoveries of radiation releases, evidence of mismanagement, attempts to sabotage or disrupt the facility, etc.). The second process that may trigger significant adverse impacts is that of stigmatization.

PERCEPTIONS OF RISK AND SOCIAL AMPLIFICATION OF RISK

Nuclear waste has several unique characteristics that strongly suggest the potential for a repository to have adverse effects on the region in which it is located.

1. Although technical experts are supremely confident that nuclear wastes can be transported and disposed of safely (Carter, 1987; Cohen, 1985) the technology of high-level nuclear waste disposal is complex and largely untried. There are genuine hazards associated with such a facility, and the nature of these hazards is only partly understood.
2. From the time that radioactivity was discovered shortly before 1900, nuclear energy has been unique in the power of the imagery and symbolism that has surrounded it. Weart (1988) traces the salience and persistence with which both positive and negative meanings have become attached to things nuclear. His analysis demonstrates the strength of nuclear imagery and its broad penetration into our social and cultural consciousness over the past 90 years.
3. Contemporary evaluations of nuclear power and nuclear waste could hardly be more negative. Nuclear power stands out in studies of risk perception as unknown, uncontrollable, and dreaded, with the perceived potential to produce immense numbers of fatalities, even in future generations (Slovic, 1987; Slovic, Lichtenstein, & Fischhoff, 1979). Nuclear waste tends to be perceived in a similarly negative way (Kunreuther, Desvousges, & Slovic, 1988; Slovic, Layman, & Flynn, 1991).

These public perceptions have evoked harsh reactions from experts. One noted psychiatrist wrote that "the irrational fear of nuclear plants is based on a mistaken assessment of the risks" (Dupont, 1981, p. 8). A nuclear physicist and leading advocate of nuclear power contended that "the public has been driven insane over fear of radiation [from nuclear power]. I use the word 'insane' purposefully since one of its definitions is loss of contact with reality. The public's understanding of radiation dangers has virtually lost all contact with the actual dangers as understood by scientists" (Cohen, 1983a, p. 31).

Research on risk perception paints a different picture, demonstrating that people's deep anxieties are linked to numerous realities, including the reality of radiation's unique and powerful qualities, the reality of nuclear power's links to nuclear weapons proliferation and war, the reality of many serious examples of mismanagement (e.g., the releases of radioactive material into the environment from military reactor sites), and the reality of extensive media coverage documenting major and minor problems and controversies involving nuclear technologies. Attempts to "educate" or reassure the public and bring their percep-tions in line with those of industry experts face great difficulties because industry and government lack trust and credibility and because evidence of incompetence is much more persuasive than evidence of competence.

Perceptions of risk play a key role in a process labeled "social amplification of risk" (Kasperson et al., 1988). Social amplification is triggered by the occurrence of an adverse event, which could be a major or minor accident, a discovery of pollution, an incident of sabotage, and so on. Risk amplification reflects the fact that the adverse impacts of such an event sometimes extend far beyond the direct damages to victims and property and may result in massive indirect impacts such as litigation against a company or loss of sales, increased regulation of an industry, and so on. In some cases, all companies within an industry are affected, regardless of which company was responsible for the mishap. Thus, the event can be thought of as a stone dropped in a pond. The ripples spread outward, encompassing first the directly affected victims, then the responsible company or agency, and, in the extreme, reaching other companies, agencies, or industries. Examples of events resulting in extreme higher-order impacts include the chemical manufacturing accident at Bhopal, India, the disastrous launch of the space shuttle Challenger, the nuclear-reactor accidents at Three Mile Island and Chernobyl, the adverse effects of the drug Thalidomide, the Exxon Valdez oil spill, and the adulteration of Tylenol capsules with cyanide. An important feature of social amplification is that the direct impacts need not be too large to trigger major indirect impacts. The seven deaths due to the Tylenol tampering resulted in more than 125,000 stories in the print media alone, and inflicted losses of more than $1 billion upon the Johnson & Johnson Company, due to the damaged image of the product (Mitchell, 1989).

It appears likely that multiple mechanisms contribute to the social amplification of risk. First, extensive media coverage of an event can contribute to heightened perceptions of risk and amplified impacts (Burns et al., 1993). Second, a particular risk or risk event may enter into the agenda of social groups, or what Mazur (1981) terms the partisans, within the community or nation. The attack on the apple growth-regulator "Alar" by the Natural Resources Defense Council demonstrates the important impacts that special-interest groups can trigger (Moore, 1989).

A third mechanism of amplification arises out of the interpretation of unfortunate events as clues or signals regarding the magnitude of the risk and the adequacy of the risk-management process (Burns et al., 1993; Slovic, 1987). The informativeness or signal potential of a mishap, and thus its potential social impact, appears to be systematically related to the perceived characteristics of the hazard. An accident that takes many lives may produce relatively little social disturbance (beyond that caused to the victims' families and friends) if it occurs as part of a familiar and well-understood system (e.g., a train wreck). However, a small accident in an unfamiliar system (or one perceived as poorly understood), such as a nuclear waste repository or a recombinant DNA laboratory, may have immense social consequences if it is perceived as a harbinger of future and possibly catastrophic mishaps.

The concept of accidents as signals helps explain our society's strong response to mishaps involving nuclear power and nuclear wastes. Because the risks associated with nuclear energy are seen as poorly understood and catastrophic, accidents anywhere in the world may be seen as omens of disaster everywhere there are nuclear reactors and wastes, thus producing responses (e.g., increased regulation, public opposition) that carry large socioeconomic impacts.

STIGMATIZATION

Substantial socioeconomic impacts may also result from the stigma associated with a nuclear waste repository. The word *stigma* was used by the ancient Greeks to refer to bodily marks or brands that were designed to signal infamy or disgrace—to show, for example, that the bearer was a slave or a criminal. As used today, the word denotes someone "marked" as deviant, flawed, limited, spoiled, or generally undesirable in the view of some observer. When the stigmatizing characteristic is observed, the person is denigrated or avoided. Prime targets for stigmatization are members of minority groups, the aged, homosexuals, drug addicts, alcoholics, and persons afflicted with physical deformities or mental disabilities.

Although the sociological and psychological treatment of stigma typically pertains to interpersonal contexts far removed from that of radioactive waste disposal, the concept of stigma can clearly be generalized from persons to environments (Edelstein, 1988). Times Beach, Missouri, and Love Canal, New York, come quickly to mind as examples of stigmatized environments.

A dramatic example of stigmatization involving radiation occurred in September 1987, in Goiania, Brazil, where two men searching for scrap metal dismantled a cancer therapy device in an abandoned clinic. In doing so, they sawed open a capsule containing 28 grams of cesium chloride. Children and workers nearby were attracted to the glowing material and began playing with it. Before the danger was realized, several hundred people became contaminated and four persons eventually died from acute radiation poisoning. Publicity about the incident led to stigmatization of the region and its residents (Petterson, 1988). Hotels in other parts of the country refused to allow Goiania residents to register; airline pilots refused to fly with Goiania residents on board; automobiles driven by Goianians were stoned; hotel occupancy in the region dropped 60% for six weeks following the incident and virtually all conventions were canceled during this period. The sale prices of clothing and other products manufactured in Goiania dropped by 40% after the first news reports and remained depressed for a period of 30–45 days, despite the fact that none of these items was ever shown to have been contaminated.

Rationale and Method

Building on the theoretical concepts and research described above, we designed a series of studies to determine the potential for a nuclear-waste repository at Yucca Mountain to have adverse effects on tourism, migration, and business location decisions.

Our first efforts followed the direct approach of asking people in a national survey to indicate whether a nuclear waste repository located 100 miles from a site would reduce the desirability of that site as a place to attend a convention, vacation, raise a family, retire, or locate a new business. Depending on which of these activities was targeted in the question, between 41% (attend a convention) and 73% (raise a family) said that a repository would reduce the desirability of the region. It appeared that the more time people thought they would be spending in an area, the more likely they were to assert that the repository would make it a less desirable place in which to be.

In response to direct questions, interviewees consistently anticipated that a repository would decrease the attractiveness of a place. However, in light of the aforementioned problems with projecting impacts far into the future on the basis of answers to hypothetical questions, such data are suspect. Therefore, the present studies employed an indirect strategy, based on the notion of environmental imagery. Studies of environmental imagery appear to have the potential to provide a sound and defensible theoretical framework from which to understand and project possible impacts of a nuclear-waste repository on tourism and other important behaviors. Accordingly, the present studies were designed to:

- Demonstrate the concept of environmental imagery and show how it can be measured.
- Assess the relationship between imagery and choice behavior.
- Describe economic impacts that might occur as a result of altered images and choices.

The concept of imagery is not new to the study of environment and behavior. Geographers, cognitive and environmental psychologists, marketing strategists, and consumer theorists have written at length about the importance of images in our environmental consciousness and our behavior (Boulding, 1956; Kearsley, 1985; MacInnis & Price, 1987; Paivio, 1979; Saarinen & Sell, 1980; Weart, 1988). However, to our knowledge, no one has used a design such as ours to link imagery to the behaviors of concern here.

Our research was designed to test the following three propositions:

1. Images associated with environments have diverse positive and negative affective meanings that influence preferences (e.g., in this case, preferences for sites in which to vacation, retire, find a job, or start a new business).
2. A nuclear-waste repository evokes a wide variety of strongly negative images, consistent with extreme perceptions of risk and stigmatization.
3. The repository at Yucca Mountain and the negative images it evokes will, over time, become increasingly salient in the images of Nevada and of Las Vegas.

If these three propositions are true, it seems quite plausible that, as the imagery of Las Vegas and of Nevada becomes increasingly associated with the repository and things nuclear, the attractiveness of these places to tourists, job seekers, retirees, and business developers will decrease and their choices of Las Vegas and Nevada within sets of competing sites will decrease.

Support for these three propositions, therefore, would demonstrate the mechanism whereby the repository could produce adverse effects upon tourism, migration, and business development in Nevada and this demonstration would occur without having to ask people to make introspective judgments about their future behaviors.

SURVEY DESIGN

In order to test the propositions described above, we first conducted three extensive studies of imagery and preference. Studies 1 and 2 surveyed representative samples of residents in Phoenix, Arizona. Study 1 elicited images for four cities and asked people to indicate their preferences among these cities as places to vacation, take a new job, or retire. Study 2 did the same for four states. Study 3 surveyed a national sample of business executives, asking for their images of each of four cities and their preferences among these cities as places to open a new business or expand an existing business. All three surveys were conducted by telephone. Each survey had a sample size of about 400 persons.

The survey questions in Studies 1 and 2 were nearly identical. The cities questionnaire asked respondents to provide images for San Diego, Las Vegas, Denver, and Los Angeles. The states questionnaire elicited imagery for California, Nevada, Colorado, and New Mexico. These cities and states, in addition to Las Vegas and Nevada, were chosen for the study because they are important vacation destinations for residents of Phoenix.

The images were elicited using a version of the "method of continued associations" (Szalay & Deese, 1978), adapted for use in a telephone interview.[3] Image elicitation was always the first task in the survey. In the cities survey, the elicitation interview proceeded as follows:

> My first question involves word association. For example, when I mention the word baseball, you might think of the World Series, Reggie Jackson, summertime, or even hot dogs. Today, I am interested in the first SIX thoughts or images that come to mind when you hear the name of a PLACE.
>
> Think about _______ [CITY] for a minute. When you think about _____ [CITY] what is the first thought or image that comes to mind?
>
> What is the next thought or image you have when I say _____ [CITY]?
>
> Your next thought or image?
>
> What is another thought or image you have about _____ [CITY]?

This continued until six associations were produced or the respondent drew a blank. Then the procedure was repeated for the next city. The order of the cities was rotated across respondents. The procedure was identical for the states and business location surveys.

Following the elicitation of images, respondents were asked to rate each image they gave on a scale ranging from very positive (+2), somewhat positive (+1), neutral (0), somewhat negative (–1), or very negative (–2).

Respondents in Studies 1 and 2 were then asked to rank the cities/states according to their preference for a vacation site (long weekend vacation for

---

[3]The study of associations has a long history in psychology, going back to Galton (1880), Wundt (1883), and Freud (1924). Szalay and Deese argue that word-association techniques are easy and efficient ways of determining the contents and representational systems of human minds without requiring those contents to be expressed in the full discursive structure of human language. In fact, they argue, we may reveal ourselves in associations in ways we might find difficult to do if we were required to spell out the full propositions behind these associations through answers to questions.

cities; week or longer vacation for states). Subsequent questions asked for a preference ranking among these cities or states as retirement sites or places to move to assuming equally attractive job offers in each place, much in the same manner as vacation preferences were elicited. Additional questions assessed the extent of previous visits or living experiences in each of the cities or states, and the existence of family or close friends in each of those places.

Next, up to six images were elicited to the stimulus "underground nuclear waste storage facility" and the stimulus "nuclear test site."

The survey also asked, "In which state has the federal government proposed to build an underground facility for storing radioactive wastes?" and "In which state is the nuclear test site located?"[4]

The survey of corporate decision-makers first elicited images for each of four cities—Phoenix, Las Vegas, Denver, and Albuquerque—and then asked the respondents to evaluate these images on the –2 to +2 rating scale, as in the other surveys. These individuals were then asked to rank these cities in order of preference as a location for opening or expanding a business, assuming that market conditions and cost conditions were about equal.

SURVEY SAMPLES

Adults 18 years of age and older in Phoenix were surveyed with the cities questionnaire during the period April 13 through May 4, 1988. The states telephone survey was conducted in Phoenix between May 16 and June 8, 1988. The survey of corporate decision-makers took place between June 9 and July 29, 1988.[5] In each survey, more than 70% of the initial contacts resulted in completed interviews.

Results

CITIES SURVEY

Respondent characteristics. When asked, "Who in your household makes the final decision on vacations?" about 80% of the respondents said either that they did this themselves, or did so jointly with their spouse or partner. The percentage of respondents that had visited the target cities during the past five years ranged from 40% (Denver) to 65% (Los Angeles). Of those persons, the percentage that had spent a weekend in these cities during the past two years ranged between 50% (Denver) and 76% (San Diego). These self-reports support the choice of Phoenix residents as an appropriate target population for Las Vegas tourism and support the selection of San Diego, Los Angeles, and Denver as appropriate competing sites for vacations.

Images. In response to the stimulus word "Las Vegas," images associated with gambling, casinos, hotels, bright lights, and entertainment were dominant, followed by imagery pertaining to the climate and physical landscape, money, crime, and immorality. Imagery related to nuclear waste and the nuclear test site was very infrequent (only two images out of more than 1500). Table 6.1 presents the hierarchy of images elicited by the stimulus phrase "underground nuclear waste storage facility." The imagery was overwhelmingly nega-

---

[4]The nation's nuclear weapons test site is located in Nye County, Nevada, adjacent to the proposed repository site at Yucca Mountain.

[5]The sample was selected from the 1988 edition of *Who's Who in Corporate Real Estate.*

Table 6.1 Images Associated with an "Underground Nuclear Waste Storage Facility"[a]

Category	Frequency	Images included in category
1. Dangerous	179	Dangerous, danger, hazardous, toxic, unsafe, harmful, disaster
2. Death/sickness	107	Death, dying, sickness, cancer
3. Negative	99	Negative, wrong, bad, unpleasant, terrible, gross, undesirable, awful, dislike, ugly, horrible
4. Pollution	97	Pollution, contamination, leakage, spills, Love Canal
5. War	62	Warm, bombs, nuclear war, holocaust
6. Radiation	59	Radiation, nuclear, radioactive glowing
7. Scary	55	Scary, frightening, concern, worried, fear, horror
8. Somewhere else	49	Wouldn't want to live near one, not where I live, far away as possible
9. Unnecessary	44	Unnecessary, bad idea, waste of land
10. Problems	39	Problems, trouble
11. Desert	37	Desert, barren, desolate
12. Non-Nevada locations	35	Utah, Arizona, Denver
13. Nevada/Las Vegas	34	Nevada (25), Las Vegas (9)
14. Storage location	32	Caverns, underground salt mine
15. Government/industry	23	Government, politics, big business

[a]Basis: $N = 402$ respondents in Phoenix, Arizona.

tive. By far, the most frequent associations were dangerousness and death and their synonyms, followed by pollution, negative concepts, and radiation. Although we did not ask people to score these images, it seems likely that most of them would have been judged "very negative," a –2 on our five-point scale. Although some images pertaining to "necessity" came at the 17th position, they were very few in number (17) and included the phrase "necessary evil" given by two respondents. The words "Nevada" and "Las Vegas" were weakly associated with the repository, which was not surprising, given the low level of awareness of where the site is proposed to be located.[6]

Images of the nuclear test site were similarly negative and exhibited considerable overlap in content with the images of a nuclear-waste storage facility. Major test-site images included radiation, death, danger, cancer, destruction, and Nevada. More people associated Nevada with the test site (82 mentions) than with the repository.

Predicting preferences from images. To predict preferences among cities from images, we developed a scoring rule, the summation model, which simply sums the ratings for all the images a respondent produced for each city. A person's preferences among cities are hypothesized to be predictable from these sums.

[6]Only 19.6% of the cities sample knew that Nevada had been selected as the leading candidate for an underground facility for disposing of radioactive wastes and 46.8% knew that the nuclear weapons test site is in Nevada.

An example, illustrating the application of the summation model to the data of one respondent, is given in Table 6.2. For this respondent, the rank order of summation scores exactly matched the preference order for vacation sites.

When ranks generated by the summation model were compared to the actual ranks generated by the respondents when they stated their preferences, the model did quite well, correctly predicting 55% of the number 1 ranked vacation cities and 56% of the fourth ranked cities, with somewhat less accuracy in predicting intermediate ranks (if the model lacked predictive validity, we would expect a 25% hit rate by chance). The exact rank order of four cities generated by the summation model matched the exact rank order of the respondent 26.4% of the time (perfect matching of ranks would be expected by chance only 4.2% of the time).

A second set of tests was conducted with the summation model. Each of the four cities was paired with every other city—making six pairs in all. For

Table 6.2 Images, Ratings, and Summation Scores for Respondent 132[a]

Sample subject	Image no.		Image rating
San Diego	1	2	Very nice
San Diego	2	2	Good beaches
San Diego	3	2	Zoo
San Diego	4	1	Busy freeway
San Diego	5	1	Easy to find way
San Diego	6	2	Pretty town
		Sum = 10	
Las Vegas	1	−2	Rowdy town
Las Vegas	2	−1	Busy town
Las Vegas	3	−1	Casinos
Las Vegas	4	−1	Bright lights
Las Vegas	5	−2	Too much gambling
Las Vegas	6	0	Out of the way
		Sum = −7	
Denver	1	2	High
Denver	2	0	Crowded
Denver	3	2	Cool
Denver	4	1	Pretty
Denver	5	−2	Busy airport
Denver	6	−2	Busy streets
		Sum = 1	
Los Angeles	1	−2	Smoggy
Los Angeles	2	−2	Crowded
Los Angeles	3	−2	Dirty
Los Angeles	4	−1	Foggy
Los Angeles	5	0	Sunny
Los Angeles	6	−2	Drug place
		Sum = −9	

[a]Based on these summation scores, this person's predicted preference order for a vacation site would be: San Diego, Las Vegas, and Los Angeles.

every respondent and every pair, the image score for city B was subtracted from the image score of city A. The resulting 2,346 A–B scores across all respondents were ordered from extreme negative to extreme positive and this distribution was partitioned into five subsets, as equal in size as possible (range = 419–511 comparisons in each subset). Finally, within each subset, the percentage of respondents who ranked city A more favorably than city B as a vacation site was calculated. When the plot of the mean A–B difference was most negative (mean = –6.2), A was preferred as a vacation site for only 27.4% of the pairs. For the subset in which the mean difference was most in favor of A (mean = + 11.4), 90.7% of the preferences favored A. The best fitting regression line through these five points had a slope of .037, indicating that every one-point increase in the mean difference score was associated with a 3.7% increase in the percentage of choices favoring city A.

Figure 6.1 illustrates the performance of the summation model across all pairs of cities. The choice proportions for specific pairs of cities (e.g., Las Vegas vs. Denver) were found to be quite similar to the combined plot in Figure 6.1.

The data in Figure 6.1 show that imagery and preference for vacation cities are strongly related. If city B has a more positive set of images than city A (as indicated by simply summing the affect ratings across however many images were produced for each city), then city B is more likely to be preferred as a vacation site. If city A has more positive imagery, then city A is more likely to be preferred as a vacation site.

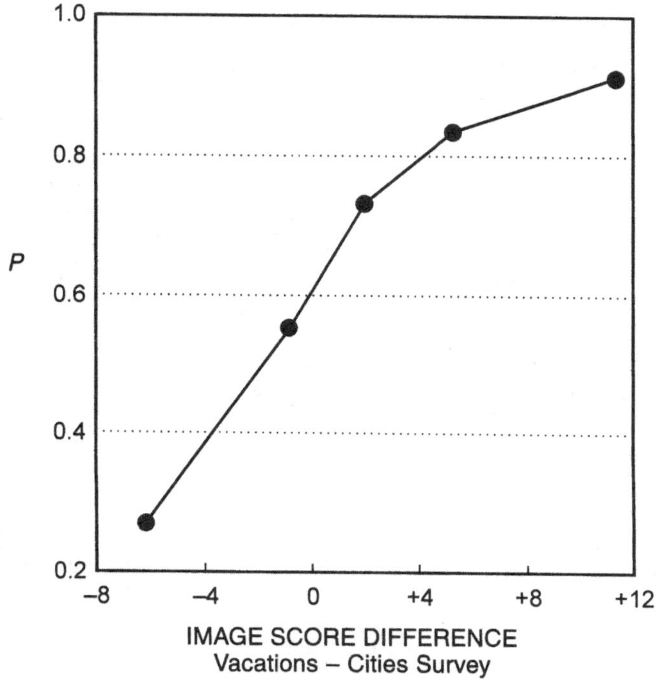

Figure 6.1 Relationshp between mean image score differences (City A – City B) and proportion of times *(P)*. City A was ranked higher than City B in the respondent's preference rankings for vacation sites. All possible pairs of cities are included in this analysis.

Predicting Job and Retirement Preferences. The summation model was applied in similar fashion to the prediction of job preferences and retirement preferences for the cities survey. The hit rates were similar to those reported earlier for vacation preferences. The functional relationships relating job and retirement preferences to image scores were almost identical to the relationship shown in Figure 6.1.

RESULTS: STATES SURVEY

As in the cities survey, more people (41.0%) knew the location of the nuclear weapons test site than knew the location being considered for the repository (24.5%). The summation model was found to be about as accurate in predicting vacation, job, and retirement preferences among states as it was for predicting preferences among cities.

Imagery associated with "a nuclear waste storage facility" and the "nuclear test site" was extremely negative for respondents in the states survey and was almost identical to the imagery obtained in the cities survey. Whereas few people in the cities survey expressed nuclear-related imagery in response to the stimulus words "Las Vegas," about 10% of respondents in the states survey produced nuclear imagery in response to the stimulus "Nevada." Such images included the terms nuclear testing, nuclear bomb, nukes, explosions, and radiation. The mean image score for Nevada for these persons was 0.18. The mean image score for persons who did not associate Nevada with things nuclear was 2.56 (a statistically significant difference; $p < .001$). As expected, persons with nuclear imagery assigned lower (poorer) preference rankings to Nevada than did persons without such images (see Table 6.3). These findings are important because they suggest that Nevada has already undergone some stigmatization as a "nuclear place."

RESULTS: CORPORATE DECISION-MAKERS SURVEY

Parallel analyses were carried out with the images and preferences of the corporate decisionmakers. The summation model correctly predicted 47% of the first-choice locations for siting a new business and the functional relationship between image scores and preferences for pairs of cities looked much like the relationship for vacation preferences in Figure 6.1.

In summary, three separate surveys totaling more than 1,200 respondents demonstrated that a simple summation model applied to sets of images did a good job of predicting expressed preferences for cities and states in which to vacation, take a new job, retire, or site a business. The slopes of the best-fitting

Table 6.3 Preference for Nevada as a Vacation Site Among Respondents Who Do and Do Not Exhibit Nuclear Imagery[a]

	Nevada preference rank				
	1	2	3	4	Mean rank
Nuclear imagery present ($N = 39$)	3	3	46	49	3.41
Nuclear imagery absent ($N = 354$)	6	16	51	27	2.98

[a]Cell entries are percentages within each row.

lines relating preferences among pairs of cities/states to differences in image values were quite steep, indicating that a change in one or two images could imply a substantial shift in preference probability.

ADDITIONAL ANALYSES

Additional studies were done on these data to test and evaluate the link between imagery and preference.

Independent raters. The predictive accuracy of the summation model was quite high. One possible criticism of the data collection method is that the high degree of predictability is an artifact of allowing respondents to rate their own images. People's ratings may have influenced their preferences, thus inflating the relationship.

To test this hypothesis, we conducted a fourth study, using samples of young adults from Eugene, Oregon as subjects. One group of subjects ($N = 150$) produced images for four cities and ranked the same cities according to their attractiveness as vacation sites, much as was done in the survey of Phoenix residents. Some respondents produced images first and then indicated their vacation preferences. Others gave their preferences first. Unlike the Phoenix surveys, however, subjects in the Oregon survey did not score their own images. Instead, the more than 2,000 different images produced by the subjects were rated by a different group of 28 subjects. The "artifact" hypothesis predicts that the summation model would have much poorer predictability in this study, because there is no possibility that image ratings can influence the preferences (or vice versa).

The results of this study were surprising. The model's hit rates were excellent (66% accuracy in predicting first choices) and the functions relating differences in image scores to preference probability were again remarkably linear, with slopes only slightly less steep than those obtained in the Phoenix survey. The high predicative accuracy of the image models in the Phoenix surveys does not appear to be an artifact of the image evaluation procedure used in those studies.

Insensitive gamblers. Another challenge to the summation model was devised in the form of a hypothesis that people who like to gamble will not be influenced much by other attributes of Las Vegas. This hypothesis was tested using 246 respondents in the cities survey who produced the term "gambling" as one of their images of Las Vegas. These individuals were separated into subgroups according to the value they assigned to the gambling image. Next, a separate analysis of the relationship between image difference scores and preference probability for Las Vegas, analogous to the analysis in Figure 6.1, was performed within each subgroup. Difference scores were computed by pairing Las Vegas with each of the other cities and subtracting the score for the other city from the score for Las Vegas.

The hypothesis predicts that those who see "gambling" as extremely positive (i.e., who rated it as a +2) would have vacation preferences for Las Vegas that are less sensitive to image differences compared to the preferences of people who are less favorable toward gambling (i.e., who rate "gambling" as intermediate or negative in value). The data did not support the hypothesis. The curves relating image score differences to preference probabilities for Las Vegas were not significantly different for groups of people who differed in their evaluation of gambling. In other words, people who viewed gambling as a very positive feature of Las Vegas were just as much influenced by other positive and negative images as were people who had less positive views of gambling.

Effects of repository knowledge and test-site knowledge. Additional analyses were conducted using the states survey data to determine the impact of knowledge about the state being considered for the nuclear waste repository and knowledge about the state in which the nuclear-test site is located upon images and preferences for Nevada as a vacation site. These two types of knowledge were found to be related. Persons who knew that the repository was being considered for Nevada were somewhat more likely to know that the test site is in Nevada (71%) as compared to those who lacked knowledge of the repository (55% knew the test-site location). Similar results were obtained in the cities survey, where the corresponding values were 70% and 41%.

Additional analyses showed that the presence of a nuclear image in one's image set for Nevada was determined more by knowledge of the test-site location than by knowledge of the repository location. Nuclear imagery was produced by 15% of those persons who knew the test-site location compared to 2% of those who did not know the location. Corresponding figures associated with knowledge and lack of knowledge of the proposed repository were 12% and 9%.

Summarizing the results from these analyses, we see that the proposed Yucca Mountain repository has not yet infiltrated people's images of Nevada and has not yet had much effect on their stated vacation preferences. The test site, which has been a feature of Nevada for many years, has had a stronger influence on images and preferences. Knowledge that the weapons test site is in Nevada appears to have led to an increase in nuclear-related imagery for Nevada and nuclear imagery is associated with decreased preference for Nevada as a vacation site.

Imagery and vacation behavior. The previous analyses demonstrated that images could predict expressed preferences for vacation sites. Can image scores also predict actual vacation trips? To address this question we attempted to resurvey the 802 respondents from our 1988 Phoenix surveys some 16–18 months later (October–December 1989). We were successful in reinterviewing about 130 persons in each of the two samples (cities survey and states survey) studied earlier.[7] Again, we elicited word associations to each of the same four cities or four states and asked for positive/negative ratings of each image produced. In addition, we asked the respondents to indicate in which of these cities (or states) they had vacationed since the previous survey was conducted.

The predictive capability of the word-association image scores was tested by means of logistic regression analysis using a person's 1988 image score for a state or city to estimate the probability that that person would vacation in a place during the subsequent 16–18 months (until the date of the repeat survey). The estimated probabilities for both cities and states are presented in Figures 6.2 and 6.3. These data show that the affective qualities of a person's images of a place were clearly related to the probability that the person would subsequently vacation there, with the relationship being stronger for states than for cities.

[7]Across the two repeat surveys, about 60% of the original respondents could not be reached because no one answered the phone (despite repeated callbacks), the number was no longer valid, or the answering person said that the target individual no longer lived there. Of the original respondents who were contacted, 83.3% completed the repeat survey.

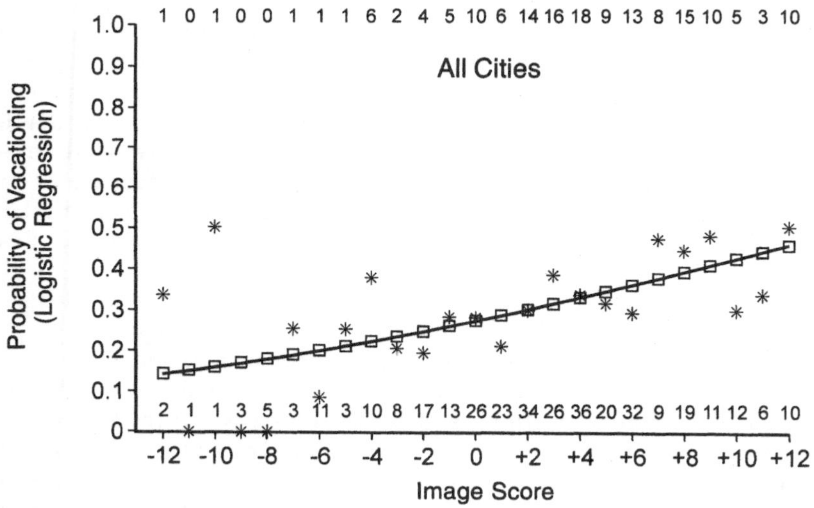

Figure 6.2 Probability of vacationing in a particular city after June 1988 as a function of image scores elicited prior to that date (Phoenix survey). Upper row of numbers indicates the number of people with that image score who vacationed in the city; lower row is the number who did not vacation in the city; * marks the proportion who vacationed. The curve is the best fit logistic function to these proportions.

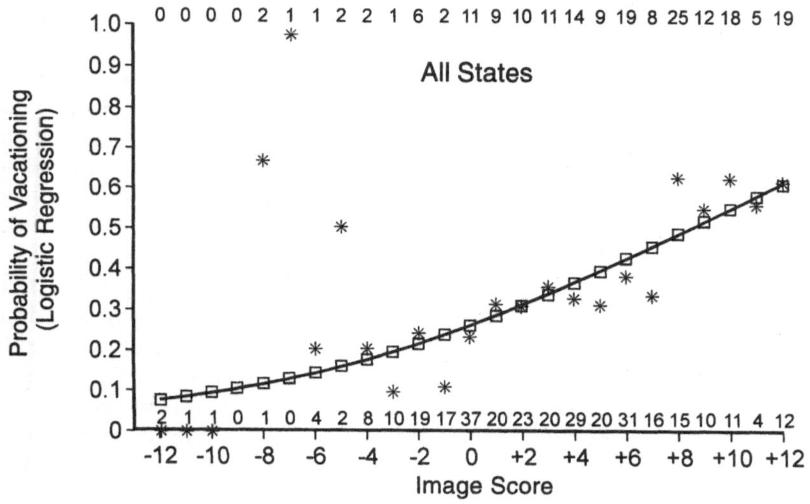

Figure 6.3 Probability of vacationing in a particular state after June 1988 as a function of image scores elicited prior to that date (Phoenix survey). Upper row of numbers indicates the number of people with that image score who vacationed in the state; lower row is the number who did not vacation in the state; * marks the proportion who vacationed. The curve is the best fit logistic function to these proportions.

Image stability. The resurveying of the Phoenix samples provided us with an opportunity to examine the stability of image scores for states and cities across a 16–18 month time span. We found moderate stability. More than 60% of the paired image sets provided by the same person had at least one identical association in both sets. When the 1988 and 1989 image scores were compared for the same person–same stimulus, they were found to correlate .52 for cities and .42 for states, across persons. The two image scores were identical in 11.4% of the cases. Almost 70% of the paired scores were within a range of ±4 points.

The changes in a person's image scores likely reflect both systematic changes over time and unreliability. The unreliability can be reduced by averaging a person's 1988 and 1989 images scores for each city or state. When this is done, the relationship between imagery and vacations taken between May 1988 and Autumn 1989 became even stronger than that shown in Figures 6.2 and 6.3. For example, when images scores were averaged, the predicted probability of vacationing in a state ranged from about .03 for the lowest mean image score (−12) to about .70 for the highest mean score (+12). This contrasts with a range between .07 and .61 when only the 1988 image score is used as a predictor.

Discussion

The present study developed and applied a methodology based on imagery in order to overcome concerns about the validity of direct questions regarding the potential influence of a nuclear waste repository at Yucca Mountain upon economically important behaviors. The results supported the three propositions that the research aimed to test: Images of cities and states, derived from a word-association technique, exhibited positive and negative affective meanings that were highly predictive of preferences for vacation sites, job and retirement locations, and business sites (Proposition 1). The concept of a nuclear-waste storage facility evoked consistent, extreme, negative imagery (Proposition 2). The nuclear-weapons test site, which has been around far longer than the Yucca Mountain nuclear-waste project, has led to a modest amount of nuclear imagery becoming associated with the state of Nevada. This provides indirect evidence for Proposition 3, which asserts that nuclear-waste related images will also become associated with Nevada and Las Vegas if the Yucca Mountain Project proceeds. Nuclear imagery, when present in a person's associative responses, was found to be linked with much lower preference for Nevada as a vacation site. The verification of these propositions implies that the repository also has the potential to cause an increase in nuclear imagery which, in turn, will produce adverse impacts on tourism and other economically important activities in Nevada.

In our opinion, these findings provide a partial answer to the question that motivated the inquiry. The mechanisms of perceived risk, social amplification, and stigma are observable in the record of past experience with nuclear and other types of hazards. In the context of the Yucca Mountain Repository, these mechanisms appear to have the potential to cause substantial losses to each of the various economic sectors at risk. We believe that it would be unwise and unfair for development of the nation's high-level nuclear waste repository to proceed without taking these potential economic impacts into consideration.

Some analysts have suggested that the nuclear weapons test site provides evidence against the above conclusions, in view of the strong expansion of the Las Vegas visitor economy during the years in which this facility has been operating. We disagree. Judging from the Phoenix survey, the test site has worked its way into the imagery of Nevada for only a small percentage of people and is rarely associated with Las Vegas. Moreover, the operations of the test site have been restricted and unavailable to full public scrutiny. Nuclear-waste transport, the operation of the waste repository, and any controversies over the safety of these activities will likely be far more visible to the public and the media. In particular, tens of thousands of nuclear-waste shipments by truck or rail throughout the United States will be a prominent reminder of the repository and its risks. As these shipments converge upon Las Vegas, nuclear associations with that city may be built to a far greater extent than has occurred with the secret, contained, underground explosions at the test site. Finally, there is no evidence that the small degree of association of the test site with the region has not actually impaired tourism and business development. Apart from the gambling industry, business development has shown little progress despite the potential attractiveness of Las Vegas for many kinds of industries.

It may also be the case that the test site and the repository will interact in a synergistic way to produce nuclear imagery to an extent that is greater than the sum of the individual contributions from each facility. Little is known about the dynamics of the process by which images become salient. It is certainly true, however, that individuals have a number of images associated with any particular place. There may be some threshold of repetition that moves a weak or unstable image from the periphery into the core image of a place. If so, Nevada's link to the nuclear weapons test site may increase its potential for stigmatization from the repository relative to a state with no existing base of nuclear imagery.

Historical analysis of major risk events has documented substantial socioeconomic impacts, but these impacts have often been transitory. Will the same impermanence hold for impacts triggered by a nuclear waste repository? In considering this question, we suggest that it is useful to distinguish two different kinds of stimuli emanating from the repository. First, there are the multitude of discrete events that are associated with the project. Second, there is the cumulative experience with the project, which reflects the characteristics of the project plus the experience across all project-related events. It is reasonable to suppose that an isolated, solitary event will generate a transitory response. It is also reasonable, however, to expect that the imagery of Nevada and Las Vegas held by the general population will reflect their cumulative experience with the repository program. Each of the discrete events that might result from the program, therefore, would have the potential to trigger two kinds of consequences—responses to the event itself, the duration of which would be related to the nature of the event, and responses based upon the cumulative image of the repository to which this event makes a contribution on the margin. Just as this cumulative image will take time to develop, it may also be more durable. In fact, to the extent that strong nuclear imagery became associated with the repository, the host region, or major communities along waste-transportation routes, the stigmatization could remain for a long time. The sort of associative material that builds an image is illustrated by the magazine photograph and excerpt shown in Figure 6.4.

NUCLEAR TESTING

Nevada, U.S.S.R.

By PETER ZHEUTLIN

A new Soviet antinuclear group called Nevada—a name chosen to attract the attention of U.S. antinuclear activists—has had an impressive first year. . .

Figure 6.4 Many kinds of associations contribute to the development of an image. Source: *The Bulletin of the Atomic Scientists*, 1990, March, p. 10. Copyright 1990 International Physicians for the Prevention of Nuclear War, Inc. Reprinted with permission.

Although this research has clarified the mechanisms by which adverse economic impacts can be generated, predicting the precise magnitude and duration of those impacts is impossible. The uncertainties involved in repository development make it inevitable that the actual impacts—physical, biological, social, and economic—will differ from the best of impact projections. There are at least four categories of uncertainty. First, the DOE plans are still largely unspecified on crucial matters; for example, it is impossible to know at this time whether waste shipments will be made by truck or rail, over which routes, and with what frequency or safeguards. Second, the risk-management policies to be followed by state and local governments are largely unknown, and could have a powerful influence on impacts. Third, there will certainly be external perturbations and surprises that may cause the repository development to differ from anything that can be foreseen at this time. Fourth, economically relevant decisions are always made in the context of alternatives; quantitative prediction requires currently unavailable knowledge of the alternatives that individuals and society will be able to choose among in the future.

In sum, our analysis indicates that the development of the Yucca Mountain Repository will, in effect, force Nevadans to gamble with their future economy. The nature of that gamble cannot be specified precisely, but it appears to include credible possibilities (with unknown probabilities) of substantial losses to the visitor economy, the migrant economy, and the business economy.

As the potential for the repository to have adverse economic impacts becomes recognized by citizens of Nevada and their government officials, the already strong political opposition to the site can be expected to intensify, making it extremely difficult for the federal government to proceed with the project.[8]

BEYOND YUCCA MOUNTAIN

The present inquiry has implications for social-impact analysis that transcend the conflicts and concerns surrounding the proposed Yucca Mountain repository. The processes of social amplification and stigma appear relevant, as well, to the analysis of impacts from any major facility that produces, uses, transports, or disposes of hazardous materials. The numerous proposed sites for disposal of low-level radioactive wastes and the many sites being considered for chemical-waste incinerators and landfills will face similar problems of perceived risk and its impacts, though probably to a lesser degree than the problems posed for Nevadans by a Yucca Mountain repository. The present study has demonstrated that the so-called "standard effects" of large engineering projects on local employment, housing, and transportation have the potential to be dwarfed by the "special effects" of risk perception and stigma. However, just as physical or technical risks can be mitigated by proper safety design and management, effects of perceived risk may be mitigated by means of management processes that instill and maintain trust and that work to protect the economic base of those individuals and communities whom the facility puts at risk.

Acknowledgments

This material is based upon work supported in part by a contract between Decision Research and the Nevada NWPO with federal funds granted pursuant to the provisions of Public Law 97-425. Any opinions, findings, and conclusions or recommendations expressed in this material are those of the authors and do not necessarily reflect the views of the federal government or the Nevada NWPO.

---

[8]In June 1989, the Nevada State Legislature passed a bill, AB 222, making it unlawful for any person or governmental entity to store high-level radioactive waste in Nevada. Concern that a high-level radioactive waste repository could severely damage the economy and environment of the state was cited by the legislature as a major factor in the passage of this bill. In litigation between the State of Nevada and the DOE, lawyers from the state's Attorney General's Office have repeatedly cited potential adverse economic impacts to support the state's opposition to the repository program.

7 Modeling Stigma: An Empirical Analysis of Nuclear Images of Nevada

Hank C. Jenkins-Smith

I. Introduction

This chapter presents the results of an analysis that builds on recent research regarding the potentially stigmatizing effects of a proposed high-level nuclear waste repository at Yucca Mountain, Nevada. Prior research (Slovic, Flynn, & Layman, 1991; Slovic, Layman, & Flynn, 1991; Slovic, Layman, Kraus, et al., 1991; Kunreuther & Easterling, 1992) examined the extent to which existing nuclear images affected perceptions of Nevada as a place to vacation, relocate or retire. Using an innovative approach for measuring images and the valences (negative or positive valuations) attached to those images, this research demonstrated that (a) nuclear images tended to be quite negative, and (b) those with nuclear images of Nevada tended to express less preference for vacationing in that state. Complementary research has argued that, due to the special nature of public perceptions of nuclear risks (Slovic, 1987), coupled with the likelihood of extensive media coverage of accidents involving things nuclear and the roles of activist groups with agendas tied to nuclear issues, signals about even modest accidents involving the proposed nuclear facility are likely to be amplified as they are transmitted to the public (Kasperson et al., 1988; Burns et al., 1990). Thus the proposed nuclear waste repository has the potential to generate a large volume of signals that attach negative nuclear images to Nevada. As the stock of nuclear images grows, Nevada may become stigmatized, resulting in behavioral change as people vacation, relocate, and retire in less stigmatized places. In sum, the prior research has hypothesized a pattern of signal generation, image formation, stigmatization, preference change, and resultant behavioral change that could result in significant losses for the stigmatized community.

This complex argument includes a sequence of important hypotheses about how images are signaled, acquired by individuals, given value, and used to generate preferences and (ultimately) behavior. Furthermore, the argument is of substantial importance for a wide array of decisions made in modern societies confronted with the necessity of managing potentially stigmatizing materials.[1] Nuclear power plants, nuclear waste repositories, chemical manufacturing plants, petroleum refineries, hospitals, and other facilities that are part and parcel of modern industrial societies have bundles of attributes with the potential to generate stigma. It is of considerable importance, therefore, that we understand as much as possible about how potential stigmatization works not just for the proposed repository at Yucca Mountain, but for attempts to site nuclear waste repositories elsewhere and for other potentially stigmatizing facilities.

This chapter focuses on the processes by which individuals acquire images of different kinds, give value to them, and rely on them in development of preferences. Using data from a set of regional and national telephone surveys, hypotheses about who gets what kinds of images, how different kinds of people value those images, and how images are translated into preferences are tested. Underlying the specific hypotheses is a more general proposition that images of places are not randomly distributed, but are best understood as bundles of images that are systematically related, and that different kinds of individuals are quite likely to acquire and use distinct bundles of images. If this proposition is correct, when new kinds of images (e.g., nuclear ones) are introduced about that place, they are likely to be more readily acquired by some people than others, and once acquired are likely to be valued differently. If so, whether a new image will stigmatize to a place depends on how readily that image is acquired, how it is valued, and how it is attached to preferences for the place by individuals who would otherwise be attracted to that place.

A. NUCLEAR IMAGERY AND THE STIGMA MODEL

Stigma research related to Yucca Mountain (Slovic, Layman, Kraus, et al., 1991) has argued that, should a high-level nuclear waste storage facility be constructed in southern Nevada, perceptions of Nevada and its cities are increasingly likely to include images of nuclear waste. Furthermore, these images will be extremely negative, and the association may contaminate general perceptions of Nevada. The resulting stigma may have behavioral consequences, as people find other places to vacation, retire, or relocate their firms. In short, Nevada could suffer serious economic consequences due to the "special effects" of construction of the high-level nuclear waste storage facility.

The importance of the model of risk perception for present purposes is that it signifies how risk perception associated with a nuclear storage facility in Nevada is expected to have its future affect. Since the dread risk mental construct is understood to be quite general among the lay public (Slovic, 1987), exposure to information (via risk amplification) about the high-level waste storage facility is likely to uniformly increase the linkage of Nevada with imagery about the nuclear waste facility and, in turn, reduce the attractiveness of Nevada as a place to vacation, retire or relocate.[2] The available data, collected for Slovic, Layman, Kraus, et al. (1991), indicate that (a) the "underground nuclear waste facility" is indeed perceived to be dreadful, (b) rela-

---

[1]Implications of the stigma argument for broader public policy decisions are discussed in Flynn, Kasperson, Kunreuther, and Slovic (1992), Slovic, Layman, and Flynn (1991), and Gregory et al. (1995, see chapter 1).

[2]This chapter does not directly address the argument that nuclear images will be magnified in transmission via a process dubbed the "social amplification of risk." See Kasperson et al. (1988) and Burns et al. (1990).

tively small numbers of people link Nevada with anything nuclear, but (c) those who do have much less favorable images of Nevada than those who do not. These empirical findings provide support for the argument that serious and costly "special effects" may result from siting the Yucca Mountain facility in Nevada.

B. THE POSSIBLE ROLE OF IDEOLOGY AND CULTURE IN STIGMA

Three aspects of the stigma model deserve more thorough analysis. First, it is not immediately apparent that people's mental images about a place will be a simple function of exposure to information. To the contrary, is seems intuitively plausible that individuals may have fairly complex cognitive filters that effectively screen out some kinds of data while screening in others.[3] Thus it may be that some individuals are more likely than others to recognize the Nevada/ nuclear link, or to retain it in their set of mental imagery once it is recognized.

Second, once people attach an image to a place, it seems likely that some images will be accorded higher salience than others. Thus, when asked to associate a limited number of images with a place, respondents are likely to identify those that are most salient to them. Given the massive information to which individuals are daily subjected, it is probable that—if prompted under congenial circumstances—the resulting image list attached to a place like "Nevada" could be quite extensive. Thus it is likely that many individuals have quite a few possible images to draw upon and will tend to restrict expression to those most salient in a specific context. And, as with the cognitive filters that may shape the content of an image set, the relative salience of particular images within that set are likely to be different across individuals.

Third, systematic differences across individuals are likely to shape both the content of their image sets and the valences they attach to images of particular places. For example, prior research suggests that ideological attitudes tend to be correlated with concern about, and preferences for, nuclear energy (Kuklinski, Metlay, & Kay, 1982; Rothman & Lichter, 1987) and nuclear waste (Nealey & Hebert, 1983; Jenkins-Smith, Espey, Rouse, & Molund, 1991). Self-described "liberals" might be more concerned about, and more opposed to, nuclear facilities than are self-described "conservatives." Similarly, the images attached to the presence or absence of attributes of a particular place (e.g., gambling or legal prostitution in Nevada) might also be significantly associated with ideological positions. If so, ideological positions could be expected to affect both (a) whether nuclear images about Nevada are particularly salient and how they are valued, and (b) how other images of Nevada (gambling, prostitution, etc.) are valued. What appears at face value to be a straightforward link between nuclear images and preferences for vacationing in Nevada might result from the relationship between ideology and an entire set of images about Nevada.

Beyond ideological characteristics, the cultural attributes of individuals might influence the kinds of images people acquire and how they are valued. One variant of cultural theory, as specified by Douglas and Wildavsky (1982), can be taken to suggest that preferences for patterns of social relationships,

[3]See, e.g., Irving Janis' (1983) discussion of "group-think" among decision makers. The more general literature on cognitive screening is vast. See Eysenck, 1990, for an overview.

and the biases that flow from these preferences, will influence both the acquisition and valence of images of a place like Nevada (also see Wildavsky & Dake, 1990; Dake, 1991; Thompson, Ellis, & Wildavsky, 1990). Within this formulation of cultural theory, an "egalitarian" might give particular attention to information about nuclear facilities because nuclear technologies are seen to be the outgrowth of large corporations and central governments—both of which are held by egalitarians to be exemplars of unresponsive and concentrated power. Once acquired, nuclear images are likely to have high salience and very negative valences. And an egalitarian might also take a dim view of gambling, as an outgrowth of greed and a source of inequality. At the other extreme, an "hierarch" might find nuclear facilities to be far less salient and dreadful (because they have faith in the experts that run and regulate them). At the same time, because hierarchs are *not* particularly concerned about inequality and its trappings, they would probably take less umbrage than would egalitarians at such activities as gambling. In both of these cases, the link between nuclear images of Nevada and evaluating Nevada as a place to vacation may express a third variable—culture—rather than a straight-forward linkage between images and vacation preferences as the extant research on stigma has suggested.

An alternative specification of culture, advanced by Inglehart (1971, 1981, 1990), argues that individuals can be usefully characterized as "materialists" or "post-materialists." Post-materialists would be expected to be more concerned about nuclear issues (and environmental quality issues generally) and more likely to put negative valences on such images. At the same time, these individuals would be likely to have less interest in, and place lower value on, such money-based recreations as gambling. Hence, all other things being equal, one would expect post-materialists to have less preference for vacationing in Nevada. Materialists, on the other hand, would be likely to have less concern about nuclear things. In addition, they could be expected to take greater enjoyment in the material glitz and glamour of Nevada's casinos. Thus being a materialist might simultaneously make one less likely to have a nuclear image *and* to have a greater preference for vacationing in Nevada.

If hypotheses derived from ideological or cultural theories are correct, the association that Slovic, Layman, Kraus, et al. (1991) find between nuclear images and negative perceptions of Nevada may be at least partially the result of a third variable (culture or ideology). Having omitted that third variable, previous research has not specifically tested for the effects of these factors. It may well be that cultures or ideologies explain *both* images of Nevada and perceptions of Nevada as a place to vacation, retire or relocate. The model underlying the prior research on stigma (shown with dashed lines) is compared with a model including cultural values and ideology (shown with solid lines) in Figure 7.1.

If the hypotheses underlying the alternative stigma model are correct, we should find significant relationships between measures of an individual's cultural attributes or ideology and (a) the contents of their image sets about a place, and (b) the valences (and associations among the valences) of the images that they attach to that place. If such associations are evident, this does not imply that mental images of a place don't matter; rather, the implication is that the presence and valence of images of a place are not simply a function of available information. They are, instead, mediated by other attributes of the

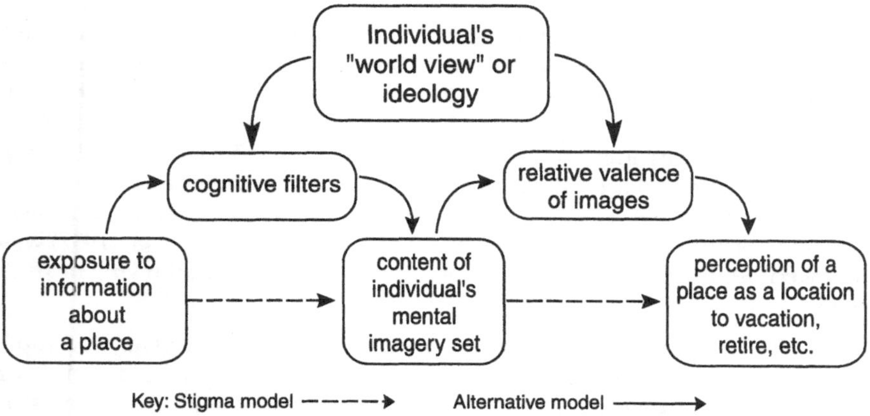

Figure 7.1 Alternative models of mental imagery and stigma.

individual. Information transmitted in the media about a high-level waste facility in Nevada may readily enter the mental image sets of certain individuals, but for others the information may tend to "bounce off" unreceptive cognitive screens. In addition, should nuclear images of Nevada be acquired, the valence attached to those images could vary systematically across different types of individuals. If so, in assessing the effects of imagery, identification of the relevant "receptive" and "unreceptive" populations may be critical.

C. DATA AND METHOD

In order to test elements of the stigma model, survey data containing images of Nevada were collected along with appropriate items to tap the respondents' ideological and cultural attitudes. Following the work done by Slovic, Layman, Kraus, et al. (1991), random household samples of respondents were collected by the University of New Mexico's Survey Research Center from residents of (a) Phoenix, Arizona[4] and (b) the entire United States. The data were collected between December 1992 and February 1993. The Phoenix-area sample consists of two parts: the first includes extensive interviews (including about 130 questions) with over 400 respondents. This survey collected images and their valences for Nevada, Colorado, New Mexico, a nuclear power plant, and a high level nuclear waste repository. The second part of the Phoenix-area sample, consisting of slightly shorter interviews with over 600 residents, collected images of Nevada, New Mexico, and of a high level nuclear waste repository. The national sample included over 800 respondents, from whom images of a high-level nuclear waste repository were collected.[5] For each sample, in addition to the imagery questions, items were included for mea-

[4]This sample is of particular importance because this is where the most substantial connection of Nevada with nuclear imagery was detected by Slovic, Layman, Kraus, et al. (1991).

[5]Images of Nevada were omitted from this sample because Slovic et al. discovered that there was almost no association between Nevada and images of things nuclear among respondents to their national sample.

surement of ideology and culture,[6] perceptions of the degree of risk associated with an array of potentially hazardous facilities, measures of support or opposition for local siting of nuclear facilities, measures of trust in relevant policy actors, and a full demographic profile. The cooperation rates (completed interviews divided by identified valid respondents) were 54.3% for the Phoenix area sample and 71.1% for the (somewhat shorter) nation-wide sample.

To measure images attached to states, nuclear power plants, or nuclear waste repositories, the technique utilized in Slovic, Layman, Kraus, et al. (1991) was followed (Szalay & Deese, 1978). Respondents were read the following instructions:

> The first few questions involve word association. I will give you a topic and ask you to tell me the first word or image that comes to mind. For example, I might say "education," to which you might respond "student," "learning," or "books." Today I am interested in the first four words or images that come to mind when you hear the name of a particular state.
>
> Think about [First STATE] for a moment—when you think about [First STATE], what is the first word or image that comes to mind?
>
> What is the next word or image you have when I say [First STATE]
>
> [First STATE] Your next word or image?
>
> [First STATE] ?

After each elicitation, the interviewer recorded each image on a computer interviewing terminal. Once the images were collected, a second set of questions followed:

> Next, I want to return to the words you associated with the three states to be sure that I understand them. When I say your word, please tell me how it relates to your overall image of the state.
>
> Let's begin with the words you gave me for [First STATE]. The [FIRST] word or image you listed was [FIRST IMAGE FOR FIRST STATE]. Please tell me how this word or image relates to your overall view of [First STATE]. Would you say it is very positive, somewhat positive, neutral, somewhat negative, or very negative?

Each image was coded with a value of +2 for "very positive" through –2 for "very negative." Once the images were collected, images for Nevada and Colorado were assigned to categories using two independent coders and a referee to resolve differences among coders.

In order to develop measures of culture, specific sets of questionnaire items were used to develop indices of cultural "types." For the cultural theory based on the work of Douglas and Wildavsky (1982), a set of items measuring aspects of cultural bias were used to calculate an average score (ranging from 1

[6]The survey items were developed in collaboration with Dr. Karl Dake. These items, used previously and validated in surveys in the United States and the United Kingdom, are designed to tap sets of values—or "cultural biases"—that characterize each of the four grid/group types.

to 4) in which higher scores indicate greater agreement with the cultural biases of each of the four (egalitarian, hierarch, individualist, and fatalist) cultural types. In order to isolate those individuals who may be most characteristic of each grid/group cultural type, those individuals with the highest 100 scores on each measurement of cultural type were identified and coded. For measures of Inglehart's concept of culture, respondents were presented with a set of four distinct types of general social goals and asked to select the two that they believed were most important. The listed goals can be characterized as material (maintaining order, fighting inflation) or post-material (protecting free speech, giving greater access to government). Individuals who chose the two material goals as their top two priorities were classified as materialists, while those who chose the two post-material goals as their top priorities were designated post-materialists. Those who chose one of each were classified as "mixed" types. Finally, for measurement of political ideology, a seven-point Likert-type scale was used for self-placement on a liberal/conservative scale.

II. Who Has Nuclear Images of Nevada?

Given equal exposure to information about Nevada, is it the case that individuals of all sorts are equally likely to acquire a nuclear image of Nevada? If not, what kinds of individuals are most and least likely to acquire a nuclear image of the State of Nevada?

Among the 1004 respondents to our Phoenix area survey, 9.3% (or 93 individuals) gave an image of Nevada that included a nuclear element. Such images included "atomic testing," "nuclear bombs," "nuclear test site," "nuclear waste" and (in one case) "Yucca Mountain high-level nuclear waste repository." Overall, then, the results confirm the frequency of nuclear images among residents of the Phoenix-area reported in Slovic, Layman, Kraus, et al. (1991).

What are the characteristics of those who do—and don't—have nuclear images? As described in Section 1, it might be the case that ideological and cultural predispositions lead to greater (or lesser) propensities to pay attention to specific kinds of images, or to give salience to those images once acquired. To statistically test for these relationships, logit regression was used, both because it has desirable modeling properties (it permits non-linear estimation and allows for dichotomous dependent variables) and because it permits inclusion of multiple explanatory variables in the same model.[7] Using a logit model, the estimated probability that a nuclear image of Nevada will be present can be calculated, given an individual's placement on an ideological or cultural measure.

Table 7.1 shows the results of a logit model using measures of political ideology and the two cultural theories to predict the presence of a nuclear image of Nevada. The ideology measure (IDEOL), employing the seven-point liberal/conservative scale, proved to be statistically insignificant, indicating that liberals are no more likely to have a nuclear image than conservatives. Among the cultural bias measures, the Hierarch measure (HIER) was statistically significant with a negative coefficient. Thus, having a Hierarch cultural bias was associated with a lower probability of attaching nuclear images of

---

[7]For a technical discussion of logit models, see Hanushek and Jackson (1977), pp. 190–199.

Table 7.1 Ideology, Cultural Bias, and Post-Materialism as Predictors of Nuclear Image Acquisition

Count	952
# Missing	52
# Response Levels	2
# Fit Parameters	7
Log Likelihood	–269.98
Intercept Log Likelihood	–286.44
R Squared	.06

	Coef	Std. error	Coef/ SE	Chi-square	P-value	R	Exp (coef)
1: constant	0.06	1.22	0.05	2.12E-3	0.9633	0.00	1.06
IDEOL	–0.04	0.08	–0.44	0.19	0.6590	0.00	0.97
HIER	– 0.86	0.25	–3.40	11.59	0.0007	–0.12	0.42
INDIV	0.32	0.26	1.20	1.45	0.2284	0.00	1.37
FATAL	0.07	0.25	0.28	0.08	0.7765	0.00	1.07
EGAL	–0.48	0.25	–1.97	3.87	0.0490	–0.06	0.62
POSTMAT	0.80	0.25	3.25	10.54	0.0012	0.12	2.23

Nevada. Those who scored high on the Egalitarian cultural bias measure (EGAL), on the other hand, were significantly less likely to have a nuclear image of Nevada. None of the other cultural bias measures proved to be statistically significant. Finally, the post-materialist measure (POSTMAT) was statistically significant, with the positive coefficient indicating that those with post-materialist values are more likely to have acquired nuclear images of Nevada.

The results of the hypothesis tests regarding the presence of a nuclear image in an individual's image set are mixed, but provide some support for the proposition that systematic cognitive filters lead some individuals to screen out (and allow in) certain kinds of images. In particular, among the Phoenix-area survey respondents, people who tended to agree with Hierarch or Egalitarian cultural biases, and who gave priority to materialist values appear to have been systematically less likely to acquire nuclear images about Nevada. Those who gave emphasis to post-materialist values were more likely to acquire nuclear images of Nevada. Thus it is evident that images of particular types are *not* uniformly distributed across a population, but tend to be filtered into, and out of, specific subgroups of that population.

These results suggest that, should a nuclear waste repository be located in Nevada, any increase in the prevalence of nuclear images will be most pronounced for certain kinds of individuals. If having a nuclear image leads to reduced vacation preferences for Nevada, then it would appear that certain kinds of people will be more likely than others to have nuclear images erode their desire to vacation in Nevada. In other words, some individuals are more (and less) susceptible to image-induced vacation preference changes due to the presence of a nuclear repository. Therefore, in order to understand the potential impact of a nuclear repository on tourism in its host state, it is important to learn who this susceptible population is. In the worst case, those who are currently most likely to vacation in that state would be the most susceptible, such that the presence of a repository would lead to greatest reduction in tourism among that state's most likely tourists. On the other hand, it might be the case that those who are attracted to the potential host state are relatively immune to acquisition of nuclear images, and therefore will not suffer a significant vacation preference change.

III. Image Valences and Preferences

What is the relationship between imagery and preferences? Slovic, Layman, Kraus, et al. (1991) showed that the negative and positive valuations—or valences—that individuals attached to their images of Nevada significantly affected the attractiveness of Nevada as a place to vacation. This relationship, in combination with the finding that nuclear images tended to have extremely negative valences, provided the empirical basis for the argument that the promulgation of nuclear images would be likely to result in lost tourism for hosts of nuclear waste repositories.

But the connection between image valences and preference formation should extend beyond vacation sites; if the measurement technique works properly, the valences that an individual attaches to images of a nuclear waste repository should be correlated with their level of fear about, and opposition to, such a facility. More important for this analysis, if there is a strong relationship between the valences given to images of a nuclear repository and the valences given to nuclear images of Nevada, then the nuclear repository images can be used as a proxy for nuclear images of Nevada in tests of the relationships between culture and ideology and nuclear image valences. In general, such a relationship would mean that one could infer from variations in the valences of nuclear waste images to the valences that would have been attached to a nuclear image of Nevada, if a person had had one. Since over 90% of the sample had no nuclear images of Nevada, this is of no small importance for this analysis.

Slovic, Layman, Kraus, et al. (1991) found a strong positive relationship between the valences that respondents attached to their Nevada images and their expressed preferences for visiting Nevada, based on data collected in 1988. The more "positive" the image valences, the more likely the individual was to prefer vacationing in Nevada. This analysis follows the earlier studies by calculating the sum of each individual's Nevada image valences (here called "NVSumScore"). These scores can range in value from –8 (for someone who had four images, all valued at –2) to +8 (for someone who had four images, all valued at +2). An individual with off-setting images (e.g., two images with valences of –2 and two images of +2) would obtain a summary valence score of zero, as would an individual who rated all images at zero (or "neutral").[8] In addition, each respondent was asked to rank their vacation preferences for four states—Nevada, California, Colorado, and New Mexico. Statistical analysis was then used to test for a relationship between the NVSumScore and Nevada's ranking among the four states. The results are shown in Table 7.2.

---

[8]Note that the extreme scores (–8 and +8) are possible only for individuals who are able to provide at least four images and who give them consistently high (+2) or low (–2) valence scores. Thus the NVSumScore variable indicates both the valuation and the "depth" of an individuals image set about Nevada.

Table 7.2 Average Valence Scores for Nevada Images by Nevada's Ranking as a Vacation Destination

Means table for NVSumScore
Effect: NVVacCho

	Count	Mean	Std. dev.	Std. err.
First choice	65	4.14	2.29	.28
Second choice	179	3.20	2.99	.22
Third choice	313	2.29	2.84	.16
Last choice	426	.98	3.05	.15

As shown in Table 7.2, there is a strong relationship between the NVSumScores and Nevada's ranking as a vacation destination. Those who ranked Nevada as their first choice among the four states (6% of respondents) had, on average, image valences that summed to 4.14. Those who ranked Nevada last (46% of respondents) had average NVSumScores of less than 1.0. Thus, these findings confirm the results of Slovic, Layman, Kraus, et al. (1991) regarding the strong positive association between image valences and vacation preferences.

A second test of the association between image valences and vacation preferences was based on an alternative measure of vacation preferences. In this case, respondents were asked the following question:

> We'd like to get your opinions about a short list of states in which many people take vacations. Assuming that there were no significant differences in the cost of the trip, please tell me how attractive each state is to you as a place to vacation. Please use a scale where ten means extremely attractive, zero means not at all attractive, and you can pick any point in between.

The question was asked for Nevada, California, Colorado, and New Mexico.[9] The vacation rating for Nevada is here referred to as "VacNV." The association between the vacation rankings and VacNV was extremely strong, with a positive linear relationship between vacation rankings and the VacNV score.[10]

How well do the Nevada image valences predict the VacNV scores? Using linear regression analysis, the NVSumScores alone predicted 27% of the variation in the respondents' VacNV ratings for Nevada—a very strong relationship for data of these kind. The model is highly statistically significant. These results are shown in Table 7.3. Clearly, the valences of the Nevada images are strongly related to vacation preferences.

To summarize, there is a very powerful relationship between Nevada image valence scores and the overall rating of Nevada as a place to vacation. This relationship holds as well for the other states for which image scores were collected. These results confirm one of the central findings of Slovic, Layman,

[9]The questions for the four states were presented in a random order to assure that there was no bias due to an ordering effect.

[10]Analysis of variance of VacNV by vacation ranking was extremely statistically significant, with a chi-square of 90.90 with 3 degrees of freedom.

Table 7.3 Predicting Nevada Vacation Preference Scores with Summed Nevada Image Valences

Regression summary
VacNV vs. NVSumScore

Count	990
Num. missing	14
R	.52
R squared	.27
Adjusted R squared	.27
RMS residual	2.17

	Coefficient	Std. error	Std. coeff.	t-value	P-value
Intercept	4.56	.08	4.56	55.53	<.0001
NVSumScore	.42	.02	.52	18.96	<.0001

Kraus, et al. (1991) and emphasize the importance of images in shaping (at least stated) preferences.

Is there also a relationship between preferences for vacationing in Nevada and the ideological or cultural attributes of an individual? As discussed earlier, it is possible that due to the strong association of Nevada with particular images, ideology and/or cultural attributes will have an overall linkage to preferences for vacationing in Nevada. To test for this relationship, a linear regression model was tested in which post-materialism, ideological self-placement, and the four cultural bias measures were used to predict vacation preferences for Nevada (VacNV). The results are shown in Table 7.4.

As shown in Table 7.4, several of the culture and ideology measures proved to be statistically significant predictors of preferences for vacationing in Nevada.

Table 7.4 Predicting Nevada Vacation Preference Scores with Culture and Ideology

Regression Summary
VacNV vs. 6 Independents

Count	950
Num. missing	54
R	.14
R squared	.02
Adjusted R squared	.01
RMS residual	2.51

	Coefficient	Std. error	Std. coeff.	t-value	P-value
Intercept	4.03	.83	4.03	4.87	<.0001
PostMDummy	–.34	.20	–.06	–1.69	.0921
Ideology	–.13	.06	–.08	–2.38	.0177
Hierarchy	.24	.17	.05	1.39	.1652
Individualist	.41	.19	.07	2.20	.0281
Fatalist	–.21	.18	–.04	–1.19	.2360
Egalit	.20	.17	.04	1.18	.2380

Those who expressed post-materialist values (represented by the dummy variable "PostMDummy") were significantly less likely to want vacation in Nevada (0.34 points lower on the VacNV scale). Those who were more ideologically conservative (high values on the Ideology measure) were significantly less likely to want to vacation in Nevada. And those who score high on the individualist cultural bias measure were more likely to want to vacation in Nevada.

While these measures do not provide anything like a complete picture of why people would chose to vacation in Nevada (only 2% of the variation in VacNV is accounted for), they do indicate that more general cultural and ideological constructs play a part in shaping preferences for vacation destinations. Our expectation, as discussed earlier, is that a more complete explanation for the link between ideology/culture and vacation preferences will derive from the relationships between an individual's ideology/culture attributes and the larger sets of images that the person has for particular places. Understanding how a particular image affects vacation preferences requires sorting out how that image fits into the larger image set of an individual, and how valences for images in that set are related to more general ideological and cultural attributes.[11]

Given that less than ten percent of the Arizona respondents (and far fewer of the national respondents) had a nuclear image of Nevada, how can tests of the relationship between nuclear images and vacation preference be performed? One test for these relationships is to see how vacation preferences correlate with the nuclear images of Nevada, among the 93 Phoenix-area respondents who had such images. It seems reasonable to expect that the values that an individual attaches to nuclear images of Nevada should be positively associated with their images of a nuclear waste repository.[12] Since there is reasonable assurance that the Nevada image valences are valid measures, a strong association lends validity to the nuclear waste images as well. Furthermore, a strong association will give us some confidence that we can use the nuclear waste image valences (that were obtained from all respondents) to test for the kinds of differences in valence that we would expect for nuclear images of Nevada.

The correlation between the valences for the Nevada nuclear images and those for the nuclear waste repository was 0.44, which was highly statistically significant ($p < 0.0001$). Thus the correlation is strong despite the fact that many of the Nevada nuclear images were of nuclear weapons tests or the weapons test site. Again, this adds to the validity of the nuclear waste repository image valences. Of equal importance, the results indicate that the image valences for the nuclear waste repository are a reasonable proxy for the kinds of valences that individuals might attach to nuclear images of Nevada, should they acquire

---

[11]As with the studies by Slovic, Layman, and Flynn (1990a; Slovic, Flynn, et al., 1991; Slovic, Layman, Kraus, et al., 1991), this study leans heavily on the valence scores attached to images to test hypotheses drawn from the models of stigmatization. For that reason, it is important to assess the validity of the measures of image valence; if the measures themselves are invalid, they will undermine the results of any hypothesis tests that are performed. These tests were performed in Jenkins-Smith (1994). The statistical evidence provides substantial support for the validity of the measures.

[12]At the same time, I would not expect the correlation to be perfect; the images of nuclear bomb testing in Nevada (and the related national security implications) might invoke a quite different set of valuations than would the storage of (primarily civilian) nuclear waste from power plants.

them. Furthermore, should an onslaught of nuclear images result from location of a nuclear waste repository in Nevada (as hypothesized by Slovic, Layman, Kraus, et al., 1991), this correlation would probably grow stronger over time. For that reason, subsequent tests in this report will rely on the valences attached to images obtained in response to the nuclear waste repository word associations.

IV. Nuclear Images and Vacation Preferences

Recent scholarship has argued that nuclear imagery in the United States and other Western societies is widely and deeply negative, primarily focused on the dread attached to visions of nuclear holocaust and fear of accidents at nuclear plants (Weart, 1988; Slovic, Lichtenstein, et al., 1979). But are nuclear images uniformly negative? Evidence from the samples suggests that they are not. Using responses to the image valences for nuclear power taken from over 400 residents of the Phoenix-area in which individuals could score the images from "very negative" (–2) through "neutral" (0) to "very positive" (+2), the sum of each individual's nuclear energy images was calculated (here called "SumNPIm"). The distribution is shown in Figure 7.2.

Figure 7.2 shows that the nuclear power image valences are far from uniformly negative; in fact, they are tilted toward the positive end of the valence scale. Furthermore, there is considerable variation across the valence scale, with significant numbers of very positive and very negative valence scores. These results indicate that, contrary to expectations, it is not reasonable to assume that nuclear images are uniformly negative.

Of course, nuclear power has some positive connotations (e.g., energy production) that are less likely to be attached to nuclear waste storage (Nealey &

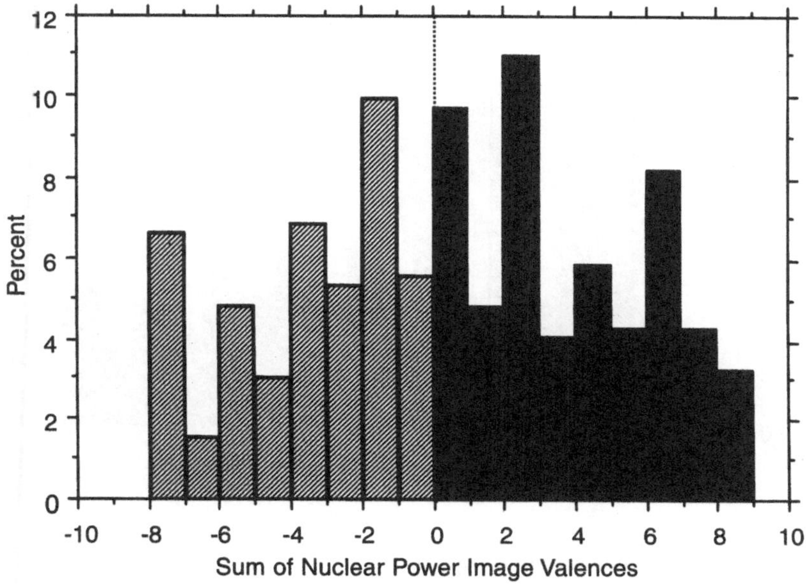

Figure 7.2 Distribution of nuclear power image valences among Phoenix-area respondents.

Hebert, 1983), so we could expect that the valences attached to the images of a high level nuclear waste repository will be considerably more negative. Using the complete set of data from the Phoenix-area sample (over 1000 observations), this expectation proves correct; the average value for the sums of the nuclear image valences is –1.48, significantly below the "neutral" value of zero. This distribution is shown in Figure 7.3.

As Figure 7.3 indicates, there is a significant bulge on the negative side of the valence scale for the images of the nuclear waste repository. Nevertheless, there is considerable variation in valences, with a substantial fraction of the sample—over 35%—having scores *above* the neutral value of zero.[13] Again, it is not correct to assume that nuclear images will be consistently negative.

Given that the valences for nuclear facilities have considerable variance, how much do the valences attached to nuclear images of Nevada vary? Are these images, perhaps, more significantly negative than the other kinds of nuclear images? This might be true if, for example, the kinds of individuals who have these images tend also to be the kinds of people who take an unusually negative view of things nuclear. In addition, if images attached to nuclear weapons are particularly nasty (as argued by Weart, 1988), then the prevalence of the attachment of nuclear weapons images to Nevada might also drive these images down.

Again using the data from the Phoenix-area sample, the mean value and distribution of the valences of the nuclear images of Nevada were evaluated. In this case, since virtually none of the respondents gave more than one nuclear image, the scale runs from –2 ("very negative") to +2 ("very positive"). The

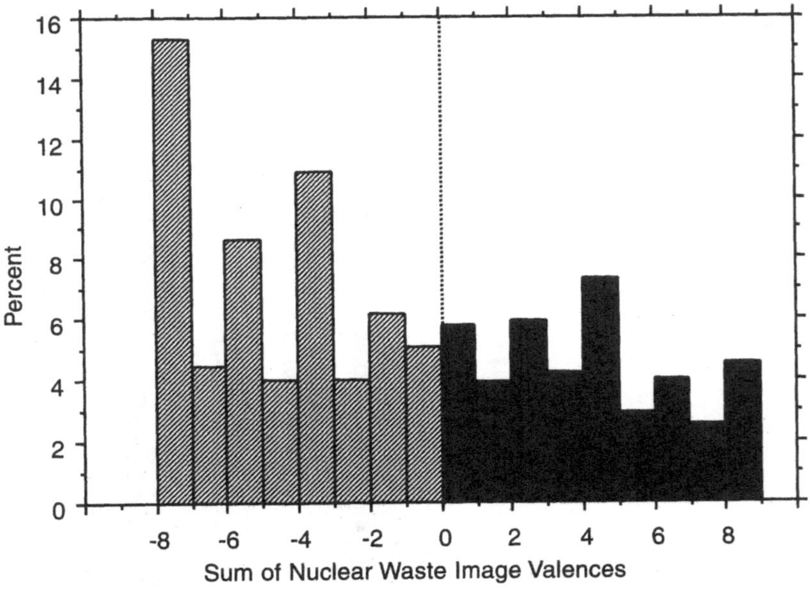

Figure 7.3 Distribution of valence scores for images of a high-level nuclear waste repository among Phoenix-area respondents.

[13]In the national sample, about 32% of the respondents provided image valence scores of above zero for the nuclear waste repository images.

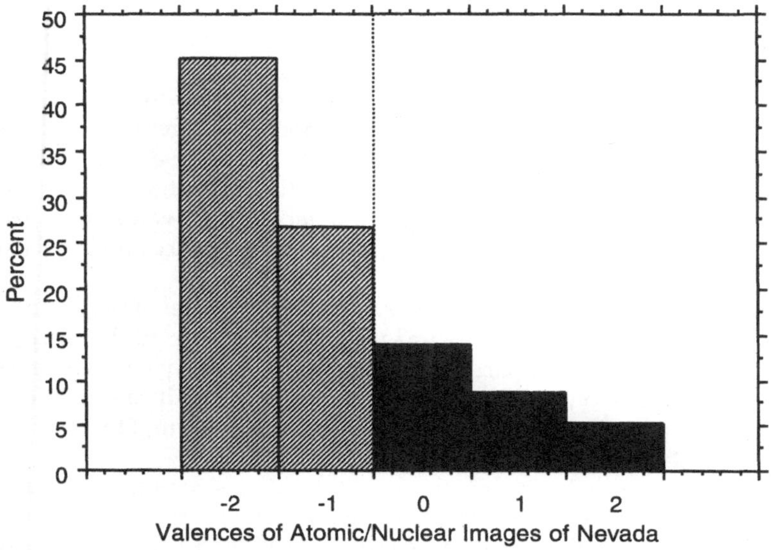

Figure 7.4 Distribution of valences for nuclear images of Nevada among Phoenix-area respondents.

average value of the valences attached to nuclear images of Nevada is –0.97 and is statistically significantly below zero. However, as seen in Figure 7.4, the distribution of valences shows significant variation. While a strong majority of those who had nuclear images gave negative valences, about 28% had valences of zero ("neutral") or above, and 14% had valences above zero. Again, while most of the valences attached to nuclear images of Nevada are negative, it is not the case that such images are uniformly negative.

Slovic, Layman, Kraus, et al. (1991) found that the presence of a nuclear image of Nevada significantly reduced preferences for vacationing in Nevada. Does that result hold in the more recent data? Table 7.5 shows the relationship between the rank of Nevada as a vacation destination among four states (Nevada, California, Colorado, and New Mexico) and the presence of a nuclear image of Nevada. These results confirm the findings of Slovic et al; among those with nuclear images, only 2% gave Nevada first rank among the vacation choices, while those without nuclear images of Nevada would rank it first 7% of the time. At the other extreme, those with nuclear images of Nevada would rank the state last 56% of the time, while those without a nuclear image would rank it last only 42% of the time. Clearly, there is an association

Table 7.5 Incidence of Nuclear Images of Nevada and Rank-Order of Vacation Preferences

	First choice	Second choice	Third choice	Last choice	Totals
No nuc image	7.00	18.67	32.00	42.33	100.00
Nuclear image	2.13	13.83	27.66	56.38	100.00
Totals	6.54	18.21	31.59	43.66	100.00

between those currently having nuclear images and reduced preferences for vacationing in Nevada.

The results shown in Table 7.5 raise several questions. As shown above, not all individuals are equally likely to have nuclear images of Nevada. Is it also the case that different kinds of individuals will place different valences on nuclear images attached to Nevada, should they have them? The remainder of this section tests for systematic differences in the values that people attach to nuclear images, looking specifically for differences that are associated with the ideology and culture.

The first section of this chapter presented a rival model of the acquisition of images, and valences for those images, that hypothesized that image valences would be driven in part by an individual's underlying ideological or cultural predispositions. If correct, there should be a significant relationship between the four cultural bias measures, based on the cultural theory of Douglas and Wildavsky (1982), and nuclear image valences.

To test for such a relationship, the four cultural bias measures were used to predict the sum of the valences for the images received in response to the elicitation "high-level nuclear waste repository" (NWSumScore) using linear regression analysis. The results are shown in Table 7.6.

The results in Table 7.6 show that two of the cultural bias measures are significantly linked to the valences attached to images of a nuclear waste repository, and that a third comes very close. As hypothesized, those who share the egalitarian cultural bias tend to have significantly lower valences for these nuclear images than others. For each one-point shift toward greater agreement with the egalitarian cultural bias, there is an estimated decrease of 1.14 points in the 16-point valence scale. In addition, those who share the fatalist's cultural bias tend to have significantly more positive valences for images of a nuclear waste repository. Each one-point shift toward greater agreement with the fatalist cultural bias is estimated to result in a 1.67 increase in the valences attached to the images of a nuclear waste repository. Hierarchs tend also to give higher valences to nuclear images (+0.51 points for each 1-point increase

Table 7.6 Relationship Between Grid/Group Cultural Bias Measures and Nuclear Waste Repository Image Valences

Regression summary
NWSumScore vs. 4 independents

Count	940
Num. missing	64
R	.20
R squared	.04
Adjusted R squared	.04
RMS residual	4.87

	Coefficient	Std. error	Std. coeff.	t-value	P-value
Intercept	–4.79	1.53	–4.79	–3.12	.0018
Egalit	–1.14	.31	–.12	–3.74	.0002
Hierarchy	.51	.33	.05	1.57	.1167
Individualist	.29	.35	.03	.84	.4025
Fatalist	1.67	.34	.16	4.92	<.0001

on the hierarch scale), though this result falls just short of statistical significance. These findings provide support for the revised model of stigmatization depicted in Figure 7.1.

Does political ideology also affect nuclear image valences? The existing literature on ideology and attitudes toward nuclear issues suggests that those who are more "liberal" should hold a dimmer view—and therefore lower valences—of a nuclear waste repository than would conservatives. To test this hypothesis, the ideology scale and a measure of the strength of an individual's association with their political party were used as predictors of the sum of the valences for the nuclear waste images. The results are shown in Table 7.7.

The results in Table 7.7 indicate that both political ideology and strength of attachment to the major political parties are statistically significant predictors of the valences for the nuclear waste repository images. As hypothesized, the more "conservative" the respondent, the more positive the image valences. An individual who was placed at the extreme conservative end of the scale would be predicted to have a summed valence score that is almost 2 points higher than would someone who was placed at the extreme liberal end of the scale. And the stronger the association with a major political party, the more positive the valences. Thus ideology and attachment to the political process significantly affect how individuals place valences on images of nuclear things.

To summarize the argument to this point, contrary to recent claims that nuclear images are deeply and uniformly negative, the valences attached to images of nuclear power and nuclear waste show considerable variation. Valences attached to images of a nuclear power plant among the Phoenix-area sample were distributed roughly evenly around "neutral," with a slight tilt toward the positive side. Images of a nuclear waste repository were significantly more negative, but still had a substantial percentage of positive valences. The valences attached to nuclear images appear to be in part the result of predispositions—ideological, cultural, and perhaps others—that orient how the individual attaches value to this particular kind of image.[14] Some individu-

Table 7.7 Relationship Between Ideology and Strength of Political Party Attachment and Nuclear Waste Repository Image Valences

Regression summary
NWSumScore vs. 2 independents

Count	673
Num. missing	331
R	.14
R squared	.02
Adjusted R squared	.02
RMS residual	4.97

	Coefficient	Std. error	Std. coeff.	t-value	P-value
Intercept	–4.17	.78	–4.17	–5.32	<.0001
Ideology	.33	.12	.10	2.73	.0066
Party Attach	.62	.28	.08	2.19	.0289

[14]See Dake (1991) for a particularly useful discussion of "orienting dispositions" as they apply across cultural types.

als—egalitarians and "liberals"—will tend systematically to attach more negative value to these images, while those who are more "conservative," and less egalitarian (and more hierarchic or fatalist), tend to attach more positive value. Thus, neither the acquisition of particular kinds of images (as shown in Section II) nor the values attached to an image are purely accidental or random. That means that, should a greater frequency of nuclear images about a place result from siting a nuclear waste repository in that place, those images will disproportionately stick in the minds of particular kinds of people and will be given systematically different values by different kinds of people. Instead of uniform stigma, then, nuclear images are most likely to have a negative impact on vacation preferences of particular subgroups of people, with less effect on others.

But the potential for stigmatization of a location due to hosting a nuclear facility is complicated still further by the potential for interaction among images. Addition of a negative image is unlikely to have much effect on preferences for a vacation site if *other* images of that place are already negative. If images tend to be associated with one another, then assessment of the possibility of stigmatization requires an understanding of how *new* images will fit into the preexisting image sets associated with the potential host site.

V. Image Sets and Associations Among Images

Thus far this chapter has focused primarily on the potential effects of nuclear images, and their valences, on vacation preferences. But what about the other images that people may have about a place? This section presents the categories of images that were obtained in response to the word associations with "Nevada" and compares their frequencies and average valences. In essence, these categories and their valences can be taken to represent the broader sets of images that people employ when shaping perceptions of Nevada. These image categories are then used to test hypotheses bearing on the potential interactions among images and how such interactions may affect the potential for stigmatization of a place. First, tests were run for the possibility that the valences attached to different categories of images may be associated; placing more negative valuations on nuclear images may be related to the value of other images in the Nevada image set. Second, tests were conducted for the influence of cultural and ideological predispositions on the most frequent image that people have about Nevada: gambling.

The images obtained in response to the word associations with "Nevada" were coded into categories, using two independent coders.[15] The categories of Nevada images, along with the percent of Phoenix-area respondents who gave that type of image, are shown in Table 7.8.

As Table 7.8 makes clear, certain categories dominate the image set for Nevada. Images of gambling (gambling, gaming, specific games, etc.), natural characteristics of the state (deserts, climate, lakes, mountains, wildlife, etc.), and Nevada cities were each mentioned by over half of the respondents, with the other image categories falling off rapidly in frequency. Nuclear images

[15]When coders disagreed about the categorization of an image—which was infrequent—a decision was reached through consultation among both of the coders and the principal investigator.

Table 7.8 Nevada Image Categories, Incidence, and Average Valences

Image Category	Incidence (%)	Avg. valence
Gambling	67.0%	0.40
Natural characteristics	65.5%	0.66
Nevada cities	51.0%	0.62
Man-made structures	18.3%	0.96
Casinos	11.6%	0.15
Miscellaneous–general	11.6%	0.47
Nuclear/atomic	9.3%	–0.97
Industry	9.3%	0.82
Vacation	8.8%	1.41
Geographic/demographic	7.5%	0.56
Entertainment	7.3%	1.02
Corruption/prostitution	5.6%	–0.68
Self/family/friend	5.5%	1.15
Money/losses	5.3%	0.32
Economy/taxes	4.5%	1.51
Miscellaneous–positive	3.2%	1.41
Miscellaneous–negative	3.1%	–0.90
Military/weapons	2.8%	0.56
Environmental characteristics	1.1%	0.09
Marriage/divorce	0.9%	0.40

tied with images of industries (at 9.3%) for the seventh most frequently mentioned image category.

The average valence for the images within each category are shown in Figure 7.5. Valences for the economy and (low) taxes take top billing, followed by a miscellaneous-positive category and vacation images. The gambling image falls well down the list, with a valence of 0.40. Natural characteristics, as the second most frequent image of Nevada, fares a little better at 0.66. Thus neither of the most frequent images held about Nevada received particularly high valences. Not surprisingly, nuclear images (with an average valence of –0.97) came in last, behind miscellaneous negative images and corruption/ prostitution (with an average valence of –0.68). These frequencies and average valences for the image categories provide a summary of the mental sketches that people make of Nevada.

In order to make sense of the relationship between images and vacation preferences, then, the entire image set for a place—including both frequencies and image valences—must be taken into account. It is not sufficient to focus exclusively on a particular kind of image within that image set.

The importance of characterizing the entire image set extends beyond the need to account for the relative importance and valuation of images. Images may be related to one another in ways that can mute or amplify the effect of introducing new images into an image set for a place. For example, if the valence of the new image (e.g., nuclear images) is strongly and positively correlated with existing images that are important components of a state's image set (e.g., gambling), then introduction of the new image will tend to reinforce *existing* perceptions of that state. This is because the positive correlation means that those who already tend to have lower valences for the gambling images

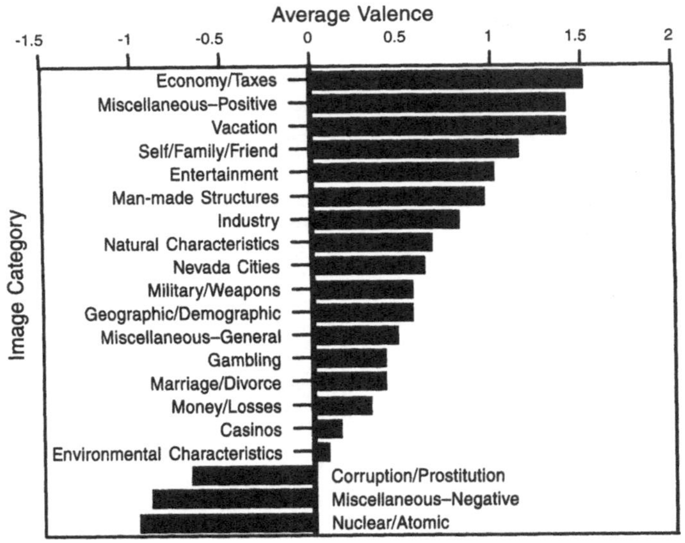

Figure 7.5 Nevada image categories by average valences.

will also assign lower valences to the nuclear images, essentially buttressing what was an already negative view of the place. For those who had more positive views of gambling, the new nuclear images will also tend to be more positive—or at least less negative. If, on the other hand, there exists a strong negative correlation between the new image and prior important images, then introduction of the new image is likely to significantly alter individual's perceptions of the place. If the nuclear image is an anathema (and hence negative valence) to those who like gambling (and give gambling positive valences), the new image will have greatest negative impact on precisely those people who previously had positive perceptions of the place.

Are there significant associations among the valences of images across the different Nevada image categories? This test is made difficult by the fact that, for some categories (e.g., the Nevada nuclear image category) the number of people having the image was quite small, making it very difficult to calculate meaningful correlations with other image valences. There were, for example, only 58 people out of over 1000 respondents who provided both a nuclear image and a gambling image. Since the correlation with the nuclear image valences are of particular importance, the valences for the images attached to a nuclear waste repository were used (since these were obtained for everyone) in lieu of nuclear images of Nevada. The strong correlation between the valences for the nuclear waste repository (as discussed earlier), and the fact that the expected change in the Nevada image set is to be derived from the introduction of a nuclear waste repository in Nevada, use of the nuclear waste repository images would appear to be a reasonable proxy for nuclear images of Nevada.[16] Table 7.9 shows the Pearson correlation coefficients, and the prob-

[16]In fact, given the strong strain of nuclear weapons references among the existing Nevada image set, use of the nuclear images of Nevada for this test may well bias the results, given that inference is being made to the kind of images that would result from presence of a nuclear waste repository.

Table 7.9 Correlation Between Valences of Nuclear Waste Images and the Valences of the Nevada Image Categories

Correlation of nuc repository valences with:	Correlation	P-value
Gambling image valence	.23	<.0001
Casino Image valence	.11	.2537
Prostitution/corruption Image Valence	.50	<.0001
Economy/taxes image valence	.10	.5433
Entertainment image valence	.38	.0013
Industries image valence	.12	.2655
Man-made structures image valence	.15	.0449
Natural characteristics image valence	.14	.0004
Nevada cities image valence	.17	.0001

abilities that these correlations occurred due to random chance, for the valences of the nuclear waste images and the Nevada image categories. Table 7.10 shows the correlation between the gambling image valences and the other Nevada image categories.

Table 7.9 shows that there are significant correlations between the valences that people attach to images of a nuclear waste repository and average valences of several of the Nevada image categories. The more positive (or less negative) the valences of the nuclear waste images, the more positive the images in the categories of gambling, prostitution, entertainment, man-made structures, natural characteristics, or of Nevada cities. Thus, these results suggest that, should nuclear waste repository images be increasingly introduced into people's Nevada image sets, those who currently give lower valences to gambling, prostitution, and entertainment images will be the most likely to attach very negative valences to these nuclear images. Those who attach more positive valuation to gambling, prostitution, and entertainment images will give the nuclear images less negative valences.

More generally, associations across the valences of images for the different categories of Nevada images suggest that there is considerable structure within image sets. Rather than being independent of one another, the valences that are attached to images within particular categories appear to be systematically related to valences attached to other image categories. For example, as shown in Table 7.10 the valences attached to gambling images are related to the valences of many of the other image sets. Those individuals who attach

Table 7.10 Correlation Between Valences of Nevada Gambling Images and Other Nevada Image Categories

Correlation of Nevada gambling valence with:	Correlation	P-value
Casino image valences	.82	<.0001
Prostitution or corruption image valences	.49	.0005
Economy/tax images valences	.57	.0001
Entertainment image valences	.26	.0431
Industries image valences	.18	.1434
Man-made structures image valences	.21	.0189
Natural characteristics image valences	.12	.0162
Nevada cities image valences	.56	<.0001

positive valences to gambling images also tend to attach more positive (or less negative) valences to images of casinos, prostitution or corruption, the Nevada economy, and Nevada's cities (among others).

These results show that the patterns in which people accord value to the images they hold about a place have considerable structure. It therefore makes little sense to think of images of a place as isolated "bits" that independently influence peoples' general perceptions of that place. It make more sense to understand them as parts of an interrelated set of images about a place. Furthermore, the valences that people currently hold about images in their image sets tells quite a lot about the values that they are likely to attach to new images that might be introduced about that place.

This analysis has shown that there are systematic relationships among the valences attached to different kinds of images, and that valences attached to nuclear images are significantly influenced by cultural predispositions. Are the valences of the other image categories influenced by these cultural predispositions as well? One test would be to assess whether valences of other images of Nevada, that are correlated with the valences nuclear waste repository images, are also significantly influenced by the culture measures. Since the valences of the gambling image category were significantly correlated with the valences of the nuclear waste image category, the gambling valences make an appropriate test. Table 7.11 shows the mean values for valences of the gambling images, broken into the four cultural bias categories.[17]

Table 7.11 shows that, on average, the egalitarians attached the lowest valence to the gambling images (0.10), and the fatalists the highest (0.60). The difference between these groups is statistically significant ($p < 0.05$). Thus it appears that egalitarians have more negative images of gambling than others, and fatalists more positive images of gambling. At the same time, egalitarians tended to give the lowest valences for the nuclear waste images, and fatalists the most positive (see Table 7.6). This suggests that general predispositions—like the cultural biases measured here—provide at least part of the organizational framework for the images that people have about a state.

VI. Images and Hazardous Facility Siting

This chapter has presented the results of a series of tests of hypotheses bearing on the ways in which images of a place, and nuclear images in particular, might

Table 7.11 Mean Values of Gambling Valences by Grid/Group Individuals in Cultural Categories

	Count	Mean	Std. dev.	Std. err.
Other	402	.47	1.25	.06
Egalitarian	62	.10	1.30	.17
Hierarch	69	.30	1.28	.15
Individualist	65	.22	1.19	.15
Fatalist	65	.60	1.43	.18

[17]Each group contains those 100 individuals who most strongly agreed with the cultural bias measures for each cultural type. The mean values are shown for those members of each group who had a gambling image.

be related to vacation preferences. Beginning with the pioneering work of Slovic, Layman, Kraus, et al. (1991), the process by which signals about a place are acquired as images by individuals and given valuation has been explored. The particular concerns have been who is most likely to acquire nuclear images about a place, how image valences are related to vacation preferences, and how valences are attached to nuclear images. More generally, the analysis has sought to explain how particular images fit into the broader image sets that individuals hold about a place, how they are associated with other images, and what kind of framework underlies those associations.

The more important findings in this chapter include:

- Some people are more likely to acquire nuclear images of Nevada than others. Those who have post-materialist values are significantly more likely to have nuclear images, while those who share either hierarch or egalitarian cultural biases are less likely to have nuclear images.
- The results confirmed the findings of Slovic, Layman, Kraus, et al. (1991) that the valences attached to images about a place are very strong predictors of vacation preferences for that place. Thus, the more positive the valence of one's images about a place, the more likely it is that one will want to vacation there.
- The valences attached to images of a "high level nuclear waste repository" appear to be reasonably valid measures of the positive and negative affect that people hold about a nuclear waste repository.
- Despite the implication of some scholars (e.g., Weart, 1988) that nuclear imagery is overwhelmingly dread-filled, the valences that people attach to nuclear images of nuclear facilities have considerable variation, ranging from quite positive to quite negative.
- The valences that people attach to nuclear images are related to their cultural and ideological predispositions. Egalitarians and self-described liberals tend to have more negative nuclear image valences, and "conservatives" and fatalists tend to have more positive ones.
- Nuclear images are part of a broader set of images about Nevada, and the valences of nuclear images are correlated with the valences of other Nevada image categories. Those with more negative valences for nuclear images also tend to have more negative images about gambling, prostitution and entertainment.
- Valences of both nuclear and gambling images appear to be influenced by cultural biases. Egalitarians tend to give more negative valences to both gambling and things nuclear, while fatalists give more positive valences.

Taken as a whole, these findings suggest that the stigma model developed by Slovic, Layman, Kraus, et al. (1991) should be elaborated. This stigma model suggests that, should a high-level nuclear waste repository be built in Nevada, the process of "risk amplification" is likely to lead to increasing numbers of people having nuclear images of Nevada. Presumably these images would be widely distributed, and about equally likely to afflict people who are attracted to Nevada as a place to vacation and those who are not. These images will be overwhelmingly negative. The negative images will reduce people's preferences for vacationing in Nevada, and ultimately alter behavior as fewer people vacation, relocate, and retire in Nevada.

The findings presented in this chapter suggest that the stigma model should be revised to account for who gets images and how they attach valences to them. If certain people are more receptive to nuclear images, and others more resistant, it is likely that—even given the process of risk amplification—nuclear images are more likely to enter the image sets and become salient for certain kinds of people. Those with post-materialist values, placing emphasis on protecting individual rights and environmental preservation, are more likely to acquire nuclear images, while those with hierarchic cultural biases (holding greater respect for authority and tradition, and a concern about maintenance of social order) are significantly less likely to acquire such images. If, as Inglehart argues, post-materialists are less concerned about material things and the trappings of material society, then it may be that those who are most likely to acquire nuclear images of Nevada are also less likely to be attracted by Nevada's gaming industry. This appears to be the case, as post-materialists and self-designated conservatives tend to have lower preferences for visiting Nevada, while those with individualist biases have higher preferences for vacationing in Nevada (as shown in Table 7.4).

Beyond image acquisition, the model should specifically address the processes by which valences are attached to images. This analysis has shown that the valences of both nuclear images and gambling are affected by broader predispositions such as ideology and culture. These predispositions seem to act both as screens and as filters that lend particular value to images depending on how they are related to the predisposition. Egalitarians and liberals, for example, tend to attach more negative valences to images of both gambling and nuclear waste repositories. Thus negative nuclear images and less willingness to vacation in Nevada may both result from a broader predisposition that also encompasses a more negative view of gambling. Looking simply at nuclear images and their relationship to vacation preferences risks identifying a relationship that is at least partly spurious.

The results of the analysis support a model of imagery and stigma that looks more like the revised model depicted in Figure 7.1. Broad predispositions, including (but not limited to) ideology and culture, tend to attract and repel different kinds of signals that are encountered, excluding some and accepting others for inclusion in individuals' image sets. Once acquired, the predispositions act to give differential salience and valences to images, which determines how the images affect overall views of a place and vacation preferences. Such a model would explicitly account for the differential propensity of certain kinds of images to be present in the image sets of different individuals, and to have systematically differing values.

The revised model should also account for the importance of the sets of images that exist prior to the introduction of the potentially stigmatizing image. As shown in Section V, images interact. If a new and negative type image is widely introduced into the images sets of a place, the effect of that image on such activities as vacationing, relocating, and retiring will be in part dependent on how the new image is associated with images in the pre-existing image sets. If the new image (e.g., a nuclear image) is *negatively* associated with the valences of images that previously had served to attract people to the place (e.g., a pristine environment), then the nuclear image is likely to lead to greatest reduction in vacation preferences among precisely those people who used to be most attracted to the place. The wide dispersion of such an image might well result in a stigmatization *among those people who used to be attracted to that place.* If, on the other hand, the new image (e.g., a nuclear image) is *positively* associated with the valences of those images that previously had attracted people to the place (e.g., gambling), then the nuclear image will be most positive (or least negative) for those who are most likely to vacation in that place. Those who were least likely to vacation in the place before (those who assigned negative valences to gambling) are the ones for whom the new images will be most negative. In that case, people who didn't want to vacation there before will now want to vacation there even less.

Acknowledgments

Gilbert Bassett, Carol Silva, Teresa Braley, and Gil Freidman provided critical assistance in the writing of this chapter. Useful comments were also provided by Richard Barke, Karl Dake, Doug Easterling, Jim Flynn, Ross Hemphill, Bob Kimball, Howard Kunreuther, Paul Slovic, Gil St.Clair, and Pat Van Nelson. The data collection and initial analyses for this project was funded by Argonne National Laboratory and the U.S. Department of Energy. Additional funding was provided by the University of New Mexico's Institute for Public Policy. None of the findings or conclusions contained in this report represent the positions of Argonne or the DOE.

8 Fear and Loathing of Las Vegas: Will a Nuclear Waste Repository Contaminate the Imagery of Nearby Places?

Doug Easterling

Introduction

Over the past 20 years, a growing number of places have gained a sense of infamy as a result of environmental "disasters": Love Canal, New York; Three Mile Island, Pennsylvania; Times Beach, Missouri; Chernobyl, Ukraine; Institute, West Virginia; Bhopal, India; Goiania, Brazil. In each case, toxic chemicals or radiation contaminated the area, causing greater or lesser degrees of mortality, morbidity, and damage to soil, air, water, and wildlife. In some cases (e.g., Bhopal), death and destruction occurred on a massive scale, while in others (e.g., Three Mile Island), it is unclear whether the accident led to any discernible increase in illness or death. However, regardless of the scope of the direct health and environmental impacts attributable to the event, each of these places attracted international notoriety and derision; even after the *environmental* disaster passed, a *social* legacy remained. Namely, millions of people came to regard these areas as unhealthy, deviant places to be avoided. In a very real sense, they have become stigmatized (Edelstein, 1988; Petterson, 1988; Slovic, Layman, Kraus, et al., 1991; Gregory et al., 1996).

Yucca Mountain—a barren ridge in the desert 90 miles northwest of Las Vegas, Nevada—has been cited as another place that might become stigmatized on account of a risky technology. This is the proposed site for the United States' first geologic repository for high-level nuclear waste (HLNW). Because nuclear waste is viewed with dread by much of the public, there is a very real possibility that a facility designed to store this material would take on highly aversive connotations. If, in turn, an HLNW repository becomes viewed as a major threat to health, the environment, and/or deeply held societal values, the possibility would arise that places near the repository would themselves become stigmatized. Officials and citizens in Nevada have expressed concern over this very prospect, fearing that the state's substantial visitor economy could be adversely impacted by a denigration in the image of places such as Las Vegas.

This chapter considers the viability of these fears in more depth, using both a conceptual framework for stigmatization and empirical studies of visitor behavior to examine the likelihood that an HLNW repository could introduce stigma on nearby places. The analysis finds enough evidence to support the *possibility* of stigmatization effects. Whether these effects actually occur will depend on a number of critical factors—the performance of the facility, future attitudes toward nuclear power, and the degree to which populated areas of Nevada are linked with the repository in media accounts. These factors are each subject to tremendous uncertainty; indeed, at this point it is unclear whether a permanent underground repository will be built at all.

Stigmatization of Place

This chapter is concerned with the prospect of *places* (e.g., Nevada, Las Vegas) becoming stigmatized by a noxious facility. Although the dynamics of stigmatization are somewhat similar regardless of whether we are referring to a place, a technology (e.g., chemical manufacturing), a product (e.g., British beef), or a person (e.g., someone afflicted with AIDS), it is useful to look more specifically at how stigmatization manifests itself when the victim is a place. For the purposes of this chapter, a place is "stigmatized" if the following three conditions are met: (a) a large number of people feel an imperative to avoid the place, (b) this imperative stems from the sense that there is "something wrong" with that place, and (c) the sense of "something wrong" is represented by some sort of mark.[1] Stigmatization requires all three of these conditions to be met, in particular avoidance combined with a sense of "something being wrong." Avoidance behavior is not always indicative of stigmatization; people avoid visiting coastal areas such as Hilton Head and the Outer Banks during hurricane season (out of a reasoned sense of caution), but many of the "avoiders" think very highly of these areas.

A vivid example of stigmatization of place is the city of Goiania, Brazil (Petterson, 1988; Brooke, 1995). In 1987, two scrap dealers smashed open a discarded hospital irradiation machine, sawed open a small platinum capsule, extracted the radioactive cesium chloride, and distributed the blue luminescent material (regarded as "carnival glitter") to their friends and relatives. Through ingestation and physical contact, 129 individuals were contaminated, of whom 50 were hospitalized and 7 died. Throughout Brazil, Goiania was regarded as a place to be avoided, with significant drops in visitation and cancellations of virtually all conventions planned for the city. In addition, residents of Goiania were denied access to planes, buses, and hotels throughout the rest of Brazil; cars with Goiania license plates were stoned; and local agricultural products would not sell. In this instance of stigmatization—which persisted in its extreme form for about a year—most of the residents of Brazil actively avoided Goiania out of a sense that the place was an incredibly dangerous place to be in any way associated with. The issue of how Goiania was "marked" is less clear; no single feature of the city came to symbolize its contaminated nature, but rather, a series of radiation-related images (e.g., waste

[1] This definition draws directly from the prior analyses of stigmatization by Goffman (1963), Jones et al. (1984), Edelstein (1988, 1997—see chapter 4), Slovic, Layman, Kraus, et al. (1991), and Gregory et al. (1996).

drums, workers into protective clothing, radiation-warning signs, children dying of radiation illness) appeared in the media and presumably came to be linked with Goiania in the public's mind.

A number of other places have been stigmatized in the sense that people regard the place to be *dangerous* to visit or live. For example, Bhopal, Love Canal, and Three Mile Island each conjure up risk-laden imagery that make the places unattractive to visit. On the other hand, stigmatization can stem from characteristics other than danger. For example, Tacoma, Washington has long been derided on account of its foul-smelling air—sulfur from the local pulp mills. The health risks associated with this particular contaminant are relatively minor, but the sheer unpleasantness of the odor (akin to rotten eggs) is enough to generate a common perception that there is "something wrong" and to induce people to avoid the city (Egan, 1988).

Stigmatization can also occur for reasons that have little if anything to do with concern over the physical consequences of spending time in the place. For example, when local officials or residents appear to have acted unethically or immorally, outsiders tend to distance themselves from the place. Arizona was avoided by tourists and organizations planning national conventions during Evan Mecham's controversial reign as governor, in which he overturned a referendum to establish a Martin Luther King holiday and made a number of racist comments in public speeches (Reinhold, 1990).

Together these examples suggest that the sense of "something wrong" can stem from many distinct perceptions: dangerous, contaminated, unpleasant, sick, immoral, unnatural, inferior, etc. Environmental contamination and health risks are common sources of stigmatization, but many other sorts of events can stimulate this phenomenon. Regardless of the precipitating event, stigmatized places tend to share a number of characteristics. First and foremost, the place conjures up aversive imagery and triggers negative emotional reactions among a large fraction of the people who live outside that place. Avoidance is thus a natural behavioral response. The widespread public aversion of stigmatized places is typically supported by the media, with repeated stories deriding whatever is wrong with the place. Stigmatization also tends to confer negative traits (e.g., sick, immoral, dangerous) on the local residents. This sort of impugning is apt to be even more extreme in cases where the locals are somehow "to blame" for the misfortunes that have befallen them (Jones et al., 1984; Edelstein, 1997). Looking within a stigmatized community, it is not uncommon to find residents feeling victimized by forces beyond their control, leading to a pervasive sense of helplessness (Edelstein, 1988; Erikson, 1994).[2]

Stigmatization from a HLNW Repository?

In contrast to the other chapters in this volume which describe the consequences, precursors, and mechanisms associated with *historical* instances of stigmatization, this chapter considers the challenges of trying to forecast the

[2]Communities that are stigmatized "through their own actions" (i.e., because they deliberately choose to be different in some important way) may regard themselves not as victims, but rather as morally superior dissidents. For example, some of the white supremacist communities—stigmatized by mainstream America—come across not as helpless, but as defiant.

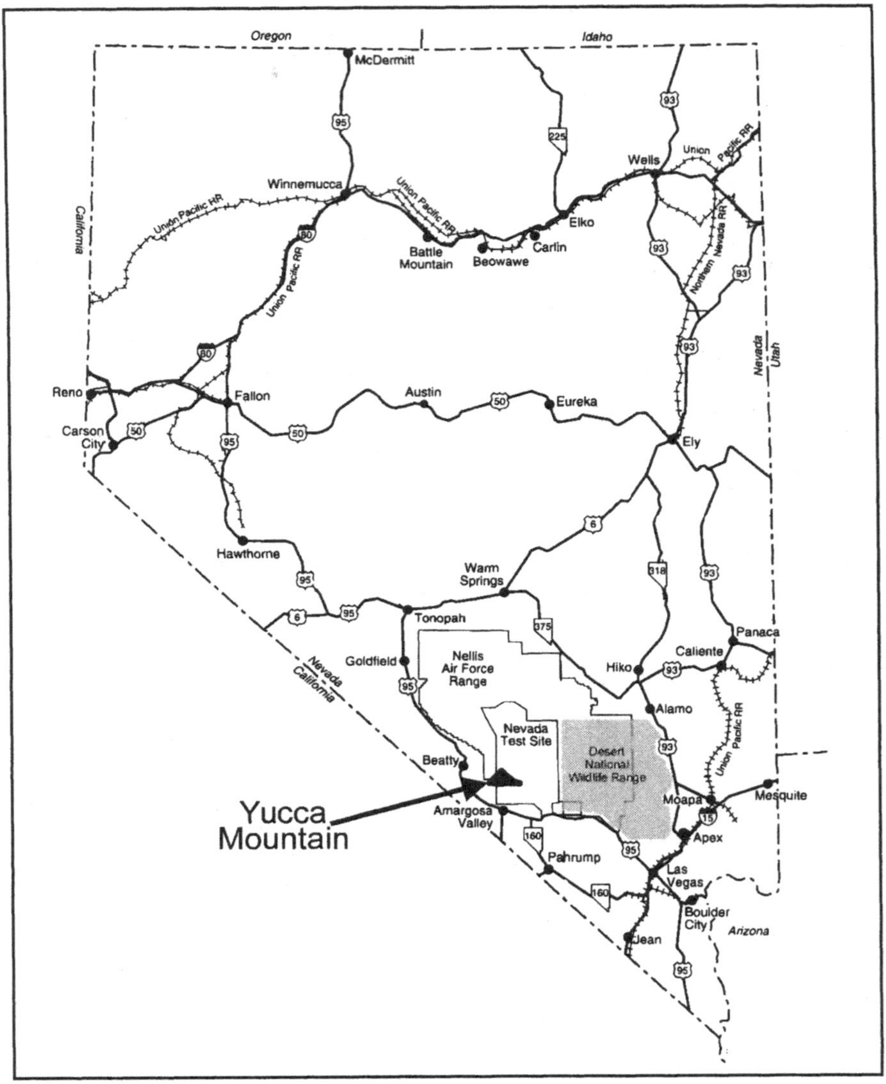

Figure 8.1 Location of the proposed high-level nuclear waste repository at Yucca Mountain, Nevada (DOE, 1999, p. 2–46).

possibility of stigmatization in response to a *future* facility—the high-level nuclear waste repository proposed for Yucca Mountain, Nevada (see Figure 8.1). The question addressed here is whether a facility situated on a barren ridge 90 miles from the nearest city—Las Vegas—could have the sort of serious socioeconomic consequences associated with stigmatization.

THE PROPOSED REPOSITORY

The DOE has spent over $2 billion over the past 15 years studying the suitability of Yucca Mountain as a site for the nation's first repository for high-level nuclear wastes. The waste will consist of spent fuel rods from commercial nuclear reactors and solidified liquid wastes from nuclear-weapons production. These wastes contain a number of highly radioactive isotopes such as strontium-90, cesium-137, uranium-238, and plutonium-239, with half-lives ranging from 239 to 24,400 years. Federal regulations call for these wastes to be isolated for 10,000 years, although other countries have adopted more conservative safety standards requiring 100,000 years of isolation (Lenssen, 1992). At present, spent fuel is stored on-site at the reactors where they were generated—either in pools or above-ground dry-cask storage units—while military wastes are stored at federal weapons-production facilities.

Under the siting approach specified in the NWPA, Yucca Mountain was identified as one of nine candidate sites for the first repository in 1983. Along with the Hanford nuclear weapons reservation in Washington State and Deaf Smith County in Texas, Yucca Mountain emerged as one of the top three finalists under DOE's search process. In 1987, Congress amended NWPA and declared Yucca Mountain to be the site-apparent and terminated all site-characterization activities at any other location. Following a series of court cases in which the State of Nevada unsuccessfully sued to block DOE from conducting on-site examinations, Yucca Mountain has been subject to a series of geologic, hydrologic, and seismic studies, including the construction of a five-mile long tunnel through the mountain. (See Carter, 1987; Jacob, 1990; Lenssen, 1992; Easterling & Kunreuther, 1995; and Flynn et al., 1995, 1997 for more details on the highly complex and contentious siting process.)

If built, the repository will consist of a 100-mile long network of storage tunnels dug 1000 feet beneath a mountain of welded tuff (see Figure 8.2).[3] Approximately 70,000 metric tons of high-level nuclear waste will be entombed in these tunnels using a "reverse-mining" technology. The wastes will be sealed in corrosion-resistant canisters, which will be emplaced in bore holes drilled into the tunnels. Once the repository is filled to capacity, the tunnels will be backfilled.[4] The combination of engineered and geologic barriers is designed to isolate the wastes from the outside environment for a period of 10,000 years.

The prospects for actually constructing the repository are unclear, given both the scientific uncertainty surrounding the ability of the site to meet the Nuclear Regulatory Commission's licensing test and the potential for political and legal challenges to the facility (Easterling & Kunreuther, 1995; Flynn et al., 1997). In response to this possibility, the nuclear industry has actively promoted legislation that would authorize a temporary above-ground storage facility at the Nevada Test Site (close enough to Yucca Mountain that the wastes could be conveniently transferred to the permanent repository if and when it is constructed). This legislation has been repeatedly vetoed by President Clinton

[3]The welded tuff—chosen as a repository site because it is geologically very stable and forms a "solid" base for containment—derives from ancient volcanic eruptions. The question of whether volcanic and/or other substantial seismic activity could disrupt a repository over its operational lifespan has been the source of much scientific and political debate (Nevada Agency for Nuclear Projects (NANP), 1995a).

[4]Under a DOE proposal currently under consideration (the Viability Assessment Plan), the repository would remain open for monitoring purposes for approximately 100 years. The intent of this proposal is to defer the licensing process until more definitive containment data have been collected, while still allowing wastes to be transferred from their current storage location to a "more permanent" disposal facility (Easterling & Kunreuther, 1995).

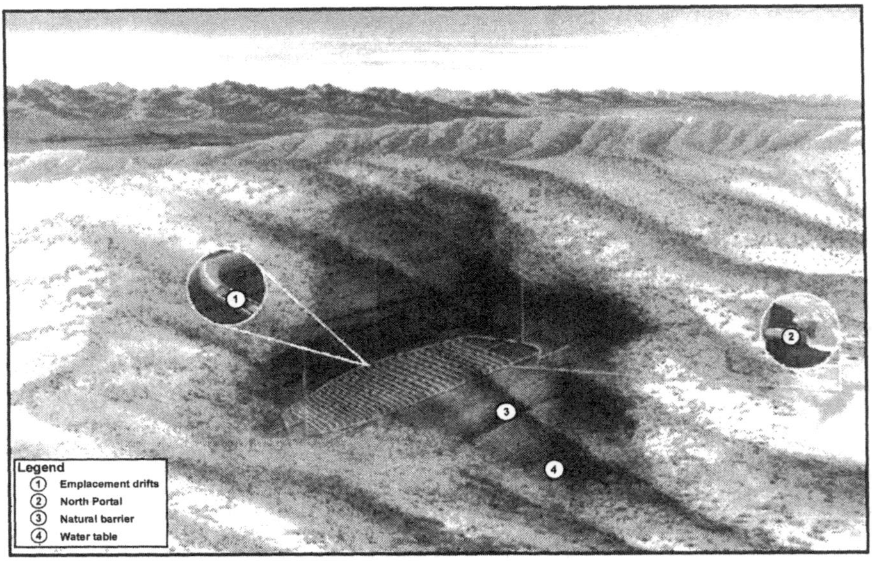

Figure 8.2 Artist's conception of proposed repository subsurface layout (DOE, 1999, p. 2–14).

and proponents have been unable to gain the requisite number of votes to override the veto in the Senate. However, it is entirely possible that a temporary above-ground storage facility for high-level nuclear wastes will be constructed in southern Nevada. If that occurs, the original concern about the socio-economic impact of a *geologic repository* will be extended to the cover the potential consequences associated with a *temporary storage* facility.

CONCERNS OVER THE STIGMATIZATION OF NEVADA

State officials in Nevada have long pointed to the possibility that the repository could adversely impact the visitor industry. For example, Senator Richard Bryan cited this as a major concern in a 1987 journal article (at a time when he was the governor):

> When one considers that tourist-generated revenues account for much of the tax base in Nevada and that tourism-related industries are the state's largest employers, any reduction in visitors could have catastrophic consequences (Bryan, 1987, p. 36).

Likewise, in a lawsuit brought against the secretary of energy in January 1990 (*State of Nevada v. Watkins*, 1990), the Nevada attorney general argued that the potential for economic losses provided the state with a legal basis for preventing a repository from being built in the state.[5] Nevada residents also endorse the potential for economic dislocation: In a 1994 survey of 800 state residents, 62% thought that it was either "likely" or "very likely" that "a repository at Yucca Mountain would cause tourists and other visitors to avoid coming to nearby communities" (Mertz, Flynn, & Slovic, 1994). In addition, the Nevada

[5]This argument was rejected by the Appellate Court.

Resort Association acknowledged the potential for visitor losses in approving a resolution that opposes the repository program (Morrison, 1991).

While Nevada officials have repeatedly highlighted the threat to the visitor industry, DOE officials have consistently downplayed this possibility. For example, Carl Gertz, DOE's former project manager for Yucca Mountain, argued that a repository would have a negligible impact on the decisions of people who would otherwise visit Nevada (Kerr, 1990). Bill Metz (1992), an economist at Argonne National Laboratory under contract with DOE, points to the continuing boom in hotel construction as an indication that investors downplay any possible consequences of a repository on the Las Vegas visitor industry.

LOOKING CRITICALLY AT THE PROSPECT OF STIGMATIZATION

The controversy over whether or not communities in Nevada will become stigmatized by a repository is unlikely to be fully resolved unless and until the facility is actually built. In the meantime, however, it is important to develop well-informed assessments of the likelihood of stigmatization. This will allow for better estimates of the full range of costs associated with the repository, as well as for the design of systems to mitigate the social and economic consequences of the facility.

For Nevada to be stigmatized in response to an HLNW repository at Yucca Mountain, it would seem that two critical conditions need to occur. First, the repository itself must be a stigmatized facility once it is built. In other words, large numbers of people will feel an imperative to avoid the repository out of a sense that there is "something wrong" with the facility. Second, this pervasive sense of "something wrong" must spread to places beyond the repository (e.g., Las Vegas). The next two sections of the paper will examine the plausibility of these two conditions.

Condition 1: Stigmatization of the HLNW Repository

There is at least some reason to believe that an HLNW repository would become a stigmatized facility. In a series of imagery studies, Slovic, Layman, and Flynn (1991) found that the term, "underground nuclear waste repository" tended to elicit responses such as *dangerous, contaminated, dirty,* and *immoral* (see Table 8.1). These images are the basic ingredients in creating stigma, suggesting that if a repository were built, much of the public would regard the facility with disdain and would actively avoid it.

The noxious images observed by Slovic, Layman, and Flynn (1991) are not surprising when one considers the sense of dread that is commonly associated with nuclear waste. Since the beginning of the nuclear age, radioactive substances have provoked special anxiety in the general public (Weart, 1988; Erikson, 1990; Slovic, Layman, & Flynn, 1991). In part, this reflects the dangerous effect that fission products (alpha particles, beta particles, and gamma rays) have on the human body. However, many other materials are just as deadly (e.g., lead, arsenic) but do not instill the same degree of dread. The difference seems to be in the way that radioactive substances harm. Conditions such as radiation sickness, cancer, physical deformities, and genetic mutations come to mind when people think about radioactive waste; radiation seems to generate "unnatural" attacks to the human body.

Table 8.1 Stigma-Related Images Elicited in Response to "Underground Nuclear Waste Repository"

A. DANGER AND HAZARDS (General Images) Danger, Hazard, Unsafe, Scary, Poison, Toxic, Problems	1,722
B. HARM TO HUMANS Death, Cancer, Sickness, Health hazard, Mutations, Birth defects	841
C. POLLUTION, CONTAMINATION, HARM (General Images) Pollution, Contamination, Harmful, Damage	497
D. DESTRUCTION AND DISASTER Destruction, Devastation, Disaster, Tragedy, End of world	312
E. HARM TO THE ENVIRONMENT Harmful to environment; Contaminates water, soil, food	232
F. LONG-TERM HARM Long-term effects, Fear for new generations	22
G. OTHER RISK-RELATED IMAGERY Love Canal, Ticking time bomb	12
H. TERRIBLE, HORRIBLE Terrible, Horrible, Horror, Terror	109
I. WRONG Wrong, Unethical, Evil, Against nature, Don't bury it	79
J. STUPID Stupid, Dumb, Folly	85
K. DIRTY AND DISGUSTING Disgusting, Gross, Dirty, Messy, Ugly, Smelly, Spooky	157
L. DUMP Dump, Garbage dump, Sewer	45
M. TRASH Garbage, Trash, Sewage, Junk	41
N. NASTY STUFF Decay, Slime, Sludge, Crud, Shit	31
O. NOT HERE Not near me, Not in my backyard, Far away, Somewhere else	242

Note. Data are from surveys conducted by Slovic, Layman, and Flynn (1990b). Responses have been recoded according to a slightly different categorization scheme.

In addition, radiation threats seem "special" because they involve an invisible, "energized" agent. This makes radiation appear all the more potent, unpredictable, and uncontrollable. In fact, Weart (1988) goes so far as to argue that radiation is regarded as a "death force" continually seeking to rob the world of its positive, life-affirming nature. This attribution is fostered in large part by the images that people associate with nuclear weapons (e.g., destruction, mushroom clouds, fallout). These associations have generalized to such an extent that apocalyptic images are elicited even for nuclear reactors and radioactive waste (both high-level and low-level).

The link between radioactive waste and nuclear weapons suggests that we are dealing not only with perceptions of *danger*, but also perceptions of *immo-*

rality when dealing with an HLNW repository. A sense of immorality can also be fostered by the nature of the technology that generates radioactive waste. For some persons, the basic notion of generating electricity through nuclear fission seems to violate the "laws of nature." Under some world views, the atomic force that binds together the nucleus is sacred (Black Elk & Lyon, 1990); splitting the atom and transmuting matter is thus viewed as an intrusion into the realm of God (Weart, 1988).

Given all these noxious connotations that are associated with nuclear waste, there is tremendous potential for a place that stores nuclear waste to also be viewed as dangerous and immoral. By design, a geologic repository is intended to isolate HLNW from the environment, and thus to mitigate the risks inherent in the wastes. However, much of the public harbors doubts about the integrity of geologic disposal (Easterling & Kunreuther, 1995). Because of the novelty of the technology, the time requirements for containment, the track record of federal agencies, and the presence of scientific disputes, the public finds reason to question whether a repository would live up to its claims. As a result, the public tends to view an HLNW repository as a particularly risky technology (Kunreuther, Desvousges, et al., 1988; Flynn, Slovic, Mertz, & Toma, 1990).

Furthermore, even if one were convinced that a repository would safely contain the hazards associated with radioactive waste, the facility might still become stigmatized. Namely, nuclear waste is perceived not only as *risky* but also as *immoral*; a repository cannot undo this trait. In fact, to some people, geologic disposal compounds the immorality they associate with radioactive waste because it involves the introduction of corrupt substances into the earth. According to the traditionalists among the Western Shoshone Tribe (whose ancestral lands include the Yucca Mountain site), burial of nuclear waste is an extreme violation of our collective responsibility. Their world view conceives of the earth as a living, active being—Mother Earth, or *Sogobia* in the Western Shoshone language (Fowler, Hamby, Rusco, & Rusco, 1990). Mother Earth is sustained by the rivers that run through her, including the ones that flow beneath Yucca Mountain. Under this world view, burying nuclear waste could conceivably contaminate the life blood of the earth and thus threaten future existence.

Ethical concerns with geologic disposal are exacerbated by the special qualities that some people ascribe to radioactive materials—traits such as "active" or "alive" (Weart, 1988). For such an individual, a repository might be envisioned as allowing radioactive waste to fester and infect the planet. Impressions such as this would account for some of the more value-laden images ("wrong," "evil," "unethical," and "against nature") that Slovic, Layman, and Flynn (1991) observed when respondents were asked to free-associate with a repository. Thus, even if a repository can contain the hazards associated with spent fuel, it cannot necessarily mitigate the moral violation that many people feel when they think about burying radioactive waste.

Overall, a nuclear-waste repository has a number of connotations that seem destined to produce strong aversion on the part of the general public. The relative depth of this aversion is apparent from the results of a pair of surveys conducted by Flynn et al. (1990)—one national and the other of Nevada residents. In these surveys, respondents were asked to specify the *minimum permissible distance* for each of a number of noxious facilities (e.g., landfill for chemical wastes, pesticide manufacturing plant, oil refinery). Of the eight facilities tested, the HLNW repository was regarded as by far the most aversive. As shown in Table 8.2, for each sample the median response for the repository was 200 miles, indicating that half of the sample reported that a proposed HLNW repository would need to be more than 200 miles from their home before they would cease to "actively protest" its construction. By comparison, the other seven facilities had median responses of between 5 and 100 miles.

The proposed HLNW repository clearly stands out as an "extreme" facility among the general public. All the available data indicate that if one of these facilities were currently operating in the United States, people would actively avoid it out of a sense of danger, dread, and/or immorality. However, it is still an open question as to whether these perceptions will persist if and when an HLNW repository is actually constructed. Is the *prospect* of this facility more aversive than will be true of the *actual* facility? The answer will depend on whether people grow accustomed to the repository (i.e., become desensitized to the current connotations) once the facility becomes a reality. This, in turn, will likely hinge on the track record of the repository once it becomes operational; an accident-prone facility would reinforce the pre-existing attributions, whereas an uneventful track record may defuse the fears that are currently associated with a repository. On the other hand, it may take a very long absence of events to counter the current perception of the repository; benignity appears to be a distant prospect.

Condition 2: Stigmatization of Nevada

Given the very real potential of an HLNW repository to become stigmatized among the general public, it is also possible that "nearby" places will be stigmatized. Simply by being physically linked with a hazardous facility, it is possible for a place to take on a "courtesy stigma" (Goffman, 1963; Edelstein, 1988). This raises the question of how close is close enough for stigmatization to occur. In particular, would people avoid the places in southern Nevada that are currently visitor destinations (e.g., Las Vegas, Laughlin, Hoover Dam)?

Table 8.2 Minimum Permissible Distance for Various Noxious Facilities

	Median response (in miles)	
Facility	National sample	Nevada sample
Underground repository for high-level nuclear waste	200	200
Landfill for chemical wastes	100	100
Nuclear power plant	60	100
Pesticide manufacturing plant	50	70
Oil refinery	30	50
Coal-fired power plant	25	50
Garbage dump or landfill	10	15
10-story office building	5	5

Note. From Flynn et al. (1990).

Figure 8.3 Stigmatization of Yucca Mountain (*Las Vegas Sun*). Copyright 1986 by Mike Smith. Reprinted with permission.

DIFFUSION OF STIGMATIZATION

The strongest "courtesy stigma" effects are apt to occur on-site; Yucca Mountain is likely to become a stigmatized place. Indeed, given the nature of the repository's design (i.e., burial within a stable geologic structure), it will be difficult to distinguish between the facility and the place. Thus, if the repository is built according to the current plan, Yucca Mountain will inevitably be identified with the facility and will correspondingly take on whatever connotations are associated with the repository. The possibility of Yucca Mountain becoming a stigmatized place is shown rather vividly in the cartoon in Figure 8.3.

Even if Yucca Mountain becomes stigmatized by the repository, this by itself will not have major economic or social consequences. People already avoid Yucca Mountain. The mountain is located on the Nevada Test Site and Nellis Air Force Range, which are off limits to the general public. Even if Yucca Mountain were accessible, most people would find the stark, barren landscape to be an unattractive vacation spot. Thus, if stigmatization were confined to the repository site, we would see essentially no change in visitor behavior.

Moving out from Yucca Mountain, the first potential for substantive stigmatization effects occurs with the unincorporated towns of Beatty and Amargosa Valley which are located approximately 20 miles from the proposed repository site (see Figure 8.1). However, even if these two communities are stigmatized (i.e., actively avoided by persons outside the area), the potential consequences of such stigmatization are relatively small. The combined population of Beatty and Amargosa Valley is approximately 2,000, and only a minor portion of the local economy is dependent on visitors (i.e., a few gas stations, restaurants, mini-marts and local-history museums along U.S. Highway 95). In fact, any stigma-induced economic losses are likely to be more than offset by the direct economic benefits of the repository—due to in-migration (i.e., repository workers moving into the area) and the associated increase in revenues to local businesses (Little & Krannich, 1990).

The real concern over stigmatization relates not to the towns closest to Yucca Mountain, but rather to the Las Vegas metropolitan area (Clark County). The economy of this area is highly dependent on visitors. In 1995, the Las Vegas area received 28.6 million visitors (including tourists, convention delegates, and gamblers) who contributed approximately $20 billion to the local economy (Las Vegas Convention and Visitors Authority, 1996). If this area were stigmatized by a repository, the socioeconomic dislocation would be tremendous.

IS LAS VEGAS CLOSE ENOUGH TO BE STIGMATIZED?

Predicting how far a stigma will spread is highly speculative at the current time; because the repository has not been built, we cannot observe how people actually respond to the facility. An imperfect substitute is provided through surveys of *intended behavior*. These sorts of surveys, common in marketing research, present respondents with some sort of scenario (e.g., a new product on the market) and ask them to predict how they would respond, in either a deterministic sense (yes/no) or by reporting a probability of emitting a particular behavior (Fishbein & Ajzen, 1975; Pickering & Isherwood, 1974; Morrison, 1979; Sheppard, Hartwick, & Warshaw, 1988). The intended-behavior methodology was employed by the Nevada NWPO as one piece of a large-scale research effort designed to predict the socioeconomic consequences of an HLNW repository (Yucca Mountain Socioeconomic Study Team, 1993).

A pair of public opinion surveys conducted in 1987 by Kunreuther, Desvousges, et al. (1988) provided the first test of whether an HLNW repository could lead people to avoid places as far away as Las Vegas. These surveys, one of Nevada residents and one of persons living in other states, contained a set of questions regarding changes in behavior (visiting on vacation, attending a convention, and locating a new business) following the introduction of a repository "near" the target community (either 50 or 100 miles away, depending upon the experimental condition). As shown in Table 8.3, the majority of each sample (57% of the national sample, 51% of the Nevada sample) indicated that a repository would make it "less desirable" to vacation in a place located 100 miles away (about the distance between Yucca Mountain and Las Vegas). Compared to vacationing behavior, convention attendance was somewhat less influenced by the repository, while locating a new business was more influenced by the presence of a repository. These data suggest that avoidance behavior could extend well beyond the actual site of the repository.

Table 8.3 Reported Effect of Repository on Behavior (1987 Surveys of General Public)

Percent indicating that repository would make place less desirable for Visiting on a Vacation

Distance to repository	National sample (n = 1201)	Nevada sample (n = 804)
50 miles	61.3%	60.8%
100 miles	57.0%	51.3%

Percent indicating that repository would make place less desirable for Attending a Convention

Distance to repository	National sample (n = 1201)	Nevada sample (n = 804)
50 miles	47.5%	50.3%
100 miles	42.7%	43.1%

Percent indicating that repository would make place less desirable for Locating a New Business

Distance to repository	National sample (n = 1201)	Nevada sample (n = 804)
50 miles	72.1%	64.4%
100 miles	68.4%	59.3%

Note. From Kunreuther, Desvousges, et al. (1988).

The usefulness of the Kunreuther, Desvousges, et al. (1988) study is limited by the wording of the behavioral-intent question: rather than asking about expected changes in behavior, the survey asked whether the repository would make a place "less desirable" as far as vacationing, conventions, or new businesses. It might be argued that this approach to questioning produced overestimates of avoidance behavior since we often end up doing things even when they are not completely "desirable." This limitation was remedied in a subsequent survey of convention attendees (Easterling & Kuneuther, 1993). This survey also asked about a range of noxious facilities in order to test whether the proposed HLNW repository has a special potential for producing avoidance behavior.

The convention attendees survey sampled 600 individuals who were members of one of six professional associations[6] and who regularly attended that association's annual meetings. The telephone survey included a long series of questions about the process that one goes through in deciding whether or not

[6]The six organizations included in the survey were: American Frozen Food Institute, American Orthodontic Society, Clinical Laboratory Management Association, Joint Council on Economic Education, National Purchasing Institute, and United Bus Owners of America. These organizations met the following criteria: (a) the organization is national in scope, (b) it holds one convention (not a trade show) each year, (c) the previous four conventions (1986 through 1989) were held in four separate cities, (d) one of these meetings was held in Las Vegas, (e) the organization was willing to provide us with its membership list. Of the 11 organizations that met all five of these criteria, the study included those that maximized the diversity of the sample.

to attend a convention. At the end of the survey, respondents were told to assume that they had made a tentative decision to attend a convention, and then found out that a particular facility was located 100 miles away from the host city. Five different facilities were specified: a prison, a nuclear reactor, a hazardous waste incinerator, a low-level radioactive waste repository, and a high-level nuclear waste repository. For each facility, respondents reported whether they would "definitely attend," "probably attend," "probably not attend," or "definitely not attend."

The results from these questions are shown in Table 8.4. Little effect is reported for the more common facilities: only 1% of the sample indicated they probably or definitely would not attend if a *prison* were within 100 miles of the host city, while 3% reported they would not attend with a *nuclear power reactor*. The three waste disposal facilities elicited stronger reactions. The HLNW repository provoked the most extreme response, with 23% reporting they probably or definitely would not attend. These data indicate that the HLNW repository poses a much stronger risk of stigmatizing nearby places than is true of existing facilities.

Despite these studies, analysts working for DOE have expressed strong doubt that an HLNW repository, even if it were a stigmatized facility, would lead to avoidance behavior 90 miles away (Bassett & Hemphill, 1991; Metz, 1992, 1994, 1996). In particular, it is unclear that a repository at Yucca Mountain would be strongly linked with the city of Las Vegas. Indeed, imagery surveys (Slovic, Layman, & Flynn, 1990b) have found that only a small minority of U.S. residents currently associate places in Nevada with things nuclear (e.g., the proposed repository, nuclear-weapons testing). The unanswerable question is whether a repository, once built, will be a dominant enough feature of the state's cognitive landscape to become linked with major tourist destinations such as Las Vegas.

Whether or not Las Vegas becomes stigmatized may depend in large part on whether the city is linked with Yucca Mountain in media accounts. This would occur if Las Vegas typically served as the dateline for stories about the repository, or if media accounts included a map of Nevada showing Yucca

Table 8.4 Willingness of Convention Attendees to Attend a Meeting After Finding out that a Noxious Facility was Located 100 Miles Away

	Distribution of response			
Facility	Definitely would attend	Probably would attend	Probably would not attend	Definitely would not attend
Prison	78.3%	20.4%	0.8%	0.5%
Nuclear reactor	70.6%	26.2%	2.2%	1.0%
Hazardous waste incinerator	65.7%	27.9%	5.2%	1.2%
Low-level radioactive waste repository	61.3%	28.8%	7.1%	2.9%
High-level nuclear waste repository	49.1%	28.0%	16.1%	6.8%

Note. Subjects were told to assume that this was a meeting that they had decided to attend prior to learning of the facility. From Easterling and Kunreuther (1993).

Mountain and Las Vegas one-eighth of an inch from each other. Even though the repository is a 1½ hour drive from Las Vegas (about the same as going from Philadelphia to the Poconos), people unfamiliar with Nevada's geography may place Las Vegas and Yucca Mountain directly adjacent to each other on their cognitive maps, which would make it easier for Las Vegas to assume the attributes of the repository, including a sense of "something wrong."

LAS VEGAS'S SUSCEPTIBILITY TO STIGMATIZATION

It appears that Las Vegas is close enough to Yucca Mountain that repository-induced stigmatization could occur. However, distance is not the only factor to consider. We should also consider the *ripeness* of Las Vegas for stigmatization.

Some places seem riper than others to become stigmatized. If a penitentiary for habitual sex offenders were built in a small nondescript town in the Midwest, that town might very conceivably be stigmatized; a group of "deviant" individuals would be introduced into a relatively benign cognitive landscape (i.e., a *tabla rasa*). However, if that same facility were plunked down in the middle of New York City, the character of the host city might change very little. New York is already characterized by an extremely rich, complex set of images and perceptions; it is hard to imagine that any single facility could unilaterally displace the existing perception of the city.

This example raises the question of whether Las Vegas is ripe for stigmatization from a nuclear-waste repository. On the one hand, the city is clearly associated with very strong imagery related to gambling, glitz, and entertainment (Slovic, Layman, Kraus, et al., 1991; Easterling & Kunreuther, 1993; Jenkins-Smith, 1994). Because this imagery is so vivid and entrenched, the repository may have little impact on how people think about Las Vegas. On the other hand, one might argue that economically Las Vegas has become completely dependent on this escapist image. If the repository becomes a strong enough stimulus to offset the existing image (i.e., if danger and contamination displace gambling and entertainment), Las Vegas would lose its primary means of attracting visitors. So, in some ways, Las Vegas is almost uniquely vulnerable to the stigmatizing effects of a repository.

It is also important to point out that Nevada is already somewhat stigmatized, at least among some people. A local columnist remarked that:

> This state is the American home of nuclear weapons explosions, legalized gambling, upstanding gangsters, all-night carousing, legalized prostitution and the International Brotherhood of Elvis Impersonators. The Federal Government thinks so much of us it is going to build a nuclear waste dump here. All we need is a bad tie and goggle eyes and we are the Rodney Dangerfield of states (Reinhold, 1989, p. A10).

It is clear that some people associate attributes such as corruption, immorality and tackiness to places in Nevada, particularly Las Vegas. The city's existing image is somewhat tarnished, at least in some quarters. For example, in the survey of convention attendees described earlier (Easterling & Kunreuther, 1990), Las Vegas had the most negative image among the 12 cities tested; 19% of the sample rated Las Vegas's "overall image" as either "poor" or "unacceptable." Given that Las Vegas is at least part way toward stigmatization, a nearby repository might compound the effect.[7]

STIGMATIZATION WILL DEPEND ON LEVEL OF HAZARD

The track record of the repository will play a major role in determining whether places like Las Vegas become stigmatized. If the facility is plagued by highly visible contamination events or other mishaps, the public will be much more likely to infer that there is "something wrong" with the repository and with places nearby. On the other hand, stigmatization is much less likely if the repository builds a markedly benign track record.

The influence of the repository's performance on avoidance behavior is clear from a study of meeting planners (i.e., individuals who are responsible for arranging conventions and other meetings) conducted by Kunreuther, Easterling, and Kleindorfer (1988). This study, carried out in February 1988, involved structured interviews with 153 meeting planners who had chosen Las Vegas as the site for at least one convention or trade show.[8] The interview first explored the process by which planners came to choose Las Vegas for the target meeting (e.g., the number of cities initially considered, the factors that were taken into account). The remaining questions had the planner reconsider the choice of a convention city under a series of scenarios in which the repository was located at Yucca Mountain.

The scenarios presented to respondents varied with respect to the specific events associated with the repository, ranging in severity from no accidents over the first 10 years of repository operation (Event 2) to an investigative report indicating multiple mishaps and a higher than expected risk (Event 7). Each respondents was presented with seven distinct repository scenarios (see Table 8.5). These scenarios were written in the form of news stories.

In addition to an event description, the scenario consisted of a description of how much media attention had been devoted to the event, either *dampened* (the story appeared only briefly), or *amplified* (extensive coverage by national media).[9] This factor was included to test the hypothesis that planners would be more apt to avoid Las Vegas if repository accidents were publicized widely (thus possibly alarming the people who were expected to attend the meeting).[10]

---

[7]Jenkins-Smith (1994) argues that the repository will have little marginal effect in stigmatizing Las Vegas because of the correlation between views about gambling and views about nuclear waste. Namely, the type of person who holds that there is "something wrong" about a repository (i.e., environmentalists, "post-materialists") is the same sort of person who is already alienated by Las Vegas; adding the facility would simply reinforce their preexisting avoidance behavior. In contrast, the individuals who are currently attracted to Las Vegas, and thus the ones whom the city can least afford to offend (i.e., conservatives, "fatalists") do not hold such negative views about a nuclear-waste repository, and thus seem unlikely to stigmatize Las Vegas in response to a repository at Yucca Mountain.

[8]Planners were recruited from a list of upcoming meetings scheduled for Las Vegas hotels. Following an initial telephone screening, planners were sent a copy of the questionnaire, and then were called back to report their responses. The response rate was a 66%.

[9]The media attention factor was varied experimentally, so that half the sample saw a given event with dampened attention and the other half saw it with amplified attention.

[10]We were also interested in whether factors such as the price of hotel rooms and meals or providing attendees with gambling chips or free show tickets might ameliorate whatever impact the repository scenarios had on planners' preferences for Las Vegas. We thus modified the scenarios to incorporate a set of amenities and varied these factors experimentally.

Table 8.5 Repository Scenarios Used in Convention Planner Survey

I. EVENTS
1. OPENING OF REPOSITORY: Construction has been completed on Yucca Mountain repository. Facility will now begin accepting shipments. (This scenario describes the purpose and physical characteristics of the repository.)
2. BENIGN HISTORY: Repository has been accepting waste for 10 years. Operations have been according to expectations. No releases of radiation or identifiable health effects.
3. MINOR ACCIDENT AT REPOSITORY: Accident involving the offloading of a transport cannister. Small radiation release on loading dock, but no significant human exposure
4. MINOR TRANSPORT ACCIDENT: Truck hauling nuclear waste overturns near Las Vegas, but no radiation was released into the environment.
5. MODERATE ACCIDENT AT REPOSITORY: Accident involving the transfer of high-level waste from shipment cask to storage container. Three workers were exposed to radiation and required medical treatment, but contamination confined to a small area at repository
6. MODERATE TRANSPORT ACCIDENT: Truck transporting high-level waste crashed head-on into a gravel truck 40 miles from Las Vegas. Radiation escaped from a defective cask and 4 firefighters were hospitalized for radiation exposure. Traffic detoured for three days.
7. REPORT OF MULTIPLE MISHAPS: An independent consultant issues a report critical of operations at the repository (e.g., sloppy worker practices and insufficient monitoring by management). The risk of radioactive contamination is higher than previously assumed. Haulers have not abided by transportation regulations. Minor accidents at the facility have led to 15 cases of radiation exposure.
II. MEDIA ATTENTION
1. DAMPENED: Extensive coverage in Nevada, but only limited mention in national media and only for a single day.
2. AMPLIFIED: Extensive coverage by Nevada press and national media; lead story in New York Times and on network news broadcasts; week-long followup.

Note. From Kunreuther, Easterling, et al. (1988).

After reading through a particular scenario, the respondent indicated whether he or she would still choose Las Vegas for the target meeting. If the person said no, they he or she indicated how Las Vegas would rank among the possible cities. One of the possible responses to this question was "Las Vegas would no longer be considered for this meeting." By comparing Las Vegas's post-scenario ranking to the ranking that the planner initially assigned for the target meeting, one can tell whether the scenario influenced the planner's preference for Las Vegas.[11]

[11]The vast majority of the planners (116 out of 153) assigned a ranking of 1 to Las Vegas in the no-repository case. The remainder had booked their meeting for Las Vegas even though this was not their top choice. Las Vegas was chosen in these cases either because the preferred city was not available for the meeting dates, or because the planner had been overruled by the officers of the organization.

The repository brings about substantial reports of avoidance behavior on the part of the sample. Table 8.6 shows that under all scenarios, at least 30% of the planners lowered their rank for Las Vegas compared to their initial ranking for the target meeting. Moreover, the size of the effect was highly dependent on the particular repository scenario.[12] In the most extreme scenario (the repository was plagued by recurrent accidents and safety lapses, accompanied by amplified media coverage), 75% of the sample lowered their ranking of Las Vegas and 43% indicated that they "would no longer consider Las Vegas" for the meeting.

The data in Table 8.6, like those from the convention attendees survey described above, provide only an imprecise estimate of the actual impact of a repository on visitor behavior; the link between stated intent and subsequent behavior is far from perfect (Fishbein & Ajzen, 1975; Pickering & Isherwood, 1974; Morrison, 1979; Sheppard et al., 1988; Easterling, Morwitz, & Kunreuther, 1991). Much of the imperfection is due to the long latency between the surveys and the opening of the repository. In fact, the individuals who are providing the intention data are only partially representative of the population of individuals who will be making decisions when the repository comes on line. The long time horizon that applies in the case of the repository allows for the possibility that actual avoidance of Nevada could be either much greater or much less than what economic agents currently anticipate.

The most important thing to take away from the convention planner survey is not a specific estimate of convention losses, but rather a recognition that the stigmatizing potential of the repository depends strongly on what actually occurs with respect to the facility if and when it comes on line. In particular, the likelihood of Las Vegas becoming stigmatized by a repository at Yucca Mountain will increase if:

1. The repository is plagued by accidents, particularly accidents that result in people being contaminated by radiation;
2. Those accidents occur along transportation routes (which are closer to Las Vegas) rather than on site; and
3. Repository-related mishaps attract significant media attention.

Until the performance of the repository is known more precisely, the prospects for stigmatization will remain highly uncertain.

Concluding Thoughts

STIGMATIZATION AND AVOIDANCE OF NEVADA

Stigmatization has become a very real threat to the social and economic well-being of communities (Edelstein, 1988). This threat provides a strong impetus for political action. For example, Sen. Wayne Allard (R–Colo.) petitioned the

[12]An analysis of variance based on a second outcome measure (likelihood of choosing Las Vegas) found highly significant effects for the repository-scenario factor and the media attention factor [$F(6,1204) = 89.30$, $p < .0001$ and $F(1,1204) = 65.44$, $p < .0001$, respectively]. Of the seven amenity factors, only two (hotel room discount and free gambling chips) significantly influenced the planner's decision, and only slightly (Kunreuther, Easterling, et al., 1988).

Table 8.6 Changes in Planners' Ranking of Las Vegas Under Repository Scenarios

Scenario	Sub-Sample	N	% who lower their ranking of of Las Vegas	% who no longer consider Las Vegas
1A. Opening of repository (base case: dampened media attention & no amenities)	Both	153	32.0	7.8
1B. Opening of repository (replication)				
a. Dampened	A	75	32.0	8.0
b. Amplified	B	78	37.2	7.7
2. Benign 10 year history				
a. Dampened	A	75	30.7	4.0
b. Amplified	B	78	30.8	9.0
3. Minor accident on site				
a. Dampened	A	75	40.0	6.7
b. Amplified	B	78	46.2	14.1
4. Minor transport accident				
a. Dampened	B	78	38.5	12.8
b. Amplified	A	75	49.	10.7
5. Moderate accident on site				
a. Dampened	B	78	41.0	14.1
b. Amplified	A	75	49.3	21.3
6. Moderate transport accident				
a. Dampened	A	75	53.3	32.0
b. Amplified	B	78	64.	30.8
7A. Report—recurrent accidents and new risk				
a. Dampened	B	78	55.1	32.1
b. Amplified	A	75	74.6	42.7
7B. Report—recurrent accidents and new risk (replication)				
a. Dampened	B	78	55.1	30.8
b. Amplified	A	75	74.6	48.0

Note. For each repository event, half the sample was assigned to the *dampened* media attention condition and half was assigned to the *amplified* condition. The specific assignment is shown with the "sub-sample" column. Each respondent saw Event 7 twice. The two replications (7A and 7B) differed from each other with respect to a number of experimental factors related to the "amenities" associated with choosing Las Vegas (e.g., free gambling chips, reduced hotel costs). From Easterling and Kunreuther (1993).

EPA to remove the business district of Leadville, Colorado from the Superfund National Priority List, arguing that:

> One of the main economic activities of Leadville is tourism and recreation. The impact of having the entire city on the NPL is a drag on that activity.... They [local restaurants] are not contaminated and should not have the stigma of being a Superfund site (*Denver Post*, 1997, p.4B).

In the case of the HLNW repository, the threat posed by stigmatization is profound. Indeed, whenever a site has been identified as a candidate for the repository (or for temporary HLNW storage facilities), local residents and officials have cited stigmatization as a potential consequence. For example, farmers in eastern Washington state feared that a repository at Hanford would lead consumers to avoid fruits and wines grown in the area (Dunlap, Rosa, Baxter, & Cameron, 1993). Concerns over avoidance behavior were also raised by state officials in Tennessee when DOE proposed to build an MRS facility in Oak Ridge (Sigmon, 1987). Fox et al. (1985) surveyed two key populations to assess the economic threat posed by an MRS at Oak Ridge. In a sample of 306 persons living outside Tennessee, 47% indicated they would change their vacation plans if they learned that their destination was located "near" an MRS facility. Among a sample of 130 business executives, 55% indicated that they would be less willing to locate a business in a county that was home to an MRS facility.

The potential consequences of stigmatization are particularly high in the case of a permanent repository at Yucca Mountain. Nevada is the most tourist-dependent state in the country, with almost 40% of the state's non-farm workforce employed in tourism-related industries such as hotels and casinos. The next highest tourist-dependent state (Hawaii) has a figure of 16% and, and for the United States as a whole, tourism employment runs at 4 to 5% (Edmonston, 1994).

Although there are huge economic stakes associated with building an HLNW repository at Yucca Mountain, it remains unclear whether the facility will in fact be stigmatized, and if so, whether this stigmatization will spread from the site to places as far away as Las Vegas. The prospect for stigmatization will depend in large part on the events associated with that facility. However, these events are subject to substantial uncertainty (Flynn et al., 1995). At a rudimentary level, the technical design of the repository (which defines the "stimulus" to which visitors respond) has not fully taken shape (Planning Information Corporation, 1995). More significantly, there are innumerable controversies over the likelihood of a radiation release (NANP, 1995a). On the one hand, opposition surrounding the repository could dissipate, the facility could be constructed in short order, and spent fuel could be shipped to Yucca Mountain and buried without incident and without undue publicity. Under this scenario, the repository would likely be a benign addition to the public's image of Nevada—either ignored or regarded as a largely neutral piece of the landscape. On the other hand, if one assumes a severe repository scenario (e.g., with a set of high-publicity accidents and controversies), there is a very real potential for significant visitor impacts, in the extreme stigmatizing Nevada as a contaminated place to be avoided.

LESSER FORMS OF AVOIDANCE IN THE SHADOW OF A REPOSITORY

Even if repository fails to produce *stigmatization* of Las Vegas and other places in Nevada, visitor losses and economic impacts could still occur. Stigmatization is an extreme process; there are other, less dramatic mechanisms through which people may decide not to visit the state (Easterling, 1997). Namely, the choice not to visit a place often is made "on the margin," not as a result of an imperative to avoid the place. Taking a vacation typically involves a choice between a set of relatively attractive options. For example, a family in California with young children may consider going either to Disneyland in Anaheim, Seaworld in San Diego, or one of the theme hotels in Las Vegas (e.g., Aladdins,

MGM Grand, Mirage). The eventual choice is apt to be only slightly preferred to alternatives. Thus, if a place like Las Vegas becomes even slightly less attractive, some tourists who would have otherwise chosen to visit there will go elsewhere instead. This same sort of "tipping the balance" process occurs regularly with convention planners; cities aggressively compete with another to host conventions, so when any one candidate city becomes less attractive, there tend to be a number of attractive challengers waiting in the wings (Kunreuther, Easterling, et al., 1988).

The marginal nature of the visitation decision process is demonstrated by a series of studies that test Slovic, Layman, Kraus, et al.'s (1991) *imagery* model. Under this theory, people naturally associate a set of images with any particular place they think of. For example, "mountains" come naturally to mind when most people think of Colorado and "gambling" comes up for Las Vegas (Slovic, Layman, Kraus, et al., 1991; Jenkins-Smith, 1994). A number of studies have shown that the set of images that come up when we think about a place influences the probability that we will visit there.

For example, Slovic, Layman and Flynn (1990b) and Slovic, Layman, Kraus, et al. (1991) conducted a set of surveys in which respondents were asked to report the images they associated with various places—cities such as Las Vegas and Denver, states such as Nevada and New Mexico. Respondents then rated these images in terms of positive versus negative connotations; summing across the different image ratings generated an imagery score for the place. These imagery scores were strongly related to the preferences that the person expressed for visiting the places: cities with higher imagery scores were ranked higher in terms of desire to visit. In addition, a longitudinal analysis found that imagery scores collected at one point in time could predict actual visitation behavior over a subsequent 18-month period (see Figure 8.4).

The predictive ability of imagery data was replicated in the convention attendees survey described above (Easterling & Kunreuther, 1993). In this survey, respondents provided images (and the associated imagery ratings) for the four cities in which their professional association had held its last four meetings. A set of logistic regression analyses tested whether the imagery scores could predict the individual's pattern of attendance at those four meetings.[13] As shown in Table 8.7, the imagery scores were significant at $p < .05$ for four of the six organizations, while in one other case, the imagery effect had a p-value of .07. The only organization in which imagery was clearly nonsignificant was the American Frozen Food Institute. In general then, attendance at past meetings seems to vary systematically according to the imagery associated with the host city: the more positive a city's imagery, the more likely the individual was to have attended a convention there.

Jenkins-Smith (1994) employed a similar methodology and found highly similar effects of imagery on visitor preferences. Namely, respondents provided images for four states (Nevada, California, Colorado, and New Mexico), along with a rating (0-to-10) of how "attractive the state is to you as a place to vacation." The imagery scores accounted for 27% of the variation in reported

[13]The analysis included a meeting only if the respondent had been a member of the association at the time of the meeting. To account for individual differences in propensity, we entered as a control variable the percentage of meetings that the individual had attended since becoming a member.

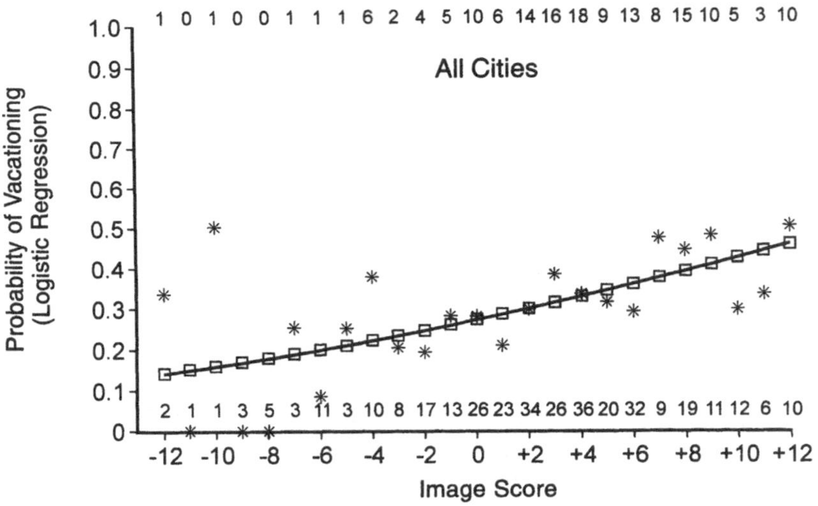

Figure 8.4 Relation between imagery and subsequent visitation behavior. Probability of vacationing in a particular city after June 1988 as a function of image scores elicited prior to that date (Phoenix survey). Upper row of numbers indicates the number of people with that image score who vacationed in the city; lower row is the number who did not vacation in the city; * marks the proportion who vacationed. The curve is the best fit logistic function to these proportions. From Slovic, Layman, Kraus, et al. (1991). Copyright 1991 Society for Risk Analysis. Reprinted with permission.

vacation preference ($t = 55.53$, $p < .0001$). Once again, we find evidence that people are more inclined to visit a place as its imagery becomes more positive.

All of these imagery studies found that the effect of imagery on visitation behavior was *linear*. In other words, an *incremental* change in the imagery of a place leads to an *incremental* change in the number of people who visit there. This means that some losses to the Nevada's visitor economy could occur through fairly minor diminishment in the state's imagery.[14] Stigmatization would clearly produce much more drastic impacts, but the prospect of losses does not hinge solely on stigmatization; even if Las Vegas is not tainted with a pervasive sense of "something wrong," a repository could lower its attractiveness as a place to visit.

STIGMATIZATION UNDER VOLUNTARY SITING

As we examine the potential for stigmatization of places in Nevada, it is important to point out that elected officials and residents in the state have actively opposed the siting of a geologic repository at Yucca Mountain (along with the temporary HLNW storage facility proposed for the Nevada Test Site). Thus, stigmatization is a viable prospect in cases where the affected communities reject the offending facility. It is plausible to assume that stigmatization would be even more pronounced under a voluntary-siting model.

[14]Incremental changes in the perceived risk associated with a place—through crime, environmental contamination, or natural hazards—can also produce incremental losses in visitor behavior (Easterling, 1997).

Table 8.7 Relation Between Host City's Imagery Score and the Decision to Attend Past Meetings: Results from Logistic Regression Analysis

Association	b	*SE*	G^2
Amer Frozen Food Institute	.035	.048	0.53
Amer Orthodontic Society	.177	.055	10.14**
Clinical Lab Management Assn.	.066	.037	3.29+
Joint Council on Economic Education	.132	.041	10.42**
National Purchasing Institute	.079	.040	3.85*
United Bus Owners	.106	.039	7.28**

Note. "Imagery score" is computed as the sum of the ratings associated with the images elicited by the city. Respondents could report up to six images for each city. Each image was assigned a score from –2 ("very negative") to +2 ("very positive"). Thus, a city's imagery score can range from –12 to 12. From Easterling and Kunreuther (1993).
$^{+}p = .07$. $^{*}p = .05$. $^{**}p < .01$.

According to Jones et al. (1984) and Edelstein (1997), the "origin" of a stigma is crucial to determining the size of the effect: stigmatization is more potent in cases where whatever is "wrong" with a place is self-induced, or at least appears self-induced to people outside the place. Just as stigmatization of people is more pronounced among diseases such as AIDS that have a behavioral component to their etiology (Jones et al., 1984), stigmatization of places is aggravated by volition in the siting of deviant facilities. In effect, the "contamination" aspect of stigmatization is exacerbated by the "immorality" aspect; a community is seen as profoundly deviant if the place is flawed or marked through the deliberate actions of its residents.

The "origin" aspect of stigmatization suggests that voluntary approaches to siting may be more apt to stigmatize places that host noxious facilities. In particular, a nuclear-waste storage facility may be more likely to produce stigma when the host state "invites" the facility than when the facility is sited over the objections of the host state (as is the case with Yucca Mountain).

The potential for stigmatization in fact emerged as an issue under the federal government's attempt to use voluntary siting for an MRS facility. The Office of the Nuclear Waste Negotiator (ONWN; operational from 1990 through 1994) canvassed states, counties, and Indian tribes in an attempt to find a place willing to consider hosting an MRS facility in return for significant financial and other compensation (ONWN, 1993; Easterling & Kunreuther, 1995). A small number of counties entered into negotiations, but were forced out of the process by the governors of their respective states. Most of the jurisdictions that actively participated in the process were Indian tribes, with four tribes still interested when the Office was terminated by Congress.

Counties and tribes opted out of the voluntary siting process for a number of reasons, including health and environmental risks, concerns over inter-generational equity (i.e., selling off future generations in order to increase the wealth of current residents), and the prospect of stigma-induced economic effects (Gowda & Easterling, 1998). Those involved in the process were often concerned that the act of "accepting" a facility that is as suspect as a nuclear-waste storage facility would cause outsiders to view the host community as deviant and/or unethical.

One exception was the Mescalero Apache tribe in New Mexico which ultimately broke off from the federal process to negotiate with a private consortium of utilities to build an MRS on its lands. Even though the Mescaleros have a highly tourist-dependent economy (with the Mountain Gods casino complex and the Ski Apache resort), tribal officials believed that the MRS would only serve to enhance its economic standing (Satchell, 1996). On the other hand, a sizeable number of tribal members questioned the judgment of the council in soliciting the facility (Linthicum, 1995).

MITIGATION OF STIGMA

Assuming that stigmatization is an authentic threat with a nuclear-waste repository, the question arises as to whether anything might be done to mitigate the effects (i.e., to minimize the avoidance and ostracization of places near the facility). The answer to this question, in turn, hinges primarily on the question of whether a nuclear-waste repository can be made to appear less risky, deviant, unethical, onerous, etc. To date, the obvious remedies—an elaborate licensing process, years of scientific study, a combination of engineered and geologic containment barriers, concerted media campaigns—have failed to overturn the public's skepticism and animosity toward the repository (Flynn, Slovic, & Mertz, 1993; Easterling & Kunreuther, 1995).

It is entirely possible that if a repository is actually built, the public will come to find it a more benign "mark" than they currently envision. This would argue in favor of simply going ahead and developing the repository despite concerns over stigmatization and other potential consequences to people and the environment. On the other hand, the public may never grow comfortable with this facility regardless of how familiar they become with it. As with many other forms of nuclear technology (Weart, 1988), the repository seems to stir up a level of dread that defies mitigation or "education." The most important question seems to be not, "How do we mitigate the stigma associated with a nuclear-waste repository?" but rather, "What sort of violation do people experience when they are confronted with this facility?" Once we more fully understand the origins of the deep emotional reactions triggered by a repository, we will have more insight into the question of whether or not this facility is fundamentally appropriate.

9 The Effects of the Rocky Flats Nuclear Weapons Plant on Neighboring Property Values

Wayne L. Hunsperger, MAI, SRA

Introduction

The Rocky Flats Nuclear Weapons Facility, located 16 miles northwest of downtown Denver, Colorado, has been the source of thousands of jobs and the consequent economic impacts for more than four decades. In this sense it has contributed to the economic base of the metropolitan area. At the same time, it has processed some extremely dangerous materials and thereby existed as a source of risk to nearby residents and properties.

Historically, there have been a number of on-site accidents and off-site releases of hazardous substances. In June of 1989, the plant was raided by the Federal Bureau of Investigation (FBI), and its operator, Rockwell International, subsequently pleaded guilty to environmental crimes and paid an $18.5 million fine. Given the conflicting economic, social, and risk-related issues associated with Rocky Flats, any study of property value impacts due to the plant and its operations is complex.

One result of the FBI raid was a class action suit against the facility contractors claiming compensation for health and property claims. The motion for class certification was filed in June, 1993. The population claimed for the health effects was about 40,000 people (in 1990) and the property value class area contained about 15,000 properties. The studies reported here were conducted on behalf of the plaintiffs.

One essential obligation for the plaintiffs was to determine property values impacts due to Rocky Flats and, if found, to quantify the effect. Because these impacts could be subtle, a research program was designed to examine the problem from a number of perspectives. Consequently, experts were engaged from a number of disciplines—survey research, risk analysis, statistics, economics, and real estate appraisal.

Rocky Flats History

Rocky Flats was established in 1951 on a 10 square mile tract in Jefferson County, Colorado, as one of 17 major facilities that make up what is known as the

"Nuclear Weapons Complex." The plant opened in 1953 with Dow Chemical Company as the first Operating Contractor. Dow was succeeded in 1975 by Rockwell International Corporation, who operated the plant until EG&G took over in 1990. Until 1990, plutonium triggers for nuclear weapons were processed, purified, machined, and prepared for assembly. These triggers, called "pits," also include uranium, beryllium, and other materials provided by various facilities in the DOE's Nuclear Weapons Complex. The completed pits were shipped to the Pantex plant near Amarillo, Texas, where they were assembled with other components to build nuclear weapons.

Due to the way hazardous materials were handled, stored, and disposed of over the years, the facility experienced a number of accidents and releases involving toxic substances. In 1989, Rocky Flats was added to the National Priorities List for cleanup under the federal Superfund program. As part of an investigation into possible environmental mismanagement, FBI and EPA agents raided the plant in June, 1989. Subsequently, a federal grand jury was hearing testimony about the operation of the plant and possible violation of environmental law when the U.S. Justice Department agreed to dismiss criminal charges and allow Rockwell to pay a fine. In an unusual response, the grand jury strongly opposed this plea bargain agreement and has since sued the Federal Government for the right to tell its story publicly. These events were covered extensively by the Denver area news media.

Environmental Conditions and Their Effect on Property Value

Once the public becomes aware that a community has been (or may be) exposed to radioactive contamination, real property in the area can be stigmatized (Gregory et al., 1995). In the case of Rocky Flats, the stigma issue was exacerbated by the 1989 FBI raid. Since that date, the investigations and subsequent closure of the plant have received considerable coverage by the news media and a number of past, present and future risks have been enumerated (Flynn et al., 1998). While not all documents have been released by the DOE, it is known that: (a) drinking water may have been tainted, (b) soils on-site are highly contaminated, (c) nearby residents have been subjected to releases of hazardous materials, (d) 14½ tons of plutonium remain on site, (e) barrels used to store plutonium and uranium waste have leaked, and (f) the community does not know if or when hazardous substances will be removed from the site.

There have been numerous efforts to define stigma in a real estate context, although as yet no uniform and precise definition exists within the real estate literature. Generally, stigma is considered to be a condition of real property that negatively affects the current value of the property and possibly its future value as well. This may be separate from loss of use.

The hazard to real property is likely to be greater for on-site contamination than for properties proximate to contamination because loss of use and/or cost of remediation may represent servitudes or liens against the property. Properties proximate to contamination may be adversely affected if a plume is moving in their direction or if they are downwind from contaminants that can be transported through the air. In the case of the class action in this case, properties near Rocky Flats were included because they fell within a plutonium contamination plume. This contamination, if known and understood, reinforces adverse public perceptions even when no dramatic health impacts are documented. In effect, "because stigmatization is based on perceptions of risk, places can suffer stigma in advance of any demonstrated physical impacts" (Gregory et al., 1995).

Methodology

In order to study this problem, we developed a model consisting of five components designed to determine if the area is stigmatized, and quantify the effect (if any). The five components are summarized in Table 9.1.

Data Collection and Findings

REAL ESTATE MARKET RESEARCH

The *Real Estate Market Research* included interviews with representatives from a variety of public and quasi-public entities, as well as individual market participants, during a 2½-year period (1994 through mid-1996). In addition, numerous public documents were collected and reviewed, including resolutions passed by the various municipal and county jurisdictions, zoning regulations and other policy documents. The findings are summarized as follows.

Municipalities

The City of Broomfield concerns focused on the public drinking water supply. Great Western Reservoir, located immediately east of the Rocky Flats buffer

Table 9.1 Valuation Model

Component	Definition/purpose
Real Estate Market Research	Review public policies and documents as well as interview public and private market participants in order to understand market reaction to Rocky Flats. This is the first step in determining the likelihood (or extent) of stigma.
Analogous Case Studies	Examine the effects of other environmental disamenities to: (a) understand how real estate markets in other settings react to, or perceive, risk, (b) study how these reactions translate into overall value, (c) test the reasonableness of other valuation or evaluation techniques, and (d) apply these findings to the Rocky Flats context.
Market Sales Information	Review real estate sales data in order to identify trends or any direct effects to prices.
Multiple Regression Analysis	Compare sales data within the Class Area to sample data elsewhere in order to illustrate and quantify the influence of the plant and the 1989 FBI raid.
Public Opinion Surveys	Scientifically survey market participants in order to determine the attitudes of potential purchasers and quantify any reluctance to purchase within the Rocky Flats Class Area.

zone and subject to drainage and runoff from Rocky Flats, had been the source of Broomfield's water supply. Due to contamination, Great Western Reservoir was closed as a source of drinking water, and Broomfield is developing an alternative water supply, in part financed by $80 million from the federal government and the DOE. Even so, the city is concerned that the land around the reservoir no longer has viable commercial, residential, or public uses beyond being held as open space.

Like Broomfield, the City of Westminster, as well as the Cities of Thornton and Northglenn, had problems with their public water supply. The primary source of drinking water for these towns is Standley Lake, which past environmental studies reported as contaminated by surface water and airborne contaminants from Rocky Flats. This condition prompted the Standley Lake Protection Project to intercept runoff from Rocky Flats. The $25 million project was funded by the Department of Energy.

Soil contamination is another problem. For example, in the early 1980s an owner of land west of Simms Street approximately two miles east of the Rocky Flats plant submitted a plan to the City of Westminster for a residential development to be known as Colorado Hills. Soil testing showed contamination by Rocky Flats. The city rejected the plan and prohibited residential development west of Simms. Eventually the city purchased the land and converted it to open space. Since that time, virtually all the government jurisdictions in the area, including Jefferson County, Broomfield and Westminster, have prohibited residential development west of Simms Street.

Counties

Because the Rocky Flats Nuclear Weapons Plant is located in Jefferson County, the County Commissioners and Health Department have been active in Rocky Flats oversight activities. On September 8, 1994, the County Board of Commissioners passed Resolution No. CC94-654 that continues an undeveloped buffer zone around Rocky Flats. Similarly, the Jefferson County Board of Health passed Resolution No. B0H94-4, on September 14, 1994, in part recognizing that:

> There have been discharges of both radioactive and non-radioactive hazardous substances into the environment surrounding the plant. The board believes that the plant poses a potentially significant and long-term risk of harm to public health, safety and the environment due to, among other things, the past releases and potential future releases of hazardous substance into the environment and the past, present, and future storage of large amounts of plutonium and other hazardous substances.

In the Fall, 1996, Brown, Barton, and Bjork (1996) released their report, *Rocky Flats Community Needs Assessment*, which, among other things, recommended *a moratorium on building and other development in the area adjacent to the buffer zone, until such time as the land is proven to be suitable for unrestricted use.*

The Rocky Flats Future Site Use Working Group (1995), of which Jefferson County is part, noting that approximately 14½ tons of plutonium remain at the site, forwarded to the Honorable Hazel O'Leary, then Secretary of Energy, a statement that, *"These (radioactive) materials pose an on and off-site hazard as long as they are on the site."* Further, according to the group, *"These uncertainties affect use and development of adjacent land."*

As part of their zoning regulations, Jefferson County adopted a "radiation map" which included the geographic area of Rocky Flats and the surrounding lands. For any development proposed within this geographic area, the developer is required to prepare a "radiation report" and "radiation plans," which must be approved by the Colorado Department of Health (unless it delegates its authority to the Jefferson County Health Department) prior to plat approval. Within recent years, there has been almost no subdivision activity within the radiation map area.

In a proactive move, the Jefferson County Economic Council (1996) recently petitioned the State of Colorado to create an Enterprise Zone for lands just south and east of Rocky Flats. This would give special tax considerations to developers in the area. The Council believes such a step is necessary to induce development in an area that has been blighted by Rocky Flats.

Market Participants

Because financing is critical to real estate value, we interviewed a number of residential lenders regarding their policies toward making loans in the Rocky Flats area. All those interviewed do make loans in the area, but said they would not finance property with a demonstrated health hazard. Lenders entrust the appraiser to identify areas of concern and adequately reflect the impact on value, if any.

Realtor opinions of value and marketability within the Class Area are extremely diverse. Some believe property is impacted, some believe it is not. Some report that potential purchasers inquire about Rocky Flats, while others indicate that they rarely are asked. It was apparent from our interviews, however, that prospective purchasers obtain little or no negative information with respect to Rocky Flats from realtors. This lack of information is a point of concern expressed by respondents to researchers who conducted the *Rocky Flats Community Needs Assessment* (Brown et al., 1996).

ANALOGOUS CASE STUDIES

The *Analogous Case Studies* were developed from a variety of sources including the economics and appraisal literature, other professional journals and publications, Appraisal Institute text materials, and court documents (i.e., expert reports and trial verdicts).

When valuing an individual property, sales of similar properties are used for comparison. When analyzing entire communities, such a technique is not possible because entire communities do not sell. Thus, the use of case studies becomes a surrogate for direct sales comparison. Case study examples are helpful in defining market reactions and quantitative impacts associated with various types of environmental conditions. In some ways, they are similar to the concept of *stare decisis*, or case law, used by the legal profession.

We reviewed the case studies for evidence of market reaction in terms of behavior and quantitative impact on value. Although some economists might be concerned about the issue of "benefits transfer," we believe the case studies provide a legitimate framework for estimating diminution in property value. Even if the case study communities and their environmental conditions differ, the impacts can be placed in context (Roddewig, 1994). The following studies document market behaviors surrounding various environmental disamenities.

McClelland, Schulze, and Hurd (1990) examined property values surrounding the Operating Industries Inc. (OII) landfill in California. The purpose of their paper was to document public perceptions of health risk and compare those perceptions to land value data. They conducted a multivariate data analysis of sales and a public sample telephone survey. Some of the more significant findings were:

1. Among the approximately 4,100 homes near the site, the estimated loss in property value was approximately 7% before the landfill was closed and 3.5% after the landfill was closed.
2. Despite local requirements regarding disclosure, 62% of the respondents were not aware of the landfill when they bought their homes. That finding suggests that with more awareness the likely losses in property values would have been even larger.
3. Even when scientific evidence suggests there is no possibility of adverse health, public perceptions often differ and may result in a negative impact on property value.

One standard reference study for analyzing housing prices due to proximity to an environmental disamenity was published by Kohlhase (1991). She analyzed the impact of EPA announcements and policy actions on housing markets surrounding ten National Priority List sites, including examination of home prices in Houston's Harris County before the superfund act and after (1976 and 1985). The multiple regression analysis of these data found:

1. Diminution in value occurred out to a limit of approximately six miles.
2. The market did not distinguish between severity of the sites. For example, diminution in value was greater near one site that was considered less severe than another.
3. Loss in value was due to perception that in some cases was contradictory to scientific evidence.
4. Cleanup efforts can enhance property values depending upon the nature, timing and certainty of cleanup.

Because Rocky Flats presents the issue of radioactive contamination, we looked at the literature on property values and nuclear risks.

One example involves the analysis of property values by Beron (1992) around the Feed Materials Production Center in Fernald, Ohio. In the context of a lawsuit, these studies were designed to determine if nearby properties had suffered losses in value. Only properties within physical sight of the facility were found to have experienced measurable impacts. Since Fernald is a large and highly visible facility, these conditions applied to approximately 5,500 households. In contrast to the Beron study, Rosen and Burke (1987) found damage to considerably more properties, including many homes not within sight of the plant.

Two studies assessing the impacts of the transport of radioactive materials and the impacts on property values apply to Rocky Flats and show market resistance to location along transportation corridors. In a New Mexico case, land was acquired from a property owner in order to construct a highway to transport nuclear waste to the Waste Isolation Pilot Project (WIPP) site near Carlsbad, New Mexico (*City of Santa Fe v. Komis*, 1992). The City of Santa Fe argued that there was no loss in value to the remainder property. On the other hand, the property owner's appraiser relied in part upon a public opinion survey. The survey was designed to measure the market's knowledge and opinions with respect to diminution in property value. Forty-one percent of the respondents believed that residential property near the road would sell for between 11% and 30% less than comparable property not in proximity to the road. The Komis family was awarded $337,815, a finding that was upheld by the New Mexico Supreme Court in 1992.

A survey of residents along a key interstate highway route in eastern Oregon by MacGregor et al. (1994) produced a number of findings about radioactive wastes that are relevant to the Rocky Flats case:

1. These respondents expressed high levels of concern about the health and safety risk associated with nuclear materials and most particularly with nuclear wastes.
2. A majority strongly agreed that accidents involving the transport of hazardous waste are inevitable.
3. Up to 70.5% of the respondents said that transportation risks were greater than storage.
4. Overall, negative images accounted for 88% of the responses elicited in response to the prompt: *nuclear waste transport*.
5. Public perception of these risks contrast with technical community beliefs that radioactive waste transportation and disposition will be safe.

Smith and Desvouges (1986) developed an economic model to test the premise that nuclear power plants and hazardous disposal sites seem to be the least acceptable "locally undesirable land uses." They estimated a dollar measure of household desire to avoid living near such sites with a hedonic property value model and demand surveys. Extrapolating their mathematics, it would appear that if diminution extends outward 10 miles (as the authors suggest) those houses nearest the source would have been damaged on the order of approximately $4,140 each in 1984 dollars.

The case studies suggest some impact on surrounding real estate due to actual contamination and/or proximity to a negative environmental condition and they indicate more negative reactions to nuclear issues than to other potential dangers or contamination.

MARKET SALES INFORMATION

The *Market Sales* research considered vacant land prices and improved residential property prices both inside and outside the Class Area. Vacant land sales for 367 parcels that occurred over an eight-year period within the Rocky Flats area and in comparable Denver metropolitan areas were examined to see if the transactions had been influenced by Rocky Flats or the 1989 FBI raid. This analysis was supplemented with a study of Boulder County open space land acquisitions in the vicinity of Rocky Flats to see if purchase prices bore any relationship to proximity to Rocky Flats.

The Metro Area Multiple Listing Service (MLS), which compiles the listing and sales data for homes, served as the basis for an analysis of the improved residential properties. Data were compiled for end of year (12-month) statistics provided by MLS for 1989, when the FBI raid took place, and 1995, the most recent available data. Comparing prices from 1989 to 1995 produced a compound appreciation/depreciation rate for each area and a compound change rate per year for the six-year period.

The Rocky Flats Class Area was compared with all submarkets within the suburban Jefferson County area and with the area surrounding the Rocky Mountain Arsenal, which like Rocky Flats, is a widely known superfund site from which contamination has migrated off-site (EPA, 1991).[1]

Table 9.2 shows within the Class Area compared to other locations. This table contains raw data not adjusted for time or other considerations. However, because the parameters are constant for each category, the data are informative. Vacant land in the class area tends to sell at lower prices than elsewhere in Jefferson County and even the area near the Rocky Mountain Arsenal.

Next, individual vacant land sale transactions in the vicinity of Rocky Flats were considered. For example, in November of 1992, 357.83 acres of land located just north of the plant were sold for $612,000 or $1,710 per acre. When it was impossible for the buyer to obtain long-term financing for the property because of its proximity to Rocky Flats, the original lien holder was unwilling to foreclose and take possession of the property, but did facilitate, in effect, the liquidation of the property by discounting the note so the net real cost was $612 per acre. In 1993, Boulder County acquired the property for open space at $2,732 per acre. While this was considerably higher than the 1992 net purchase price, it is at the extreme low end of prices paid by Boulder County for

Table 9.2 Land Sale Comparisons

Cumulative Totals (1988-1995)

	Jefferson County North: Class Area	Jefferson County North: Outside Class Area	South Boulder County	Jefferson County South	Adams County: Rocky Mountain Arsenal Area
Total acreage sold	2,245.17	2,957.31	8,100.74	3,467.73	2,678.27
# parcels sold	42	31	163	87	44
Weighted average	$10,351 per acre	$10,621 per acre	$12,189 per acre	$25,477 per acre	$16,120 per acre
Price difference w/class area	Base	$270/ac.	$1,838/ac.	$15,126/ac.	$5,769/ac.
Price difference as %	Base	+3%	+18%	+146%	+56%

[1]The 17,000-acre Rocky Mountain Arsenal site is a facility owned and operated by the U.S. Army. It is located 10 miles northeast of Downtown Denver. Over the years, the facility was used by both government and industry to manufacture and dispose of various chemical products and munitions including rocket fuels, herbicides, pesticides, nerve gases, mustards and munitions.

open space. It also implies that the highest and best use of this property was as open space.

In reviewing residential properties, resale prices for comparable residential properties in various locations within the metropolitan area were examined with a focus on overall appreciation or depreciation rates within sub-markets. Specifically, appreciation/depreciation over time within the Rocky Flats Class Area was compared to other metropolitan areas. These results are shown in Table 9.3.

While the differences are subtle, the submarkets exhibiting the lowest compound rate of change in price are generally those areas surrounding and including the Class Area. This data does not lend itself to quantification for property appraisals, but it is once again suggestive of stigmatization.

MULTIPLE REGRESSION ANALYSIS

In order to more specifically determine the impact on property values and quantify it if appropriate, Dr. John Radke of the University of California at Berkeley conducted a systematic, quantified study of the property value history of residential real estate in the Class Area. The information data base was organized using a Geographic Information System (GIS), which allowed use of location variables. All Multilist sales data in the vicinity of Rocky Flats were compared against representative sample data from comparable areas in the metro area.

Eleven years of sales records were pooled to obtain a sufficient sample of sales records in three submarkets: commercial, vacant land, and multi-residential. Monthly sales of single-family residential properties were entered into a data base. The data represented almost a 100% sample within the class area over an eight-year period and a representative sample outside the contour area over the same study period. These data were examined to make a determination of the gain or loss in value within each sub-class.

Multiple regression modeling techniques were used to address three issues: (a) to see if property values were depressed around Rocky Flats and if Rocky Flats was the cause; (b) to see if the June 1989 FBI raid of Rocky Flats played a role in property value diminution; and (c) to determine the amount of diminution in property value, assuming the influence of the Rocky Flats plant and the FBI raid.

Table 9.3 MLS—Residential Metrolist Data Summary

Area	Avg. price 1989	Avg. price 1995	Avg. annual appreciation	Rocky Flats class area?
Jefferson North	$83,127	$124,023	6.90	yes
Jefferson North Central	$91,016	$129,759	6.09	yes
Broomfield	$94,009	$138,627	6.69	
North Suburban West	$77,405	$110,319	6.08	yes
Jefferson Central	$82,656	$113,075	5.30	
Jefferson South Central	$80,659	$123,991	7.43	
Jefferson West	$106,311	$171,014	8.25	
Jefferson South	$97,196	$156,488	8.26	
North Suburban East	$43,825	$69,580	8.01	
North Suburban Central	$63,979	$104,059	8.44	

In order to address these issues, hedonic price modeling was used with property value as the dependent variable. Generally, previous property studies of this type used the physical attributes of the property as independent variables. However, by using a GIS base location, proximity and accessibility are introduced as independent variables.

Exactly 6,392 data points covering an eight-year period (almost 100% of single-family residential sales activity) within the class area were pooled and compared against a 4% stratified random sample of sales throughout the Denver metropolitan area. The confidence level was in excess of 95%. Weighted least squares estimation was used to balance the influence of samples inside and outside the Class Area because different percentage samples were used.

Two methods were used to empirically determine the effects of Rocky Flats on surrounding property values. Method 1 assumed that the effect is uniform within the Class Area and there is no effect outside. Method 2 assumes that any effect was dependent on air distance from the borders of the Rocky Flats facility, and went out to a certain distance to be determined empirically.

The following tables (Table 9.4 and 9.5) show the results of each method.

Because the specific purpose was to measure impact on value within a defined geographic area, e.g., the Class Area, we considered Method 1 to be the more appropriate model. However, it is important to recognize that some of the properties within the sample pool, although outside the Class Area, are within the cutoff distances delineated in Method 2. Thus, the percentage of property value loss calculated in Method 1 is conservative.

The estimates of loss in value for residential property after the 1989 FBI raid range from 5.45% to 9.33%; the average over time is approximately 8%. The fluctuation may be attributable to economic climate and media coverage. Expressed in 1993 dollars, the range of loss for single-family detached residential property in the class area was estimated at between $64,503,340 to $97,510,144. Table 9.6 shows the estimated loss by year for single-family detached properties within the class area (1993 constant dollars).

As might be expected, these results show that loss in value fluctuates over time. The average over the eight-year period was about 8%. The low point of diminution was in 1995, which corresponds to arguably the best housing market in Denver's history and may reflect the decline in negative media coverage of Rocky Flats since the FBI raid.

PUBLIC OPINION SURVEYS

The City of Arvada (1989) and the City of Broomfield (Talmey-Drake Research and Strategies, 1990) commissioned public opinion surveys shortly after the 1989 FBI raid. The EG&G Rocky Flats Community Relations Department conducted some interviews in November 1994.

Decision Research conducted the 1995 survey reported below and in Flynn et al. (1998, reprinted as Chapter 20, below), on behalf of the plaintiffs.

Talmey-Drake Research & Strategy, Inc. (1990), a public opinion and market research firm in Boulder, Colorado, conducted the Broomfield Issues Survey ($N = 406$) by telephone between September 7–17, 1990. The sample has a 4.9% margin of error at the 95% confidence level.

The survey asked about (a) residents' health concerns about "the Rocky Flats Nuclear Plant" and (b) community perceptions about Rocky Flats and its effects on the city's drinking water. A majority of Broomfield residents (54%)

Table 9.4 Mean Property Value Loss Within Medium Plutonium Contour: Methods 1 & 2

				Method 1		Method 2	
Year	MPC	No. of samples	Mean sale value ($	Mean loss ($)	Loss rate	Mean loss ($)	Loss rate
1988	outside	343	110,597.20	–8,158.55	—	–7,046.15	—
	inside	324	106,263.50		7.68%		6.63%
1989	outside	358	119,766.00	–7,838.04	—	–5,527.64	—
	inside	365	107,457.40		7.29%		5.14%
1990	outside	416	112,944.30	–9,474.82	—	–8,572.35	—
	inside	370	101,578.50		9.33%		8.44%
1991	outside	318	115,399.60	–8,046.32	—	–8,607.47	—
	inside	493	104,658.30		7.69%		8.22%
1992	outside	290	113,403.70	–9,854.25	—	–10,129.90	—
	inside	624	107,310.40		9.18%		9.44%
1993	outside	380	119,696.70	–9,897.10	—	–9,429.55	—
	inside	557	115,628.00		8.56%		8.16%
1994	outside	282	119,134.70	–9,194.64	—	–9,726.80	—
	inside	546	122,597.50		7.50%		7.93%
1995	outside	274	119,196.40	–6,796.14	—	–6,948.29	—
	inside	452	124,629.40		5.45%		5.58%
Total sample		6,392					

Table 9.5 Method 2: Distance Dependent Loss Estimation Up to Empirical Cutoff Distances

Year	Coefficient level	Effect per mile	Cutoff distance (miles)	Standard error of coefficient	T-statistic	Confidence level
1988	–0.024438	–2.41%	7.0	0.007960	3.070	99.79%
1989	–0.023008	–2.27%	6.5	0.008990	2.559	98.95%
1990	–0.018295	–1.81%	9.0	0.004312	4.242	100%
1991	–0.024903	–2.46%	7.5	0.005930	4.200	100%
1992	–0.020314	–2.01%	9.0	0.003989	5.093	100%
1993	–0.019414	–1.92%	8.5	0.004314	4.500	100%
1994	–0.015373	–1.53%	9.5	0.003480	4.417	100%
1995	–0.024927	–2.46%	6.5	0.008404	2.966	99.70%

Table 9.6 Berkeley Estimate of Loss in Value

Year	% loss	Avg. price*	$ Loss/ home	No. of properties	Total loss
1988	–7.68%	$106,263/prop	–8,839.9	8,979	$79,373,462
1989	–7.29%	$107,457/prop	–8,449.6	8,979	$75,868,958
1990	–9.33%	$101,578/prop	–10,452.4	8,979	$93,852,100
1991	–7.69%	$104,658/prop	–8,718.7	8,979	$78,285,207
1992	–9.18%	$107,310/prop	–10,859.8	8,979	$97,510,144
1993	–8.56%	$115,628/prop	–10,824.3	8,979	$97,191,390
1994	–7.50%	$122,597/prop	–9,940.3	8,979	$89,253,954
1995	–5.45%	$124,629/prop	–7,183.8	8,979	$64,503,340

*Because this figure represents the actual average (1993) price for properties within the impacted area, property diminution is inherent. Thus, loss per home is calculated as 7.68% x ($106,263 x 92.32%).

said they were more concerned about Rocky Flats in 1990 than one year earlier, particularly with respect to drinking water. Over a third (37%) said Rocky Flats posed a moderate health threat to the residents of Broomfield, and another 25% said the threat was very serious. Only 10% reported that Rocky Flats did not pose a health threat at all. In terms of trust and confidence in the new contractor, 40% said they trusted the new management of the Rocky Flats more than they trusted the previous management while 33% did not, and 27% did not make a selection in the comparison. Over two-thirds (67%) disagreed with the statement that "Rocky Flats has no effect on property values in Broomfield." A result of the concerns recorded by the survey was that the City of Broomfield closed the Great Western Reservoir as a source of city water.

The "Arvada Citizen Attitude Survey" was conducted by the City of Arvada (1989) several months after the FBI raid, to gather citizen opinion about city services, and other issues including concerns about Rocky Flats. Frequency responses, shown in Table 9.7, were available but the data base and back-up material for the survey were not.

Arvada residents support the removal of plutonium from Rocky Flats and strongly oppose leaving it on-site. A substantial minority (42%) believed that Rocky Flats had created a health threat to their household.

In 1995 Decision Research and The University of Maryland Survey Research Center surveyed residents 18 years of age and older, currently living in the Denver metropolitan area who: (a) had bought a home in the past five years, or (b) had plans to purchase a home in the next couple of years, or (c) had been actively in the market for a home some time during the past five years. Two sample frames were used: a list of recent home purchasers obtained from the U.S. West telephone company, and a random digit dialing (RDD) sample, which was generated by the University of Maryland Survey Research Center. An oversample of the Arvada/Westminster area resulted in 188 respondents from these two communities.

The survey had the following objectives: (a) To obtain interviews with respondents who were knowledgeable and experienced in the Denver area residential housing market; (b) to interview a random selection of qualified respondents in the Denver metropolitan area and a random selection of

Table 9.7 1989 Citizen Attitudes Survey

	Survey year	Strongly agree	Agree	Neutral	Disagree	Strongly disagree
The City should support the removal of plutonium processing operations	1989	38%	22%	22%	10%	7%
from Rocky Flats	1987	*	*	*	*	*
The City should support the removal of plutonium	1989	*	*	*	*	*
processing operations from Rocky Flats if present employment levels can be maintained	1987	46%	22%	19%	9%	5%
The City should oppose the incineration of toxic	1989	*	*	*	*	*
radioactive waste at Rocky Flats	1987	56%	17%	15%	7%	4%

Do you feel the Rocky Flats Nuclear Weapons Plant northwest of Arvada has created a health threat to your household? (Asked in 1989 only)

Yes 42% No 32% Don't Know 27%

*Not asked that year.

respondents in the communities of Arvada and Westminster; (c) to obtain evaluations of community characteristics for Arvada, Westminster, and either Highlands Ranch (located in the southern part of the Denver metropolitan area) or Ken Caryl (located in the southwestern part of the Denver metropolitan area); (d) to ask about and record respondent evaluations of the advantages and disadvantages of living in Arvada and Westminster; (e) to measure respondent perceptions of the Rocky Flats nuclear weapons facility and their evaluations of the overall effects of the facility on the desirability of homes in Arvada and Westminster; (f) to determine how changes in the distance from Rocky Flats and variability in cost might affect the desirability and the acceptability of a house.

A total of 604 persons was interviewed during the period August 31, 1995 to October 1, 1995 including 416 respondents from the Denver Metropolitan Area and 188 respondents from Arvada/Westminster. The margin of error for a total sample size of 604 respondents is 4.1% at the 95% confidence level (for 416 respondents the figure is 5.0%; for 188 respondents it is 7.2%). Several ways of calculating the response rate were used resulting in a conservative estimate of 54% and a more realistic estimate of 63%.

This survey found that public perception that Rocky Flats is considered to be an important health risk and to produce adverse impacts on nearby residential properties. A set of questions asked respondents about the acceptability of a house located close to Rocky Flats under a variety of conditions to see how people would evaluate the risks and benefits under different scenarios. A small proportion of the respondents (16.1%) said they would accept an otherwise desirable house within the 2-4 miles of Rocky Flats without a discount and 15.2% would accept a house within the 4-6 mile area without a discount. Those who rejected the existing safety buffer zone as adequate and even the added 4-6 miles distance, were offered a progressive series of discounts to the house's cost and finally asked to specify a discount of their own choosing. An additional 16.6% would accept a house in either distance category with discounts. However, almost half (46.2%) of the respondents would not trade distance and/or discounts of any size as compensation for a house located within six miles of Rocky Flats.

The average discount for the 2-4 mile category for those who required a discount for this scenario ($n = 29$) had a mean value of $30,740 and a median value of $20,000. When the calculations include those who would not require a discount for a purchase in the 2-4 mile category, the mean value is $9,270 and the median value is $0. (That is, a greater number would buy without a discount than the number who would buy only with a discount.)

The average discount for the 4-6 mile category for those who required a discount for this scenario ($n = 40$) had a mean value of $15,747 and a median value of $10,000. When the calculations include those who would not require a discount for a purchase in the 4-6 mile category, the mean value is $6,084 and the median value is $0. (That is, a greater number would buy without a discount than the number who would buy only with a discount.)

A large and important proportion of the survey respondents (46.2%) are not included in these calculations of discount values because they would not buy at any discount, even if they could fix the sales price at an amount of their own choice. This level of resistance to housing associated with Rocky Flats would be expected to result in negative marketing and price pressures on houses in the communities of Arvada and Westminster.

In a related question, respondents were asked if Rocky Flats had affected the value of property within six miles or so. More than two-thirds of the Arvada/Westminster respondents (69.7%) and 81% of the Denver metropolitan area respondents agreed that there were impacts on property values. When asked if the direction of these impacts was to increase or decrease property values, 92.4% of the Arvada/Westminster and 97.0% of the Denver metropolitan areas respondents said that Rocky Flats has reduced property values.

Conclusions

The combination of *Real Estate Market Research* and *Public Opinion Surveys* focused on the investigation of attitudes, perceptions and actions (or anticipated actions) for the public and private sector marketplace within the Rocky Flats sphere of influence. Both public and private market participants were concerned about the risks associated with Rocky Flats and these concerns were directly related to knowledge of the 1989 FBI raid. Public Opinion Surveys conducted by local municipalities following the raid led to efforts to improve or protect the quality of drinking water. Over time, the public sector has responded to concerns about public safety by restricting development in certain locations (i.e., use of a radiation map) and by designating land use in some instances to open space only. Boulder County, Jefferson County, Broomfield, and Westminster have all added land to open space that was once thought to be developable.

The 1995 Decision Research Survey documented similar reactions within the private sector. The survey supports four primary conclusions about property values associated with Rocky Flats.

The Decision Research survey showed that people who recalled the FBI raid some years later were less likely to purchase property in the nearby communities.

The association with Rocky Flats makes housing in that community less desirable and works to reduce the market size and therefore the price of housing in Arvada and Westminister. This antipathy is most strongly expressed by the 46.2% of the Denver metropolitan area residents who said they would not buy a house associated with Rocky Flats, even when they were offered distance as a protection and discounted prices as an incentive.

Respondents wanted extraordinary distances between themselves and the plant itself. Denver metropolitan area respondents said that the "closest distance to Rocky Flats" they would consider was a mean distance of 21 miles and a median distance of 15 miles.

Only 16% of the Denver metropolitan area respondents would consider a house in Arvada and Westminster when a discount was offered. The average discount including those who would buy in the 2-4 mile range without a discount (i.e., their discount price is $0) was $9,270 and in the 4-6 mile range it was $6,048.

These findings are consistent with results from the *Analogous Case Studies*, which generally show adverse impacts on neighboring property values due to an environmental disamenity. Beron (1992) found that the impact applied to only those properties from which the disamenity could be seen. Kinnard (1992), in his study of New Jersey radium contamination, concluded that the degree, significance and duration or persistence of any negative price effects depend in large part on whether the contamination is perceived by potential buyers as an isolated event or a continuing condition. Depending upon distance from the environmental disamenity, loss in value was found among the case studies to range between approximately 3.5% and 35%. Range of impact also varied from a mile or two to as much as 22 miles for the siting of a nuclear power plant (Smith & Desvouges, 1986).

The GIS-based sales data analyzed by the Berkeley researchers using multivariate regression analysis techniques concluded that the property value impact from Rocky Flats extends beyond six miles from the plant and over time has averaged approximately 8% per year for detached single family homes. We note that the amount of property value diminution spiked just following the 1989 FBI raid, which was accompanied by overwhelming media attention.

The raw data summarized in the *Market Sales Information* research also supports the opinion that property values are diminished. These results show that vacant land sales within the Class Area produced the lowest average price per acre of any of the five geographic areas studied and that Rocky Flats was a factor.

The results of each component of the study add to the evidence that Rocky Flats has produced negative impacts on nearby property values. There are a number of locational and aesthetic attractions to the areas around Rocky Flats and there is a market for housing in the area, but these research results show that the area is stigmatized for many potential buyers. Based upon these findings, it is our opinion that over the study period (1989 through 1995), vacant land and residential property values have been consistently diminished by the influences of the Rocky Flats Nuclear Weapons Facility and the 1989 FBI raid of the plant.

Place, Product, and Industry Stigma

In this section, eight chapters examine and/or comment on the broader range of risk, media, and stigma. Two chapters deal with property, four chapters deal with products, and two chapters deal with stigma at the industry level. Adams and Cantor evaluate the basis for property stigma based on litigation cases they have examined. They emphasize the special nature of home ownership. Pijawka and his colleagues present the case study of a major fire that caused health and environmental damages in an inner city South Phoenix, Arizona neighborhood inhabited mainly by low-income, minority residents. They find that the hazardous event produced disproportionate property losses to the residents.

The four chapters on products begin with Mitchell's classic essay on the economic costs of the 1982 Tylenol poisoning. Powell writes about British Mad Cow Disease (BSE, bovine spongiform encephalopathy). Chan describes the Canadian blood supply and its exposure to viral contamination. Graham addresses the risk conditions of automobile air bags, making some observations on the attenuated public response and commenting on the possible reasons.

Two chapters address stigma issues and the chemical industry. Leiss examines the process by which chemicals containing dioxin came to be stigmatized and how these compounds came to represent the potentially hazardous products of the chemical industry. Long provides an insider account of how a major chemical manufacturer struggled and changed in response to rapidly changing public attitudes toward their products over a 20-year period beginning in the 1960s.

10 Risk, Stigma, and Property Value–What Are People Afraid Of?

Gregory D. Adams and Robin Cantor

Introduction

Taking the taxonomy of Stigmatized Places, Products, and Technologies (as laid out by the editors of this collection of papers), we address a specific subset of Places—namely real property. We will argue that the value formation process for real estate is fundamentally different than for most other goods commonly purchased by households, and that this difference has profound implications for the way that risk and uncertainty may impact the value of real estate. This leads to a discussion of how diminution in property values due to perceptions regarding the risk of the property (commonly referred to as "stigma") may arise, and conditions under which it may persist.[1] The second section of the paper addresses particular issues that commonly arise when stigma impacts on property value are addressed in the legal context. These issues are both methodological and philosophical. We conclude with an informal discussion of the role of the media in the formation of stigma impacts on property value and brief thoughts on how the media may mitigate these impacts.

Before delving into the body of the paper, however, let us briefly outline our relevant education and professional background. We are economists by training, profession, and natural inclination. While we know it rarely to be strictly applicable, we find it useful to analyze behavior as if humans act rationally. Also, a majority of our consulting engagements are in the context of litigation. Therefore, we tend to be a skeptic ("prove it") and concerned about issues of causality ("whose fault is it") and empirical measurement ("how big is it").

---

[1]One of the objectives of the conference that stimulated these papers was to help clarify the concept of stigma. Here, we follow the general convention in the literature to consider property stigmatized from any number of perception-based causes. Unfortunately,, it would be preferable to have a cleaner definition that could distinguish between perceptions based on risk concerns and perceptions based on moral conflicts or social outrage.

Value Formation For Real Estate

With real estate, like any market good, value is in the eye of the beholder—perception is reality. Therefore, if someone thinks it is dangerous to live next to a nuclear power plant, eat apples, etc., the value *to that person* of a home near a nuclear power plant, an apple, etc. will be reduced, regardless of whether any real danger exists or not. This does not necessarily imply that the market value of the good will be reduced, since market value is determined based on the "average" preferences of all buyers. However, with real estate more than most goods, the preferences of a comparatively few buyers may impact market value. Therefore, real estate would seem to be more susceptible than most goods to the impacts of stigma.

"WHAT DO YOU CARE WHAT OTHER PEOPLE THINK?"

The process by which persons value real estate is fundamentally different from most other household goods. Most goods purchased by households, such as food items, are purely "consumption" goods—that is, the value of the good is solely a function of the "utility" or use value of the good. For instance, the value that someone places on an apple is solely a function of that persons (perhaps subjective) evaluation of the benefits from consuming that apple (which may include taste, alleviation of hunger, nutrition, etc.). Real estate, however, is usually both a "consumption" good and an "investment" good. While people typically purchase a home for the primary purpose of providing shelter,[2] they usually do not plan to live in the same home for eternity. Couple this with the fact that a home is the largest purchase that most of us will ever make, and it is clear why people commonly worry as much about the resale value of a home as they do about how nice a place it will be to live. Therefore, an individual's "willingness to pay" for a home is a function of both the (perhaps subjective) evaluation of the benefits from living in that home (which may include the size of the home and yard, local area characteristics such as schools and parks, and the safety of the home and area), and the anticipated resale value of the home at the (usually uncertain) point in the future when they will want (or need) to move.

While this distinction may be obvious, it has profound implications for how people value property and how uncertainty about the "riskiness" of a property may affect the value of the property. For example, consider the impact of the Alar scare on the value of apples. If I personally am convinced that the treatment of apples with Alar poses no risk, then questions regarding the safety of Alar should not affect my willingness to pay for apples (regardless of whether I am right or wrong about the safety of Alar). Basically, I do not care what other people think. With real estate, however, I care very much what other people think. If I anticipate that I will want (or need) to sell my house at some point in the future, I will be concerned about other people's perceptions on the riskiness of the property since this will affect their willingness to pay, and their willingness to pay directly affects my willingness to pay. Note that this is not a phenomena that is unique to the issue of environmental risk. For example, some people may worry that property value diminution can follow the integration of traditionally segregated neighborhoods, even though these

[2]Or, in the commercial context, for the primary purpose of providing a stream of income.

same people may not harbor explicit prejudices. The regrettable state of unenlightenment that prevails in their region, however, may provide a rational basis for these concerns.

And it really is not only what other people think that is important, but what I think that other people think, and what I think these people think that other people think, *ad infinitum*. To see this, simply consider that the value I place on a home is (in part) a function of how much I think others may be willing to pay for the home (the resale value). The value that these people place on the home is in turn a function of what they think still others may be willing to pay for the home. Because of the sensitivity of real estate values to what other people think, the potential for stigma impact based on "irrational" behavior or "collective hysteria" is greatly magnified. It is easy to imagine a situation[3] where, while there is some uncertainty regarding the risk of a piece of property, no potential buyer individually believes there to be an abnormal risk, yet at least some potential buyers believe there may be some other potential buyers who may believe there is a risk. In this case, diminution in property value may occur. Note that not only do no potential buyers believe there is a "real" risk, but no potential buyers even believe that any other potential buyer believes there is a risk (that is, I do not need to believe that someone else may be wrong in assessing the true magnitude of the risk). Rather, all that is required to generate diminution in value from the uncertainty is that some potential buyer believes that other potential buyers believe that other potential buyers believe there is a risk (that is, I only need to believe that someone else is wrong in their assessment of other peoples beliefs). Therefore, even assuming that everyone individually is a rational scientist and can correctly assess the true risks, stigma impacts can still occur if everyone is not certain that everyone else is also a rational scientist.

Institutions can also contribute to the unique susceptibility of real estate to stigma impacts. Houses are expensive and few of us can pay cash for our home. Therefore, the ability to obtain a mortgage (and a favorable interest rate) on a home is a very important factor in determining a potential buyer's willingness (and ability) to pay. If uncertainty regarding a property makes lenders unwilling to grant a mortgage (or only willing to grant a mortgage at a higher interest rate), the market value of the home will likely fall. There are several reasons why lenders may be hesitant to mortgage "risky" properties. First, since the property typically serves as collateral for the loan, properties that are more likely to suffer a decrease in value represent riskier loans (since in the event of a default, the lender may be left with an asset that is worth less than the outstanding loan balance). In addition, laws that may make lenders liable for remediation costs and/or third party tort claims will discourage lenders from entering the "chain of liability" for risky properties (particularly since lenders are often inviting "deep pocket" targets for these claims).

THE PERSISTENCE OF MEMORY

The process by which real estate values are formed also has implications for the persistence of stigma impacts. Property values can be, for many reasons, "path dependent." That is, the value of a property today is function (to some degree) of the value of that property in the past. When values are path depen-

[3]But surprisingly difficult to model mathematically.

dent, temporary uncertainty regarding the risk of a property may lead to a permanent diminution in the value of that property.

Why might property values be path dependent? The first reason that comes to mind is what economists refer to as "network" or "coordination" problems. Assume that the value of homes in an area is impacted by the perception of a possible risk. Further assume that the uncertainty is resolved and it is widely agreed that the risk did not exist. If everyone now agrees that the risk does not exist, shouldn't I be willing to pay "full price" for the home? Not unless I am confident that everyone agrees there is no risk *and* I am confident that everyone knows that everyone agrees there is no risk. After resolution of the uncertainty regarding the risk, I would be reluctant to be the first person to pay "full price" for a home in the area. However, once a number of others had bought homes at "full price," I may be more comfortable concluding the storm had passed and pay full price myself. Of course, no one wants to go first, and as a consequence no one may go at all.

Stigma impacts may also persist because temporary decreases in property value due to questions about risk may lead to more permanent changes in the "character" of the neighborhood. For instance, assume that people generally wish to live in an area where their neighbors are as financially fortunate (or more so) as themselves.[4] Also assume that there is a positive relationship between the price of homes and the wealth of persons who live in those homes. Then, when property values fall (due to temporary uncertainty), the average wealth of the area residents will fall (as homes are sold to lower wealth buyers).[5] If buyers' willingness to pay for a home is predicated to some degree on the average wealth of the area residents, then people will be willing to pay less for the homes in the impacted area even after the uncertainty is resolved, simply because "they want to live around a better class of people."

Another demographic factor that may influence home values, and thereby affect the persistence of stigma impacts, is the percentage of renters (versus owner/occupants) in a neighborhood. It is generally thought that the higher the percentage of renters in an area, the lower property values.[6] The reason for this effect may relate to the average wealth of area residents discussed immediately above, since renters may generally be less wealthy than owners. Also, renters (and/or landlords) may be less diligent in the maintenance and improvement of these properties.[7] Whatever the specific etiology, if a higher per-

---

[4]A common home buying maxim is "Buy the cheapest house in the most expensive neighborhood you can afford."

[5]The dynamic impact of stigma effects on property values, and issues of environmental justice, has been analyzed by Vicki Been. Been argues that locally undesirable land uses may lower property values, leading to a larger percentage of low income residents. As such, even if undesirable land uses are not originally sited in low income areas, market dynamics may lead all areas near these land uses to become low income areas. See Been (1994a, 1994b).

[6]Percentage of owner/occupants and average income are often used in statistical (or "hedonic") models as explanatory variables when predicting home values.

[7]Of course, inadequate maintenance may also be undertaken by owner/occupants. When home values fall due to a perception that an environmental risk makes the homes less safe, it may be rational for the owner/occupant to invest less in maintaining the asset that now has questionable long term value. An interesting philosophical questions arises when the risk turns out to be nonexistent, but the stigma persists because of inadequate maintenance. Who is to blame for the long-term diminution in value, the party responsible for the perceived risk or the party responsible for the inadequate maintenance? This issue will be discussed in more depth, below.

centage of renters in an area tends to lower property values, then stigma impacts from temporary uncertainty may persist long after the uncertainty has been resolved if the percentage of renters in the area increases during the period of the uncertainty.

Why should we think that the percentage of renters in the area may increase during the period of the uncertainty? Two reasons come to mind. First, renters tend to have lower income and so have less choice in where they live. Therefore, if property values in the area fall with the advent of the uncertainty, renters may have fewer options to live elsewhere and will be more likely to accept the risk. Also, recall the discussion above regarding the influence of what other people think on the value of property. Even people who completely discount the possible risk will be unwilling to pay "full price" for the property, since they will be concerned about the beliefs of others regarding the potential risk (and the impacts of these beliefs on the resale value of the home). When renting, however, property is a pure "consumption" good—renters are not concerned abut resale value, and therefore are not concerned about what other people think. Thus, one should expect that, so long as there are some potential renters who discount the risk, rental rates should be less impacted than selling prices. One may expect that owners who wished to relocate during the period of the uncertainty, may choose to rent their homes rather than sell. By doing so, however, they will increase the percentage of renters in the neighborhood and may lay the foundation for a longer lasting negative impact on property values.

Of course, not all temporary diminution in property value need be permanent. One could easily posit a theory of arbitrage whereby, once the risk is determined to be nonexistent, the now "under-priced" homes are sought out by bargain hunting buyers. If these buyers compete with each other to purchase the homes, then prices should return to their original level.[8] Indeed, if home values were completely path-dependent, then initiatives such as urban renewal would never be fruitful. While we know of no comprehensive, systematic study of whether property values recover from stigma impacts after the removal of the risk (or the perception of risk), the empirical data we have examined, and anecdotal experience, support a theory of recovery. However, we suspect that whether property values recover in a particular instance is an empirical question.

Stigma in the Context of Litigation

As one might expect, uncertainty regarding the safety of a property often leads directly to the steps of the courthouse. Toxic tort property damage cases are very common, and economists are often called upon both to address the magnitude (if any) of the impact on property values, as well as to offer insight into the cause of any impact that is found. Both of these types of inquiry involve one in difficult, but interesting, issues.

MEASUREMENT

Measurement of the impact of an idiosyncratic risk on a set of properties is not always straightforward. The typical approach, taken by economists and appraisers, is to compare the selling prices of the "impacted" homes with other

---

[8]Adjusted, perhaps, for any intervening changes in the area demographics or quality of housing stock.

identical, but not "impacted" homes. Of course, like snowflakes, no two homes are identical and the selection of the "control" properties is crucial. Since many sources of stigmatizing risk are also associated with other dis-amenities (aside from the possible risk of living next to a nuclear power plant, it does not present the most scenic view), one must be careful to not attribute to stigma impacts that are truly caused by other confounding factors. This may not be easy, since all areas with similar aesthetic dis-amenities may also suffer from similar risk concerns. Similarly, casual empiricism would suggest that stigma related property damage claims are more common in times of economic recession and generally stagnating home values. If so, one must be careful to segregate the impact of the perceived risk from that of overall changes in the real estate market. While we will not go into the technical details of empirical measurement of property value impacts, a large literature on this subject does exist. Interested readers are referred to the bibliography at the end of this paper.

In some situations, the comparable sales technique may not be possible at all. A common case is where the plaintiff has not sold their home. The value of the home may well change before the plaintiff sells, especially if the safety concerns are currently subject to a great deal of uncertainty.[9] In this case, how should we determine the "damage" to the plaintiff. One response is to reply that, until the plaintiff has sold the home, they have not suffered a loss (in terms of property value). Therefore, until the plaintiff actually sells the home, she should not bring suit.[10] There are two potential problems with this approach. First, plaintiffs may not have unlimited wealth and credit. Since a home represents a large part of the wealth of a typical household, and since the resolution of litigation is usually not swift, homeowners may not be financially able to sell their home, take the loss and then sue for damages.

A second consideration is that the plaintiff may be subject to a statute of limitations. These statutes specify the maximum length of time that may elapse between the time that the plaintiff knows, or reasonably could have known, of the property damage and when she files suit. Therefore, if the plaintiff waits until she sells her home, she may lose the right to sue. The underlying issue here is when does the "damage" to a property owner occur—when the value of a capital asset falls, or when the owner of that asset liquidates that asset? Judges may adopt either answer when deciding on the appropriate "trigger" date to begin the tolling of the statute of limitations. Therefore, by waiting until the home is sold and the damages are liquidated, the plaintiff may run the risk that the statute of limitations on her claim is judged to have expired. If the plaintiff sues before selling her home, however, she may be compensated for damages that they never actually incur (if the diminution in value disappears by the time she sells her home).[11,12]

[9]If the safety concerns are confirmed, the home value may fall, while if the safety concerns prove unfounded, any temporary decrease in the value of the home may (or may not) disappear.

[10]We believe attorneys refer to the plaintiff's claim as being "not ripe."

[11]Issues of statues of limitations also become complex since the march of scientific knowledge is generally better characterized as a meandering stroll. At exactly what point does one draw the line and say, "From this date everyone should have realized the magnitude of the risk and the potential impact on property values."

[12]An analogous issue arises in cases involving shareholders seeking damages from an event that decreased that value of a stock. If the shareholders do not sell, and the stock price recovers, have the shareholders been damaged? Also, how does one determine if the stock price has truly "recovered" since in the absence of the temporary drop in price, the stock may have increased in value even more?

A final issue with regard to measurement is whether plaintiff damages should be based upon "average" or "market" value, or the plaintiff's subjective value of the home. A person may value their home at a price much higher than the home would command in the market (if, for instance, they grew up in the home). As such, one may argue that the damages suffered by this person in the event that they are forced to abandon the home should be based on this subjective value. In theory we believe this to be correct, however, in practice the law generally restricts damages to the market value of the home (which we also believe to be correct). The problem is that every plaintiff will have an incentive to overstate their personal value of the home, and there is usually no practical way in which to verify these claims. However, we can imagine counterexamples where one may reasonably establish a person's subjective value. These include instances where the owner had rejected an above market offer to purchase the home, or where the owner carried an above market level of insurance on the property.

BLAME

Issues of culpability entail some of the most difficult, and therefore most interesting, questions that arise in toxic tort property damage litigation. At first glance, it may seem uncontroversial that, when an actual risk is proven, culpability for any associated property value diminution should lie with the party responsible for the risk (the power company, etc.). However, the doctrine of "coming to the nuisance" may mitigate against so broad a conclusion. Consider the example where a person buys a home next to a high voltage transmission lines in, say, 1970. At the time, homes near the power lines sold for a slight discount (due perhaps to the inferior view, or perhaps to some slight concerns about risk). Assume that a few years later a highly publicized "study" reveals that these power lines may present a health risk, and property values plummet. If sued for property value diminution, the power company may argue that the plaintiff knowingly purchased a home next to the power lines, at a discount, and thereby accepted the risk. The home owner may reply that, while they knew of the existence of the power lines, they were unaware of the true magnitude of the risks (which only became generally known after the purchase). Is the plaintiff entitled to recover? If the "true" risks were reasonably foreseeable by the plaintiff at the time she purchased her home, we would argue that she is not entitled to damages (regardless of whether she recognized the risks herself). If the true risks were known (or reasonably foreseeable) to the power company, but not to the plaintiffs, then recovery by plaintiffs seems justified. What if the risks were reasonably foreseeable to neither party?

Even more difficult questions arise when the risk is not "real." As discussed above, property value diminution can be very real, even when the perceived risk is not. A plaintiff may argue that, whether or not the risk is real, the perception of a risk has damaged their property value, and the source of the perception of the risk is the power lines. Therefore, the power company should pay. The power company may argue that, while it may be responsible for any real risk, it cannot be held responsible for (incorrect) perceptions of risk. Who should be responsible for incorrect perceptions? Do we hear any votes for the media? Our belief is that, as between homeowners and the power company, homeowners should bear the risk of the impact of incorrect perceptions on the value of their homes. Just as homeowners bear the risk that certain architectural styles may go in and out of favor, so they bear the risk that certain locations may become less favored, based purely on the subjective preferences of buyers.[13] However, we also feel that the media, and others who propagate the false perceptions, may bear (moral, if not legal) responsibility in these cases as well. Of course, it is not always bad to have pulled the fire alarm when the building was not actually on fire. After all, one may have genuinely thought they saw smoke (which was actually steam), and the costs of waiting to see flames before acting can be quite high. On the other hand, we were all taught the story of Chicken Little as children, and know the consequences of crying "the sky is falling" with no basis for the alarm. We would suggest that a test of reasonableness should apply. Were the concerns justified based on the information available *at that time,* and the relative costs of being wrong (either in concluding there is a risk when in fact there is not, or in concluding there is no risk when in fact there is a risk.)[14]

Of course, most circumstances fall into the "gray area" between the two above examples. That is, one cannot usually say with absolute certainty whether the risk is real or not. In these cases, who should bear the responsibility for a decrease in property values? The legal tradition that plaintiffs bear the burden of proof would seem to indicate that when the evidence is indeterminate we should treat the case as if the risk were not real. However, in civil cases, plaintiffs are only required to meet their burden by "a preponderance of the evidence," not "beyond a reasonable doubt." It is questionable, however, whether juries[15] represent the best panel for ascertaining the true state of scientific knowledge.

Finally, consider the (not infrequent) case where the plaintiffs themselves can be thought to have contributed to the perception of risk and the concomitant diminution in property values. If the risk is "real," plaintiffs can hardly be criticized for publicizing the risk. However, what if the risk turns out to be unfounded, or the subject of continued uncertainty? Again, a test or "reasonableness" seems appropriate. Were the plaintiffs concerns justified by the information available *at the time*? Did the plaintiffs express their concerns and

---

[13]Indeed, property value diminution cases may be based completely on a theory of changing consumer preferences. For instance, homeowners near a (long standing) landfill may sue for damages, not based on any new evidence of risk from that particular landfill, but from a (more recent) general reluctance on the part of buyers to purchase homes near landfills.

[14]These two types of errors are known in statistics as Type I (or "false negative") and Type II (or "false positive") errors. Statistical decision theory holds that the optimal decision is based on a "loss function" which calculates the probability of each type of error and the cost of that error. For instance, in the case of the possible building fire, one may be 25% sure, upon seeing smoke (or is it steam?) that the building is on fire. Should you pull the alarm, or investigate further (which will take time) and not pull the alarm until you are at least 50% sure? The costs of waiting, if the building is really on fire, are likely to be large so the optimal decision is probably to pull the alarm before you are at least 50% sure the building is on fire. On the other hand, the cost of a false alarm is not zero, so it is not optimal to pull the alarm whenever one has any inclination at all that the building may be on fire. Costs of a false alarm may include: (a) the (money) costs of the business interruption; (b) the inconvenience costs to the tenants; and, (c) and the increased risk that another real fire will break out while the fire truck is responding to the false alarm.

[15]Who, in a complex toxic tort case, will be comprised only of persons willing and able to take many months off from work.

pursue the issue in a responsible manner (did they approach the State Health Board, as well as the local newspaper and contingent fee attorney).

Recognize also that plaintiffs can contribute to property value diminution simply by instigating litigation, since buyers may be reluctant to "buy into a lawsuit" regardless of the buyers perception of the risk. In fact, the potential for stigma impacts related to litigation have been recognized longer than the potential for stigma impacts from sources such as landfills and power lines. Many common sources of stigma, such as power lines landfills, are not subject to mandatory disclosure in real estate transactions (or have only recently become so). However, the existence of litigation surrounding a property has long been subject to disclosure. The potential for confounded impacts of both litigation and other sources on property values also complicates the empirical measurement of these impacts. In many cases where a nuclear power plant, landfill, etc. are thought to have adversely affected property values, the property owners have filed suit. Therefore, while the magnitude of the total impact may be measurable with some degree of precision, it may be incorrect to attribute this impact solely to the nuclear power plant, landfill, etc.

The Impact of the Media

Clearly, the media can lay a large role in the existence, magnitude and duration of stigma impacts on property value. As we have argued above, it is not the facts regarding a risk that creates property value diminution, but rather the perception of a risk or a negative image. Therefore, it is the media, much more than the scientists and even the truth itself, that determine whether a stigma impact will exist in a particular situation. While not experts in this area, we subscribe to what we believe are widely held criticisms of media coverage of "risk situations": shallowness and sensationalism. Stories such as "Study Links Landfill to Cancer" seem much more likely to receive prominent coverage than a story on a study that found no link between the landfill and cancer. Below we discuss one (perhaps isolated) example of the imbalance in media reporting of stigma impacts on property values, and then offer a few suggestions for how media coverage may be improved.

AN EXAMPLE

A large industrial corporation was sued for personal injury and diminution in value of residential properties in the neighborhood surrounding the facility. The personal and property injuries were alleged to result from air releases of toxic gasses from the facility. Our firm was retained by the defendants to provide expert evaluation of the property damage claims. Early this year, the case was presented to an "Advisory Jury." This was essentially a nonbinding minitrial, lasting three days, that was held under the auspices of the presiding court and was open to the public. The proceeding was covered by reporters from local metropolitan newspapers. The schedule of the proceeding had the plaintiffs present their case on the first day, with the defendants presenting their case on the second. As such, one can probably guess that the testimony on the first day would tend to indicate that there had been an impact on surrounding property values and medical injury to the plaintiffs. However, all of the testimony on this day was by plaintiffs' counsel or expert witnesses retained by plaintiffs.

The next day, newspapers carried a prominent article with a large headline that read "Toxic Chemicals Reach Homes." The article quoted the plaintiffs' counsel and expert witnesses in detail. Following the second day of the proceeding, when the defendants presented their case, the newspaper carried no coverage. On the third day, the advisory jury returned a complete defense verdict; the jury found that no plaintiff had suffered any medical or property injury, and the jury recommended no damages be awarded to any plaintiff.

A few days later (in the Sunday edition) the newspaper carried a large and prominent story covering the jury's verdict. This coverage was remarkable. In the days following this news story, two people who read the story called and expressed the sentiment: "Too bad about that case, your client really got hammered. The jury must not have believed anything you said." Consider this, the jury rejected each and every plaintiff claim (for property value damage and medical injury), yet two persons who read the news coverage of the event concluded that the plaintiffs had obtained a strong verdict (and thus, that the plaintiffs' claims were determined by the jury to be true).

It is also interesting to consider the two people who mistakenly inferred the outcome of the trial after reading the newspaper coverage. The first was a Phi Beta Kappa graduate of Duke University with a degree in Engineering and an MBA from Stanford, who is scientifically/technically sophisticated. The other person was a client; a senior law partner who specializes in toxic tort litigation like the case covered by the newspaper. Neither is the type that one would expect to be confused by a newspaper article discussing the outcome of a trial (on an admittedly complex topic). However, both concluded after reading the article that the jury had accepted the plaintiffs' claims, when in fact the opposite was true.

A reading of the article easily reveals why they were confused. The article quotes the plaintiffs' counsel at length, as well as the plaintiffs' consultants and expert witnesses. Statements and testimony of the defendants' counsel and witnesses are also mentioned. However, the tone and content of the entire article is such that two very sophisticated and educated professionals drew the exact opposite conclusion regarding the outcome of the trial. It is not hard to imagine what other people in the area must have concluded when reading the article. If some of these people were contemplating the purchase of a home in the area, then property value diminution may result, but not from evidence that the hypothesized risk is real.

SUGGESTIONS

First, let us restate that we are economists. Further, we view the media as no different than the seller of any other good in many respects, and realize that sellers only respond to the demands of buyers.[16] Therefore, we view the (seemingly) increasing sensationalism of stigma issues in the media as a result, not a cause, of a (seemingly) increasing level of concern regarding these issues in the general populace. In general, we do not think that the problem (and we do think there is a problem) will be cured simply by better reporting. Rather, we think that only through a better understanding of the socio-cultural influences on these concerns will we be able to minimize the negative impacts of exces-

---

[16]When media coverage is simply viewed as another good, does this justify subjecting media coverage to product defect liability?

sive fear of stigmatized places, products, and technologies. However, the media can contribute to this effort.

We would urge the media to give balanced attention to research that shows no risk from controversial place, products, and technologies as to those studies that do indicate a risk. Also, when reporting a "new" scientific finding, determine if the finding is widely held and if not, endeavor to find and present the opposing viewpoint. Recognize that those offering opinions on the potential risk may have a vested interest in the issue. Certainly attorneys (for both plaintiffs and defendants) cannot always be considered an impartial voice. Finally, help the populace realize that there are real tangible costs to overestimating risks. While one cannot usually find a teary-eyed victim of overly cautious reactions, societal welfare is reduced when useful products are taken off the market without a sound basis for their removal.

11 Environmental Stigma and Equity in Central Cities: The Case of South Phoenix[1]

David Pijawka, Subhrajit Guhathakurta,
Sarah Lebiednik, John Blair, and Suleiman Ashur

Introduction

An equitable distribution and minimization of environmental risk are a significant part of our efforts to build sustainable communities. Achieving this goal of environmental equity is difficult given the present pattern of land uses in our central cities and the widespread stigma often associated with old industrial neighborhoods. Understanding the various dimensions of environmental equity including the role of stigma and perceived risk has taken on increasing importance at local and national levels as attempts are initiated to measure the degree of inequity, identify its nature, and seek ways of mitigating its effects.

This study focuses on one such inner city minority community in South Phoenix, Arizona, where latent feelings of inequity were intensely amplified by a contamination event. The study addresses the ways in which perceived inequities and environmental risks are interrelated and the impacts of such perceptions and risks on property values. Quality Printed Circuits (QPC) operated an electronic component plant using metals, acids, and organic compounds in its operations. A four-alarm fire occurred at the plant on August 31, 1992, chemicals ignited and dispersed in a toxic smoke plume that affected a two-square-mile area to the north and east of the plant. A half square mile zone encircling the fire (including two schools and almost 1000 residents) was evacuated for about eight hours.

Although the conflagration was of short duration, the social and political repercussions of the event lasted almost five years as a result of debates over real and perceived health risks, environmental inequity, and remedial policies. This study shows that residents' awareness of problems like the high con-

---

[1]A more extensive paper on this topic, including strategies for mitigating environmental inequity and amplification effects, was published in the *Journal of Planning Education and Research,* vol. 18, 1998, pp. 113-123.

centration of industries in the area, their concern over procedural inequities that were greatly intensified in the aftermath of the fire and the perceived risks over remedial contamination were reflected in depressed home prices in the impacted area. Four years after the event, the gap between home sale prices in the affected area and the adjacent neighborhood continues to grow. We suggest from this case study that minority communities may become disproportionately affected by environmental risk and seriously disadvantaged financially when a hazardous accident occurs in the presence of existing environmental inequities.

The study builds on the work of researchers like Kasperson et al. (1988) and Slovic, Layman, Kraus, et al. (1991) who have investigated the linkages between hazards, social amplification, and risk-induced economic consequences including property value declines following contamination events. This study is made more robust by adding two contextual factors to broaden the framework. One is the condition of the housing market before and after the event. The second is the character of the study area, the presence of minorities, distributional and procedural issues of environmental equity, and resident perceptions of those inequities. Introducing these contextual factors enables us to improve our understanding of the many dimensions of environmental equity and their role in the process of social amplification of risk.

The article is built around three anchors. First, it discusses the literature on property values and contamination and presents the concepts of stigma and social amplification of risk in an environmental equity context. Second, the resident mobilization and amplification in the study area are examined. Third, the article presents evidence of substantial property value declines in the vicinity of the smoke plume as a consequence of the social amplification of risk in the aftermath of the QPC fire. The article concludes by demonstrating that residents' perceptions of environmental inequities resulted from several interrelated phenomena such as: a) a disproportionate concentration of industry due to official encouragement of industry to locate in the study area; b) the risks associated with a poor physical environment resulting from industrial concentration; c) inadequate government enforcement of standards and regulations; and d) the specific governmental response to the fire. These sensitivities were confirmed by the results of a resident perception survey. The various unresolved equity issues surfaced as the basis for a multi-year controversy over risk management and community mobilization.

Property Values, Environmental Equity, and Stigma

ENVIRONMENTAL STIGMA AND PROPERTY VALUES

There is now a large body of work documenting the effects of environmental contamination and unwanted facilities on property values (for example, Page & Rabinowitz, 1993; Chalmers & Roehr, 1993; Elliot-Jones, 1992; Patchin, 1991a, 1991b; McClelland et al., 1990; Mundy, 1992; and Kohlhase, 1991). The majority of these studies have measured the property value impacts of contamination events according to their severity, spatial extent, or duration. Soil or groundwater contamination from poor waste management practices has often been a focus of these studies. Much of the research on land value diminution from the superfund designations falls into this category (Kohlhase, 1991).

Another category of research has addressed proposed projects, their perceived threats, and the potential for stigmatized environments. This body of work has extensively dealt with siting hazardous waste disposal facilities and transporting radioactive wastes. Such studies have called on concepts of stigma and social amplification to explain the adverse price impacts of perceived risks and helped to construct a conceptual framework for explaining land price diminution (Gregory et al., 1995; Flynn et al., 1995; MacGregor et al., 1994; Mushkatel et al., 1993; Oakes, Donnelly, Garcia, & Karvia, 1993; and Kasperson et al., 1988). However, no research has explored the implications of contamination incidents in a disadvantaged area and placed the evaluation in a broader equity framework encompassing the housing market and the socioeconomic structure of the study area.

It is clear from the numerous studies that differing impacts may result from the same or similar hazard (Page & Rabinowitz, 1993) and that more severe effects may be felt on higher rather than lower priced houses (McClelland et al., 1990). It also appears that areas with a high ratio of renters, or areas dependent on employment in the polluting industry may be significantly less sensitive than expected to environmental threats (NANP, 1995b); that diminished selling prices may recover in some cases while others display little temporal change (Colwell, 1990; Elliot-Jones, 1992; Kolhase, 1991; and McClelland et al., 1990); and that there may be price gradients which relate land value to proximity to the hazard (Nelson, 1982; Nelson, Genereaux, & Genereaux, 1992; and Smolen, Moore, & Conway, 1992). Variation in study findings may be explained by the range of methodological approaches, the litigation focus of many individual studies, and variations in the method of valuation. However, the dominant cause may be the lack of a consistent theoretical base that explains the consequences of contamination in urban areas and the role played by risk perceptions in their sociopolitical contexts.

There are two main components in price diminution. The first consists of the relatively tangible cost of remediation. The second is stigma. Chalmers and Roehr (1993) conducted a survey of relevant literature (for example, Patchin, 1991a, 1991b; Slovic et al., 1994; Slovic, 1993; and Gregory et al., 1995), which suggests that stigma represents more than the difference between the values of uncontaminated and fully remediated property. It encompasses an intangible concept of risk that lies beyond direct clean-up costs and includes further discounts to compensate investors, purchasers, or lenders with the risks associated with the formerly contaminated property. The inference is that stigma may continue to affect property values without real or observable contamination (The Appraisal Institute, 1992). Indeed, Patchin (1991a) suggests that the aggregate loss of land values as a result of stigma could exceed 10% of the total U.S. real estate value. Patchin (1991a), Hunsperger (1991), and Chalmers and Jackson (1996) indicate that mildly contaminated property and even the suspicion of pollution can produce extreme caution in buyers and lenders, thus deflating market prices.

A concept that is closely related to stigma is social amplification. Kasperson et al. (1988) view the concept of social amplification as bridging the gap between the results of a technical risk analysis and a risk evaluation which incorporates a broader range of psychological, sociological, and cultural perspectives. Scientists and government view risk as the statistical probability of exposure. The public sees risk as the consequence of exposure, so amplification processes attempt to explain a perplexing situation where risk events which have minor or uncertain physical consequences can generate much stronger public concern than conventional risk analysis would suggest. Whether the technical interpretation of risk is correct or not, it is too narrow to serve as the main framework for policy formulation.

There is a growing body of judicial support for victims of stigma-induced property value loss. Colwell (1990) and McEvoy (1994) report several electromagnetic field (EMF) lawsuits that focus on perceived risks and how they have depressed real estate values. Court rulings have generally favored the landowner; not on health grounds, since the adverse effects of EMF are far from proven, but on grounds of stigma-induced price diminution. The case *City of Santa Fe v. Komis* (1992) strongly reflects judicial support for compensation for market value loss even if the loss is based on fears that are not founded on objective standards. With the partial exception of noise pollution cases (Blair, 1994), the courts have began to recognize the problem of diminished marketability of contaminated property regardless of the technical reality of the risk.

ENVIRONMENTAL EQUITY

Concerns over environmental equity surfaced in the early to mid 1980s as environmental protection issues merged with social justice goals (Gottlieb, 1993; Norton, 1991; Cutter, 1995; Bullard, 1994b). The equity movement focused on the disproportionately high environmental burdens affecting low-income or minority populations, and the unequal enforcement of environmental regulations (Higgins, 1993; Collin, 1992).

Their concerns were legitimized by several landmark events in the 1980s, one in particular being the U.S. General Accounting Office's report in 1983 showing that a disproportionate number of hazardous waste landfills were located in minority areas (GAO, 1983; Higgins, 1993; Atkinson, 1992). In the 1990s environmental equity issues have focused on empirical case studies, definitions, measuring techniques, and preventive and remediation strategies in planning. Cutter (1995) and Zimmermman (1994) define environmental justice as a way of achieving equitable allocation of resources and benefits. Differences in the definition of equity occur with the EPA expanding its meaning to embrace three concerns. First is *procedural* equity and its formal orientation towards environmental protection. It relates to compliance with regulations and their consistent enforcement. Second is *process* equity, which has a less formal role in environmental policy making and management. Process-oriented matters tend to be nonregulatory like providing opportunities for participating in decision making. Third is *distributional* equity, a term which relates to concepts of spatial fairness and the geographical distribution of environmental benefits and burdens (Collin, 1992).

Most recent environmental equity studies have centered on distributional questions and matters like facilities distribution in relationship to demographics (Bryant & Mohai, 1992). There are also some studies, which have shown that minority populations are more vulnerable to environmental contamination than previously assumed (Bullard, 1993, 1994a). National survey data presented by Flynn, Slovic, and Mertz (1994) suggest that African-Americans and Hispanics tend to perceive higher levels of risk associated with a wide array of environmental and technological hazards than was previously recognized.

Research on environmental equity has not been free of criticism. Boerner and Lambert (1994) for example, used one case study to argue that falling real estate and housing markets have created incentives for low-income minority groups to relocate to industrial areas, implying an existing knowledge base and a voluntary risk situation rather than inequity.

The QPC case is an important addition to the equity and risk perception studies. It connects a disadvantaged and minority community experiencing high ambient levels of environmental inequity with a contamination event, the subsequent triggering of a classic case of social amplification of environmental risk, and its consequences in the form of a precipitous drop in housing prices. There is also a political importance to the research since it demonstrates the role that equity issues and equity perceptions play in the social amplification of risk. In this case, perceptions of inequity (distributional, procedural, and process) heightened the controversy over risk and mobilized area residents.

The Underlying Basis for Social Amplification and Stigma in the South Phoenix Fire Event

SOCIO-ECONOMIC CHARACTER: THE BASIS FOR DISTRIBUTIONAL INEQUITY

A general study area for this analysis consisted of eleven census tracts (Figure 11.1), one of which was the site of the QPC plant. This tract (tract 1160) contained virtually all of the homes impacted by the smoke plume. The population of the study area is a little under 41,000 (see Table 11.1). It is unevenly distributed and there is much vacant and underutilized land. Tract 1160 has a

Table 11.1 Population, Housing, and Income in the Study Area by Census Tract, 1990

Census tract	Population in 1990	Minorities as a proportion of population (%)	No. of housing units	Proportion vacant (%)	Proportion rented (%)	Households below poverty line	Mean household income: Owner occupied	Mean household income: Rented
1153	2200	88	682	16.0	48	42	24471	13618
1154	2481	82	746	12.7	52	30	22217	21668
1157	4575	78	1212	6.7	24	24	28019	18928
1158	5554	75	1924	13.7	52	38	22109	13129
1159	3357	88	975	13.2	36	32	21576	15608
1160	4305	98	1496	6.6	36	38	25142	8567
1161	2506	95	1099	36.0	57	46	18246	12846
1163	4834	84	1503	9.8	23	22	33912	15233
1164	3597	77	1048	8.7	32	27	29588	17951
1165	4205	79	1243	7.4	21	14	39499	21121
1166.02	3254	94	796	4.8	42	43	40348	8073
Study Area	40868	84	12752	12.1	Not available	Not available	Not available	Not available
City of Phoenix	983403	28	422036	12.3	37	11	45485	23825

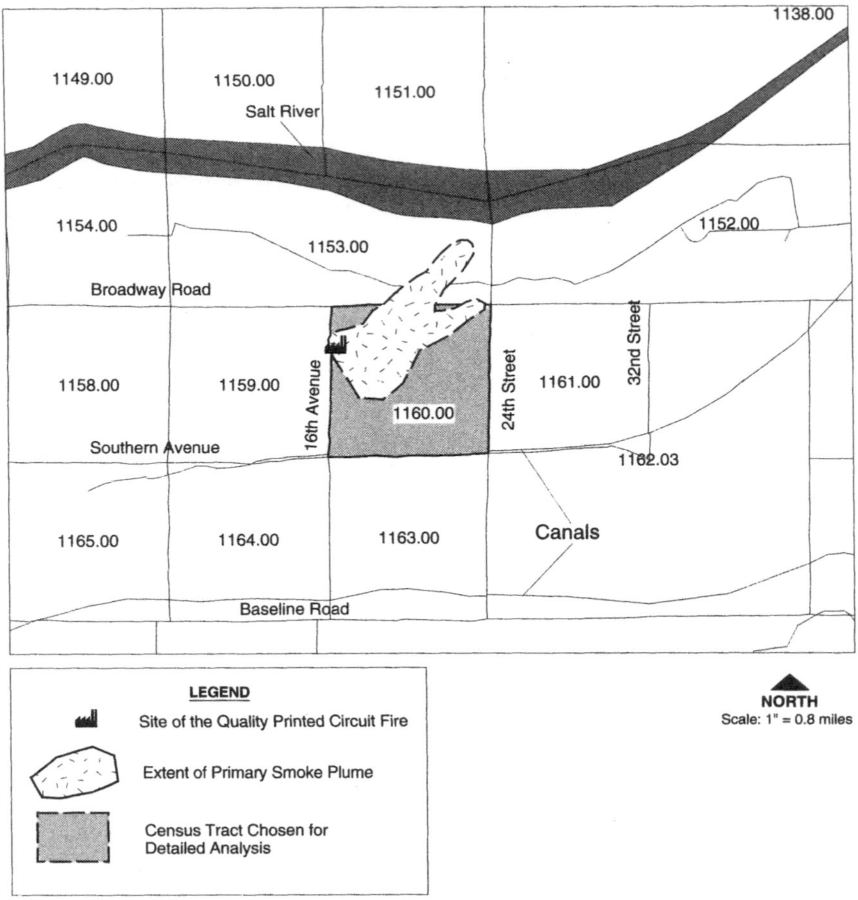

Figure 11.1 The general study area.

population of 4,305, slightly higher than the average tract in the general study area. Minorities form an exceptionally high proportion of the population, reaching 98% in 1990.

From field observation, housing in the general study area has a neglected appearance though housing is in reasonable external condition with some pockets of substandard homes. The overall vacancy rate in the housing market was 12.1% in 1990 with tract 1160 having the lowest rate at 6.4%. Tract 1160 also has a lower-than-average proportion of housing for rent, suggesting a greater degree of stability than other parts of the study area. In almost all census tracts, mean household income for owner-occupiers and renters in the study area was below the City of Phoenix average. Unemployment throughout the study area is roughly double the average for the City of Phoenix, which must be a major factor in the depressed income levels of area residents. A related trait is the high proportion of study area families living below the poverty line. Census tract 1160 records one of the highest, which at 38%, is more than three times the City of Phoenix average.

The rate of development of new commercial or industrial enterprises in the study area is very low. An average of only two permits per year were issued during 1991–1994 inclusive though there were more for extensions to existing buildings. Patrick Hendrick (Neighborhood E D Officer, Neighborhood Services Department, City of Phoenix, personal communication, July 1996) reports that building vacancy rates in the commercial and industrial sectors have declined significantly from 1990 highs of 20–25% but there is still an abundance of vacant floor space and vacant land zoned for industry. Residential resale activity is brisk but the inventory of single family dwellings is close to static with demolition almost equaling additions to stock. Hendrick (personal communication, July 1994) suggests that the negative perceptions connected with the general South Phoenix area such as high crime rates, drugs, and environmental contamination will continue to suppress demand for new housing.

One factor in the sluggishness of residential construction in the area is the patchwork of industrial and commercial zoning which often results in housing being exposed to unbuffered industrial activities such as automobile junkyards and open storage areas. These land-uses are an eyesore and may be hazardous due to leaking battery acid and engine oil. It is an important dimension to the environmental equity problem for it encourages blight and inhibits new housing. Another factor is that the cost of building even modest homes in the area is 15–20% above the average value of existing residences. New construction is discouraged by implicit over-capitalization, the gap being explained by the deteriorated state of plumbing, wiring, and other services of many houses in the area.

South Phoenix seems to be subject to higher than normal public health and environmental risks as a result of a disproportionate concentration of hazardous waste generators located in the area (ADEQ, 1995; Bolin et al., in press). The ADEQ investigation used a study area that accounts for 8.8% of the City of Phoenix's population. It has 20% of the state's hazardous waste generators (Table 11.2) and is the source of 39% of the state's shipments of toxic waste.

Table 11.2 Distribution of Hazardous Waste Generators in Central and South Phoenix, 1993

Indicator	Study area[a]	City of Phoenix	Study Area as a (%) of the City of Phoenix
No. of hazardous waste generators	105	531	20
Population (1990 Census)	86,751	983,403	8.8
Minority Population (%)	74	28	—
Hazardous waste shipments from LQGs only (tons)[b]	2,482	6,355[c]	39[d]

Source: Unpublished data from the Arizona Department of Environmental Quality (ADEQ) and the South Phoenix Waste Minimization Survey Project (ADEQ, 1995).

[a]The study area consists of Zip Codes 85034, 85040, and 85041.

[b]An LQG is a "large quantity generator" of hazardous waste producing more than 2000 kg. per month.

[c]This figure is for the entire State.

[d]This percentage is the study area in relation to the State.

These are relatively strong indicators of a distributional equity problem but there are also procedural inequities associated with the monitoring and regulation of the industry. Although 75% of large quantity generators (LQGs) of hazardous waste in ADEQ's study area file the requisite shipment plan, the rate of inspections is well under 25% of facilities. A full third of LQGs have never been inspected at all. This is another indicator that suggests that government agencies may be ignoring the environmental inequities present in South Phoenix.

SOCIAL AMPLIFICATION OF THE EVENT 1992–1996

The stored chemicals that ignited at the QPC plant on August 31, 1992 included copperfoil, sulfuric acid, and nitric acid. Immediately after the fire, many residents complained of serious health effects, and state environmental and health agencies audited the plant's stored chemicals and tested soil samples in the area. The combination of health concerns and the apparently inadequate reaction by emergency personnel led to the founding of a residents' organization called Concerned Residents of South Phoenix (CRSP) on the day after the fire.

CRSP claimed that circumstances during the emergency evacuation signaled more serious general problems related to the fire. The authorities were uncertain what types of hazardous chemicals were released because reporting procedures had been ignored. Many area residents were not warned to evacuate and despite a toxic cloud, were not informed to turn off evaporative coolers. Consequently, contaminants entered homes and became concentrated in air ducts. Without knowledge of the extent of the smoke plume or types and concentrations of chemical contaminants, residents returned to their homes three hours after evacuating.

While the fire and partial evacuation was short lived, the event was the basis for a four-year period of political and social mobilization of residents over questions of health risk, government remediation efforts, and environmental equity. Strongly held views concerning inequities hampered effective communication between residents and governmental agencies. Initial disputes over health risks and governmental actions were cautious, but they escalated in intensity over time and obscured meaningful dialogue over future management and planning of the area. Despite scientific and government reports showing no significant health risk to the public from fire-related chemical residuals in homes, residents did not trust these studies and viewed government actions as inadequate.

Three public meetings were held during the two months following the fire to resolve residents' health concerns which resulted in three studies: air duct sampling, soil sampling, and a health survey of households. The health effects survey of complaints and medical records found that there were no health risks related to residual contamination. The soil and airduct sampling studies were released four months after the fire and showed normal metal levels in soils and a slight elevation of contaminants in air ducts but concluded that those levels did not pose a health risk.

The public response to the studies amplified the question of health risks into a social and political controversy. CRSP protested that the health study was methodologically inadequate since it was extremely limited in its sampling size with only 28 households, that too few medical records were available, and that these factors prevented a definitive conclusion from being reached about health risks. The organization argued that the conclusions were not founded on fact, especially when residents complained of health problems and there appeared to be an increase in the mortality rate in the area. As a response to a perceived increase in mortality rates in the plume area, CRSP requested the state to use emergency relief funds to remove contaminants from affected homes or to move residents from the area. CRSP also argued that the soil and airduct contaminant sampling of only 11 homes was too small to be statistically significant and that the samples were not taken from homes most effected by the smoke plume.

Social Escalation and Governmental Response

Six months after the fire (January 1993), while government agencies were beginning a series of new health and sampling studies, residents of the affected area had by now concluded that serious health threats confronted the area, government-sponsored testing was inadequate and biased, and the lack of public hearings indicated government evasiveness. Previous government responses to health risk concerns stimulated feelings of procedural and process inequities. Residents expressed the view that their community was not being treated appropriately. While local and state governments saw the next step as more studies, residents were already assuming the worst and, in an atmosphere of distrust, they moved toward protest as well as political and legal solutions.

The second phase of amplification which made dispute resolution difficult to undertake was caused by a divergence of issues. For the affected community, the issue was how to remedy the existing health threats quickly. The issue for government agencies was whether there was a health risk and whether further studies were necessary. Government action during this period included sponsorship of a chronic health effects study, a self-reporting health survey, a county mortality study, and renewed soil, air and airduct contaminant sampling. Contaminant levels were found to be extremely low in the health studies, and the sampling surveys showed slightly elevated metal residue concentration, but not in excess of "state action levels." The county mortality study concluded that there were no excess deaths due to the fire.

Two Separate Directions

It is clear from ensuing incidents that community and government views of the key issue were widely divergent during the 12- to 18- month period after the fire. Two additional soil and airduct studies showed that contaminant levels did not exceed state guidelines and in January 1994, 16 months following the fire, the state closed its case. In contrast, area residents argued that sampling size was inappropriate and that there was a lack of sampling immediately after the fire as well as inaccurate interpretations of metal concentrations and mortality data. The concern over the lack of remedial actions to clean up homes and provide medical assistance resulted in a class action lawsuit filed one year after the fire seeking punitive and compensatory damages. Three months after the state closed its case a settlement was reached with residents, but it affected a limited number of households and did not resolve the continuing dispute and concern over the lack of remedial action.

Underlying the disagreements associated with health risks from the QPC event was a persistent concern over environmental equity. Two out of four public meetings that were held focused on environmental justice. Many residents in and near the affected area perceived the institutional responses as failures from manifest inequities. The debate—originally about a toxic smoke plume and health risks—moved to the perceived inequities of a politically weak minority population residing close to a disproportionate number of industries concomitant with the lack of effective government oversight.

The dispute did stimulate two inquiries by the state environmental agency on issues related to equity. Neither report is complete although they appear to be limited by their focus on short-term mitigation of risks through improved permitting, enforcement policy, and emergency planning. Neither appear to evaluate underlying causes nor propose strategies for correcting existing environmental inequities in South Phoenix.

CONFIRMING THE SENSE OF INEQUITY: SURVEY OF RESIDENTS

An important way of addressing inequities is knowing the extent to which the public perceives inequities and what their characteristics might be. Solutions to risk-based inequities may be unsuccessful if they are directed at improving siting or relocation policy and neglect the procedural and process-oriented matters which may be at the root of perceived environmental inequities. To understand how perceptions of inequity stimulated the controversy, a survey instrument was designed and distributed to 200 households located in the smoke plume. Eighty-four households returned completed questionnaires. A comparison of socioeconomic data between the sample and the larger census tract shows substantial similarities demonstrating that the sample is representative of the population.

Environmental Saliency

A series of questions were asked to measure the importance the minority community gave to environmental quality. Environmental quality was identified by 63% of respondents as a serious problem in the area. The survey population viewed only crime and employment as being more serious. In addition, 78% of the sampled population rated the environmental condition of the area as "poor or very poor" and 91% indicated that improving environmental conditions was important. Environmental quality is a salient element among residents in the affected area, confirming earlier findings by Flynn et al. (1994) that minority populations are especially sensitive to environmental risk considerations.

Distributional Equity

Based on the work of Bullard (1993) and Gelobter (1994), three questions were developed in the survey to capture perceptions of distributional equity: the extent of industrial concentration; industrial targeting and its fairness; and the costs and benefits of industrial development in the area (Table 11.3). In this context, 65% of the sample population perceived that the area in which they resided had a relatively higher concentration of industry than other areas of Phoenix and 74% agreed or strongly agreed with the statement, "This area was purposely targeted as a location for industry." Of these, almost all felt that siting industries close to residential areas in such concentrations was *unfair*. The data suggest that there are prevailing perceptions of distributional

Table 11.3 Perceived Distributional Inequity of Industrial Development by South Phoenix Residents

Level of agreement	Higher concentration of industry	Targeted for industry	Industrial targeting is unfair
Strongly agree	38%	42%	59%
Agree	27%	32%	36%
Uncertain	21%	15%	2%
Disagree	11%	8%	2%
Strongly disagree	3%	3%	1%

Responses rounded to 100%.

unevenness in the city's industrial location policy and that the outcome is unfair.

Very high and significant correlations were observed between perceptions of distributional inequity and levels of perceived risk. Survey questions asked respondents to rate levels of risk to safety and health based on the existing concentration of industry, the proximity of industry to home, and the industrial mix in the area. "Serious to very serious" levels of risk by over 75% of the population were associated with all three dimensions of distributional equity. This suggests that residents are skeptical of the benefits of industrial development in these circumstances and are concerned over the risks associated with the inequitable distribution of industries in the area.

Procedural and Process Equity

Residents' perceptions of procedural and process equity were measured along several dimensions in the survey as shown in Table 11.4. Over 77% of residents felt that areas other than South Phoenix had more control over industrial location. Residents expressed the view that they were generally not consulted in matters related to industry location or development. It was also apparent that residents' trust in industry is very low with respect to compliance with environmental regulations. Their trust in government's enforcement abilities is also low.

Table 11.4 Perceived Procedural Inequity Regarding Industrial Location in South Phoenix

Level of agreement	Other areas greater control[a]	Risk reduction by industry[b]	Trust to operate safely[c]	Government trust[d]
Strongly Agree	51%	2%	9%	16%
Agree	26%	7%	7%	21%
Uncertain	16%	27%	31%	12%
Disagree	2%	22%	23%	21%
Strong disagree	5%	42%	30%	30%

[a]Q. What is your level of agreement with the following statement? Residents n other area in Phoenix have more control over industrial location than residents in the area.
[b]Q. Industry can be expected to do everything possible to reduce risk.
[c]Q. Industry can be trusted to operate safely.
[d]Q. Government can be trusted to enforce regulatory compliance of industries in the area.

Equity Perceptions Four Years after the QPC Fire

The general view held by residents on matters related to procedural equity are even more pronounced when it comes to issues stemming from the responses to the QPC fire. Over 90% believe that governmental response to QPC-like problems would be more effective in other areas of Phoenix. Four years after the fire, and despite governmental studies showing minimal risks to health, 89% of respondents expressed concern over health risks, and 72% were concerned with declining property values. When asked how well government worked with residents to address risk-related concerns, 77% indicated the response was poor to very poor.

This survey supports the conclusion that residents of South Phoenix are conscious of distributional inequities and that the process and procedural inadequacies of the post-fire period have greatly heightened latent sensitivities to environmental risk.

The Impact of the Fire on Land Values

Studies cited above indicate that contamination incidents are often reflected in property value diminution in the affected areas. None of these studies have examined price effects in a low-income residential area where housing prices are at or near the bottom of the market, despite the national recognition given to the problems of environmental inequity in poor and minority neighborhoods. In microeconomic theory, the extent of price diminution is directly proportional to housing values. *Ceteris paribus*, cheaper housing can be expected to experience lower diminution effects. The QPC contamination incident allows the theory to be tested. Moreover, since there has been no scientific evidence of significant risk levels, the case study also permits the examination of the type of impacts resulting from social perceptions of risk and inequity.

The analysis of the extent of impacts to property values in this South Phoenix neighborhood is carried out at two geographic levels. First, trends in housing prices in the larger study area, comprised of 11 census tracts in South Phoenix, are examined. Second, a more detailed study analyzes the housing-price effects of the toxic plume from the QPC incident in census tract 1160. The toxic smoke plume was configured by ADEQ through detailed analysis of meteorological data for the day of the QPC fire. The two levels of analysis provide a framework for evaluating the intensity and the extent of property value diminution triggered by the QPC fire.

ANALYSIS OF THE GENERAL STUDY AREA, 1984–1995

Analysis of median house prices in this area for the period 1984 to 1995 reveals two distinctive features. First, median housing prices in the larger study area and Metropolitan Phoenix have diverged since 1986 (see Figure 11.2). There is a clear indication of the gradual decline in home values in the study area as a proportion of Metro Phoenix values. Trends for individual census tracts in the area are converging, suggesting that the study area is strengthening its common identity. Further, housing price trends in the general study area are similar to trends in the affected tract (tract 1160).

Second, contrary to expectations, median price trends at the level of the general study area or even the affected census tract reveal no discernible impact from the contamination incident. In fact, there was a small upward move-

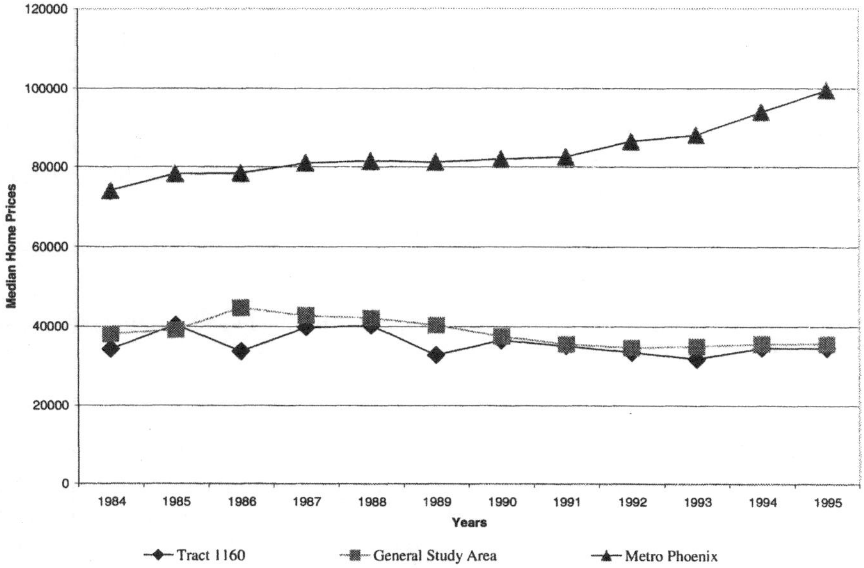

Figure 11.2 Housing price trends, Metro Phoenix and the General Study Area, 1874-1995.

ment in median housing price in the area in 1993, less than one year after the QPC incident. However, a closer look at the houses within the plume provides a different picture.

IMPACTS IN THE PLUME AREA

The plume-area analysis used housing price sales data for 1989–95 and evaluated 251 houses that were sold in census tract 1160 located both in and outside the designated plume area. About half of the single-family homes sold during this period was within the plume area. The homes sold within and outside the plume area in the tract were substantially similar in floor area, housing condition, or in access to neighborhood amenities. A final data set of 175 homes included those that sold for more than $25,000 and less than $65,000. Of the 71 homes that sold for less than $25,000 during the seven-year period, 38 were in the plume and 35 outside it. These sales were not included in the analysis since they were deemed to be less than market price and most likely the result of foreclosures. The relatively even distribution of low priced homes inside and outside the plume and subsequent inspection of the date of sale does not reveal any discernible bias of foreclosures that could affect the results of this analysis. Only 5 homes sold for more than $65,000 during that period of which two were within the plume area (one of which sold in 1993). Thus, the possibility of bias from limiting the data set is minimal.

Stratified analysis of home prices within and outside the plume disclosed a significant influence of the fire on average home prices within the plume. Homes outside the plume have appreciated by 11% over the period 1989–95 inclusive. This increase is significant at the 99% confidence level. However, the sales price of homes within the plume does not follow the upward trend that is evident in similar homes outside the plume but within the same census

Table 11.5 Significance Test for Means of Home Prices by Area and Time in Census Tract 1160

	In-plume	Out-plume	Difference	Significance
Before Fire[a]	$39,425	$37,955	4%	LOW (50%)
After Fire[b]	$37,466	$41,955	–12%	HIGH (97%)
Difference	–5%	11%	–16%	
Significance[c]	LOW (55%)	HIGH (99%)		

[a]1989–1991 inclusive.
[b]1993–1995 inclusive.
[c]Significance level indicates the probability of rejecting the null hypothesis suggesting that the difference between the two sets of home prices is actually zero.

tract. The *t*-tests reported in Table 11.5 also show that before the fire there was no statistical difference between mean home prices within and outside the plume area. After the fire, there is a statistically significant gap in the prices of homes between the two areas. The magnitude of this change is about 16% (from –5% to +11%), a fire-induced diminution in home values that is significantly larger than those reported in previous studies of similar hazardous incidents. Thus our theoretical expectations of smaller house price diminution cannot be justified from this analysis.

Conclusion

Detailed analysis of tract 1160 shows significant property value impacts while individual census tracts close to the QPC site did not show house price diminution which could be clearly attributed to the effects of the fire. The limited geographical extent of the impact of the fire can be explained by relatively strong regional demand for the low-priced segment of the housing market. An overriding conclusion is that a hazardous event in a minority area experiencing environmental inequities triggered significant social amplification of risk that resulted in serious property value impacts even in the low-income housing market. This occurred despite government reports indicating that the health risks were not significant. Diminution effects were strongly present in the QPC case and they persist almost four years later. There is little sign of attenuation despite settlement of a health lawsuit, completion of remedial actions by the government, and a steady rate of property transactions. These developments suggest that the stigmatizing effects of the accident should have been discounted by now. In addition, the plant which was the source of the conflagration no longer exists: the factory was demolished, the site cleared, and production relocated to another industrial park. The literature suggests that a major reason for risk amplification is distrust of corporate and government ability to avoid further accidents. Removing the source of the problem should have reduced perceived risks, but did not do so. It is indicative of considerable underlying concern over environmental risk in the community.

There are at least two factors that explain the failure of the in-plume housing market to rebound. First, government agencies involved in clean-up operations and health assessments were continuously and stridently criticized in the media and at public meetings so that the larger buying public was not persuaded that contamination problems had diminished enough to return to the market without risk discounts. Second, lack of confidence in government efforts was seen as indicative of larger inequities confronting the area. One might have reasoned that property values could not diminish much or for long in the moderately stable areas surrounding the fire, especially given city-wide demand for home ownership in this housing sub-market. Clearly, the stigma and degree of social amplification over environmental inequities experienced by a minority population was large enough to create and sustain a serious decline in house prices. It is especially significant in the context of out-plume price appreciation in tract 1160 and the rise in house prices in the general study area during 1992–1995.

There is both a tragedy and an irony in this case study. The tragedy is that a minority community that was already confronting environmental and social inequities experienced a significant drop in property prices at a time when the trend throughout the entire metropolitan area was for prices to rise, especially in low-income neighborhoods. This is serious enough from a general investment viewpoint since low-income and minority populations have proportionately more of their savings invested in their homes than middle- and high-income groups. Hence, the depreciation of home values in a low-income neighborhood poses serious equity problems which compensatory packages almost never acknowledge. For example the class action lawsuit noted earlier was resolved on May 19, 1995 with a settlement of $1.6 million. It only encompasses health costs for a small number of affected residents. Thus, low-income households close to industries using hazardous materials bear a disproportionate amount of physical and financial risk. Given that only two local residents actually gained employment at the QPC plant, the risk and benefit equation seems to be unduly loaded against residents of this neighborhood. We conclude that a hazardous event occurring in the context of existing environmental inequities may result in disproportionately greater losses to a minority community than was originally anticipated based on the prevailing local market economy.

12 The Impact of External Parties on Brand-Name Capital: The 1982 Tylenol® Poisonings and Subsequent Cases[1]

Mark L. Mitchell

An examination of the 1982 Tylenol® poisonings reveals stock market losses to Johnson & Johnson that far exceed direct costs and losses shared with other pain-reliever producers; this evidence provides support for the Klein and Leffler (1981) theory of brand names as quality-assuring mechanisms. Of the subsequent cases, only the 1986 Tylenol® poisonings were associated with significant stock market losses. Prior to the 1982 and 1986 Tylenol® poisonings, Tylenol® was the number one pain reliever whereas the other pain relievers that were poisoned had a much lower level of brand-name capital to lose.

I. Introduction

On September 30, 1982 Johnson & Johnson announced that three people had been killed as the result of ingesting cyanide-laced Tylenol® capsules.[2] Four more Tylenol-related deaths were reported within the next two days. Culminating in 125,000 stories in the print media alone, the poisonings were an event without precedent in American business ("Crisis Public Relations," 1982). The Tylenol® brand received over $1 billion in adverse publicity.[3] As a result, many analysts claimed the brand was dead. But the company president, James Burke, ignoring the advice of government officials and even some of his close associates, decided to spend millions to revive Tylenol®. Burke's decision will be studied in business schools for years to come. The general opinion today is

---

[1]Reprinted from *Economic Inquiry*, vol. 27, pp. 601-618. Copyright 1989 by Oxford University Press, Inc. Reprinted with permission.

[2]In researching the Tylenol® poisonings, I examined over two hundred articles in nonacademic publications including *Advertising Age*, *Drug Store News*, *Fortune*, and *Wall Street Journal* for dates, market share figures, advertising expenditures, and so forth. To save space and facilitate the flow of the paper, many of the articles used are not cited. For the interested reader, a complete list of all articles is available.

[3]According to *Drug Store News* ("The Bad News," 1982), it would have cost Johnson & Johnson over $1 billion to purchase a similar amount of air time and print space.

that Johnson & Johnson and Tylenol® made a prodigious comeback, one unparalleled in American business.

The event provides an opportunity to study the effect that an external party can have on the brand name or reputation of a firm. That is, to what degree does a firm's brand name suffer a loss in value even when the firm clearly did not intentionally lower product quality? In other words, do consumers hold firms responsible for the damaging actions of parties not associated with that firm? What is the proper gauge to use in measuring the recovery from such an event? Many would favor market share, while others might argue that the stock price is a better barometer to measure recovery. This study attempts to resolve these issues.

II. Brand Names and Product Tampering

The theory of brand names as quality-assuring devices has emerged formally in the last decade. Klein and Leffler (1981), following the arguments by Klein, Crawford, and Alchian (1978), developed a model in which the presence of firm-specific sunk capital investments, such as those incurred in developing a brand name, provide a mechanism for assuring contractual performance.[4] According to their model, if a firm cheats by reducing product quality below the expected level, the value of its brand-name capital declines to zero and the price premium which consumers were willing to pay for the firm's products is lost. Recent empirical studies by Jarrell and Peltzman (1985), Chalk (1986, 1987), Mitchell and Maloney (1989), and Benjamin and Mitchell (1989) have presented evidence in support of the theory.

Can a firm suffer a loss in brand-name capital even if management did not intentionally cheat and lower product quality or fail to prevent cheating by distributors and retailers? While the development of a brand name is largely under the control of management with firm-specific capital investments and consistent product quality, forces external to a firm may seriously damage the value of its brand name. The classic example is product tampering.

To the extent that product tampering reduces the expected safety level of a product, consumers will shy away from that product. If consumers perceive that the tampering is targeted at a firm, they will reduce their demand for similar products produced by the firm as well, at least until the tamperer is captured, the safety level is improved so that tampering cannot occur again, or the firm eliminates whatever may have triggered the tamperings. Additionally, when product tampering reveals information about the safety level of similar products produced by other firms, consumers will reduce their demand for products of those firms as well. However, as long as consumers perceive the tampering was directed at a specific firm, demand for the tampered product will exhibit a reduction relative to competing products. To summarize, the decline in the expected safety level of a product should cause consumers to revise their consumption patterns and shy away from that product, and to the extent that the product is identified with the firm, the brand-name capital of the firm itself will depreciate as well.

---

[4]See also Barzel (1982), Benjamin (1978), Shapiro (1983), and Telser (1980).

III. Johnson & Johnson and the Tylenol® Crisis

During a three-day period in late September and early October of 1982, seven Chicago-area residents died after taking Extra-Strength Tylenol® capsules contaminated with cyanide. Johnson & Johnson immediately recalled the lots from which the contaminated bottles had come and halted advertisements for the entire Tylenol® product line. Within days, it became clear to investigators that the tampering had occurred at the retail level.[5] Apparently, a few bottles of Extra-Strength Tylenol® capsules had been removed from store shelves, the cyanide added, and the bottles returned to the shelves. Learning this, Johnson & Johnson withdrew all Tylenol® capsules from the market.

In the mid-1970s, a consumer-oriented promotion campaign made Tylenol® the biggest selling item in drug, food, and mass merchandising outlets, breaking an eighteen-year dominance by Proctor & Gamble's Crest toothpaste. By 1982, the Tylenol® product line controlled 37% of the over-the-counter analgesics market, in which sales totaled $1.2 billion. Immediately following the cyanide poisonings, the market share of the entire Tylenol® line fell from 37 to 7%.[6] Although the company and its manufacturing procedures were quickly cleared of any possible direct role in the disaster, some question remained concerning the brand's survivability. Johnson & Johnson made clear its commitment to regaining Tylenol's pre-poisonings status. A month after the poisonings the company resumed regular advertising of the noncapsule products in the Tylenol® line. A few weeks later Johnson & Johnson repackaged the capsules with a triple safety seal and began an advertising campaign focusing mainly on the new packaging. The capsules gradually reappeared on the shelves in late December of 1982. Even before the return of the Tylenol® capsules, the market share of Tylenol® tablets appeared to rebound, despite heavy competition from brands never before advertised.[7] Tylenol® rebounded to a 30% market share within six months. By August 1983, Tylenol® was firmly established once again as the nation's leading pain reliever.

This evidence suggests Johnson & Johnson was successful in reestablishing confidence in the brand. Over 90% of consumers questioned in a survey felt that Johnson & Johnson was not to blame, according to the *Wall Street Journal* ("Tylenol Regains," 1982). Many analysts felt that the company had exceeded its responsibility and deserved the Comeback of the Year award. To this day, other companies' reactions to product tampering are judged relative to Johnson & Johnson's handling of the Tylenol® disaster.[8] The name Tylenol®, as well as Johnson & Johnson, seems to mean just as much as it did before the incident.

---

[5]The cyanide-laced capsules were from bottles produced in two different states, Texas and Pennsylvania, making it unlikely that contamination occurred at the plant level.

[6]Noncapsule products in the Tylenol® line were never removed from the shelves.

[7]Information Resources Inc., which surveys sales in grocery outlets in four U.S. towns, claimed Tylenol® recaptured 95% of its prior total market share by mid-December. In the first week of December alone, Tylenol's market share climbed 47%. However, much of this may be attributed to the $2.50 coupons Johnson & Johnson offered in thousands of newspapers in order to win back customers (*"Tylenol Tablets,"* 1982).

[8]According to *Advertising Age* ("Gerber Ignores," 1986, p. 3), "Johnson & Johnson is widely viewed as having written the textbook on corporate response to devastating news."

IV. Stock-Price Effects of 1982 Tylenol® Poisonings

Stock market event analysis is an often-used technique in estimating the impact of events such as the 1982 Tylenol® poisonings. It involves the identification of an event that causes investors to change their expectations concerning the discounted future cash flows of a security. The analysis is based on the theory of efficient markets, which assumes that the price of any security incorporates at each instant all currently available information and adjusts to new information as soon as the information is accessible to investors.

The objective of the empirical model is to obtain a well-specified time series of security returns for the firm; a widely accepted way of achieving this is the market model

$$R_{it} = \alpha_i + \beta_i R_{mt} + \varepsilon_{it}$$
$$\varepsilon_{it} \sim N(0,\sigma^2),$$

which assumes that the return to a security i at time t, R_{it}, is a linear function of the market return, R_{mt}, plus a random error term, e_{it}, which is uncorrelated with the market return.

By estimating the market model for a period different from the event period, the returns on security i can be forecasted, conditional on the parameter estimates (a_i,b_i) and the actual return on the market index, R_{mt}, for each day of the event period. The abnormal return

$$AR_{it} = R_{it} - (\hat{\alpha}_i + \hat{\beta}_i R_{mt})$$

measures the impact of the event on security i at time t. For events lasting more than one day, the abnormal returns are summed to obtain the cumulative abnormal return

$$CAR_{iT} = \sum_{t=1}^{T} AR_{it}$$

where T is the length of the event window.

The daily stock returns tapes from the Center for Research in Security Prices, University of Chicago, were used to estimate the market model for Johnson & Johnson for the one-year period (253 trading days) prior to the poisonings. The Center for Research in Security Prices equally-weighted market return of all New York Stock Exchange and American Stock Exchange stocks proxies R_{mt} in the empirical tests.[9] Table 12.1 displays estimates of the market model parameters with standard errors in parentheses.

Several measures were taken to test the stability of the market model estimates reported in Table 12.1. For example, the presence of nonsynchronous trading may bias the coefficient estimates of the market model parameters. To test this potential bias, the market model was replicated using the Dimson (1979) technique with $R_{m,t-1}$ and $R_{m,t+1}$, in addition to R_{mt}, as the independent variables. The abnormal returns generated from this technique are not significantly different from those reported in the text and are available on request as are the Dimson market-model estimates and all other results mentioned but

[9] The Center for Research in Security Prices value-weighted market return was also used as a proxy for the market return, but did not alter the results.

Table 12.1. Market Model Estimates[a]

Estimation period: September 30, 1981–September 29, 1982 (253 trading days)

Dependent variable	Independent variable	$\hat{\alpha}$	$\hat{\beta}$	R^2
Johnson & Johnson return	Market return	0.0007 (0.0010)	1.3544 (0.1337)	0.29
Johnson & Johnson return	Drug industry return	–0.0001 (0.0009)	1.1376 (0.0880)	0.40
Drug industry return	Market return	0.0008 (0.0003)	1.1896 (0.0463)	0.72

[a]The standard errors are in parentheses.

Data source: Center for Research in Security Prices, daily stock returns tapes 1981–82 (University of Chicago).

not reported. Also, estimation periods of 100 ($\hat{\beta}$ = 1.35), 150 ($\hat{\beta}$ = 1.46) and 200 ($\hat{\beta}$ = 1.40) trading days prior to the poisonings yielded no significant differences from the market model estimates reported.[10]

To account for factors unique to the over-the-counter drug industry, the rate of return of the drug industry has been substituted as the independent variable in place of the market return.[11] In order to discern the effects of the poisonings on the drug industry, the market model was also estimated with the drug industry return substituted as the dependent variable. Estimates from these two versions of the market model are displayed in Table 12.1 with standard errors in parentheses. The estimates of α and β from the three models shown in Table 12.1 were then used to forecast the *CAR*s over the event period.

Two event periods are examined. The first represents investors' initial reaction to the poisonings. This period should be long enough to include any substantial new information about the poisonings, and yet not be too long so as to incorporate other events which would influence the estimates. Second, after investors formed a prediction about the event, did Johnson & Johnson recover? This would occur if, after some point, investors realized that they had been wrong about their initial forecast of the response of consumers to Johnson & Johnson, especially Tylenol®, following the poisonings. In the test

[10]I also estimated the market model for post-event periods of 100 ($\hat{\beta}$ = 0.88), 150 ($\hat{\beta}$ = 0.87), 200 ($\hat{\beta}$ = 0.95) and 250 ($\hat{\beta}$ = 0.96) trading days, beginning 20 trading days after September 30, 1982 (the first day of the event window). Given the stability in the b estimates across the pre-poisoning estimation periods, the stability in the b estimates across the post-poisoning estimation periods, and the difference in the b estimates between the two periods, this evidence suggests that the difference in $\hat{\beta}$ between the pre- and post-poisonings estimation periods was due to the poisonings. A computation of moving-average betas of 50 trading days for a 500-trading-day interval symmetric around the poisonings also suggests the shift in beta resulted from the poisonings. Furthermore, there was no change in Johnson & Johnson's capital structure, asset size, and operations after the Tylenol® poisonings that would be predicted to have altered beta.

[11]The drug industry portfolio is equally weighted and excludes Johnson & Johnson. It consists of twenty-two firms that produce over-the-counter drug products, most of which are found under Standard Industrial Classification code 2834 (pharmaceutical preparations). A value-weighted drug industry portfolio was also used, but did not alter the results.

for recovery, the null hypothesis of no recovery is the estimated value of the *CAR* in investors' initial forecast of the effects of the poisonings, and recovery is measured relative to this estimate.

No one can say precisely how long it took for investors to become confident about the accuracy of the information on the magnitude of the disaster. The event began on Thursday morning, September 30, 1982, when the stock market received notice of the first two victims. All seven victims were accounted for within the next two days. By Monday, October 4, it was suspected the tampering occurred at the retail level. These suspicions were confirmed by the end of the week and no substantial new information materialized after this, based on reports from the *Wall Street Journal*, major newspapers, and trade magazines.

Table 12.2 contains three measures of the CAR for 20 trading days following the poisonings, corresponding to the three market models in Table 12.1.

The first two measures, CAR_{jj}^{m} and CAR_{jj}^{di}, are based on Johnson & Johnson's relationship with the overall market and drug industry, respectively. For both measures, the *CAR* declined markedly during the first few days in the aftermath of the poisonings, and then leveled off. In both cases, the *CAR* is statistically different from zero for all 20 trading days at the 1% level of significance using a one-tail test.[12] Hence, an abnormal decline in the value of Johnson & Johnson immediately following the Tylenol® event is indicated with a high level of statistical confidence.

The *CAR* for Johnson & Johnson declines less based on its relationship with the drug industry, CAR_{jj}^{di}, than with its relationship with the overall market, CAR_{jj}^{m} , indicating that the other over-the-counter drug companies suffered losses as well. The third measure of the *CAR* in Table 12.2, CAR_{di}^{m}, was forecasted for the drug industry (excluding Johnson & Johnson), based on its relationship with the market. During the first 6 or 7 trading days following the poisonings, there appears to be no impact on the other drug companies. However, CAR_{di}^{m} begins to decline around the 8th trading day and is statistically significant at the 1% level of significance for the 11th through 20th trading days after the poisonings, although the value of CAR_{di}^{m} is much less than for CAR_{jj}^{m} over this period.

It is not obvious why CAR_{di}^{m} does not decline immediately following the poisonings. One possible explanation is that it took a few days before it became apparent that it was not solely a Tylenol® problem, but could have occurred to other capsule makers as well. An examination of the *Wall Street Journal* and trade magazines revealed no other significant events during this period that may have contributed to the significant decline.

After the tenth trading day, the two measures of the *CAR*s for Johnson & Johnson appear to have subsided somewhat. Rather than reporting the *CAR*s

---

[12]The standard error of the *CAR* is

$$\hat{\sigma}_{car} = \{\hat{\sigma}^2_{et}\,[\Sigma_{x=1,\,Nf}\,(1 + 1/N_e + (R_{mt} - \bar{R}_m)\ /\ CSSR_{mt})] + N_f\,(N_f - 1)\ /\ N_e\}^{1/2}$$

where $\hat{\sigma}^2_{et}$ is the residual variance from the estimation period, N_e is the number of observations in the estimation period, $\bar{R}_m$ is the estimation period sample mean of the market return, $CSSR_{mt}$ is the corrected sum of squares for the market return during the estimation period, and N_f is the period over which the abnormal returns accumulate. In the past, studies generally have used only the first term in braces in calculating the standard error of the *CAR*. Cantrell, Maloney, and Mitchell (1989) show that the abnormal returns over an event window are not independent even if they are independently distributed. The second term in braces adjusts for this. This term is positive (except for the first day of the event window, when it is zero), implying that the true standard error of the *CAR* is higher than previously assumed.

Table 12.2. Cumulative Daily Abnormal Returns Following the 1982 Tylenol® Poisonings for Johnson & Johnson and the Drug Industry

Day	CAR^{m}_{jj}	t-value[a]	CAR^{di}_{jj}	t-value	CAR^{m}_{di}	t-value
1	–.060	–3.01	–.050	–2.98	–.009	–0.42
2	–.054	–2.44	–.050	–2.46	–.003	–0.42
3	–.111	–4.10	–.103	–4.16	–.006	–0.68
4	–.173	–5.54	–.162	–5.65	–.009	–0.88
5	–.129	–3.67	–.130	–4.02	.001	0.10
6	–.186	–4.83	–.186	–5.25	.000	0.02
7	–.148	–3.56	–.145	–3.79	–.003	–0.19
8	–.156	–3.50	–.139	–3.38	–.015	–0.98
9	–.206	–4.34	–.176	–4.05	–.026	–1.57
10	–.229	–4.56	–.184	–3.98	–.040	–2.30
11	–.233	–4.43	–.172	–3.54	–.054	–2.97
12	–.258	–4.69	–.194	–3.83	–.057	–2.97
13	–.219	–3.81	–.147	–2.78	–.064	–3.19
14	–.219	–3.66	–.151	–2.76	–.059	–2.85
15	–.249	–4.01	–.179	–3.14	–.061	–2.86
16	–.226	–3.52	–.149	–2.53	–.068	–3.05
17	–.253	–3.82	–.155	–2.56	–.086	–3.73
18	–.272	–3.98	–.181	–2.88	–.080	–3.39
19	–.241	–3.43	–.154	–2.39	–.077	–3.15
20	–.265	–3.67	–.176	–2.65	–.078	–3.13
$\overline{CAR}_{11,20}$ (October 14-27)	–.244	–3.85	–.166	–2.80	–.068	–3.09

[a] See Footnote 12 for computation of the *t*-statistic.
Data source: Daily stock returns tapes, Center for Research and Security Prices, University of Chicago, 1982.

for a specific date as indicative that investors were confident about the magnitude of the poisonings, the *CAR*s were averaged over a ten-day trading period, beginning on the 11th day after the event and ending on the 20th day (October 14–27). The *CAR*s were averaged over the ten-day trading period simply because it is impossible to judge the exact date when most of the information concerning the poisonings was accessible to investors. In any event, the choice of forecasts of Johnson and Johnson's losses is inconsequential for the analysis since all the *CAR*s are negative and significant following the initial decline. These averages are reported at the bottom of Table 12.2; they are all statistically significant at the 1% level, using a one-tail test. The $\overline{CAR}_{11\text{-}12}$ for

[13]I also calculated *CAR*s for Johnson & Johnson using all the market model estimates discussed earlier. Based on the other pre-poisonings estimation periods and the Dimsen market model, the resulting *CAR*s are not different from those reported in Table 12.2. *CAR*s based on the post-poisoning estimation periods are less negative than those reported in Table 12.2; the average *CAR* for $CAR_{11\text{-}20}$ is –17.8%. These less negative *CAR*s are due to the lower beta estimates from the post-poisonings estimation periods, which is arguably due to the poisonings themselves, and thus less confidence is placed in these results. Even so, these *CAR*s are statistically different from zero. Various other measures of abnormal performance were also computed. Cumulative net-of-market returns (i.e., beta assumed equal to one and alpha equal to zero) produced slightly less negative, though not significantly different results. In another test, abnormal returns were computed by substituting the mean market return over the event period for the daily market return. The resulting *CAR*s are not significantly different from those reported.

Johnson & Johnson based on its relationship with the overall market is –24.4%.[13] Based on the market price of Johnson & Johnson common stock on September 9, 1982, the day before the market received notice of the poisonings, the estimated loss to Johnson & Johnson stockholders is $2.11 billion. The $\overline{CAR}_{11\text{-}12}$ for the other over-the-counter drug companies is –6.8%; the estimated loss to shareholders of these firms is $4.06 billion. Thus, the total decline in wealth attributed to the poisonings is over $6 billion, which arguably is a lower-bound estimate considering that the poisonings likely increased consumers' expectations of product tampering in other industries as well.

After the initial reaction, did Johnson & Johnson recover? By December 1982, Johnson & Johnson had regained most of its market share. But regaining previously held market share during this period might have been temporary, due to the $2.50 coupons Johnson & Johnson offered in newspapers across the country to win back customers. Even so, the market share increase may have demonstrated that Johnson & Johnson had regained their customers' confidence. Six months after the poisonings Tylenol® had regained most of its market share, and by August of 1983, Tylenol® was once again the nation's leading pain reliever. The latter market share recovery is more qualified than the December recovery, since the December recovery was attributed in large part to the $2.50 coupons. Had Johnson & Johnson actually recovered or had it simply regained its top position in the market by way of increased advertising or price decreases?

Stated earlier, the null hypothesis of no recovery is the estimated value of $\overline{CAR}_{11\text{-}12}$ and recovery is measured relative to this estimate. An examination of the *CAR*s beyond the twentieth trading day following the poisonings does not reveal any significant recovery. The negative *CAR*s continued to mount after the period for which the initial forecasted losses were computed.[14] One year after the poisonings, the respective *CAR*s for Johnson & Johnson were: CAR_{jj}^{m} = –63.5% and CAR_{jj}^{di} = –34%. Other than the Tylenol® poisonings, there was only one event specifically related to John-son & Johnson mentioned in the *Wall Street Journal* more than once during the twelve months immediately following the poisonings. The event concerned Johnson & Johnson's removal of Zomax, a prescription pain reliever, after reports (see "Johnson & Johnson Suffers," 1983) of five people dying of allergic reactions to Zomax. The deaths were made public on March 4, 1983 and were associated with a negative abnormal return of 4.6% to Johnson & Johnson's stock price on that day. During the next four trading days, Johnson & Johnson stock declined an additional 3%. Given the grave consequences of the side effects of Zomax, a drug which Johnson & Johnson had been promoting heavily, this event may have hampered any significant recovery of Johnson & Johnson's stock price from the Tylenol® poisonings.

While the *CAR* for Johnson & Johnson (and for the drug industry as well) continued to decline after their initial drop, this should not be interpreted as evidence of continued losses due to the Tylenol® poisonings, as there was no new concrete information regarding the poisonings after the period for which Johnson & Johnson's initial wealth decline was calculated. Furthermore, an event study consisting of one event is plagued by the problem of confounding

[14]All measures of the *CAR*s were forecasted through December 1986 and are available.

events which may bias the estimated *CAR*. This does not appear to be a major problem here in that the Zomax episode was the only other event directly related to Johnson & Johnson which received considerable publicity in the post-poisonings period. However, the evidence does imply that Johnson & Johnson's stock price never recovered. This lack of recovery indicates that investors' initial forecasts of Johnson & Johnson's losses were not overestimated. Investors perceived Johnson & Johnson would attempt to recover Tylenol's lost market share and could only do so by reducing the price of Tylenol® and/or increasing its level of advertising. Since investors did not overestimate Johnson & Johnson's success with its recovery efforts, their initial forecasted losses are an unbiased estimate of Johnson & Johnson's permanent wealth decline due to the poisonings.

V. Johnson & Johnson's Brand-name Capital Loss

How much of the financial loss suffered by holders of Johnson & Johnson stock is due to the fact that the company brand name, Johnson & Johnson, and the product brand name, Tylenol®, declined in value as a result of the poisonings? The value of the brand-name capital of a firm is determined by the firm's expected quasi-rents from future sales; thus the capital loss arising from an unanticipated decline in expected product quality comes in the form of the depreciation of the brand-name capital. Johnson & Johnson may have recovered sales of Tylenol® capsules and any other of its products for which consumers have decreased their purchases; however, this recovery may be only achieved with a decrease in prices or an increase in brand-name capital investments (such as advertising) in order to revive the brand name.

In order to arrive at an estimate of the financial losses caused by brand-name capital depreciation, other losses must be netted out. Clearly, apart from the potential loss of brand-name capital, Johnson & Johnson suffered direct financial losses as a result of the out-of-pocket costs of recalling and destroying the capsules and designing a safer package. Also, during the fourth quarter of 1982 and part of the first quarter the following year, there was a loss of profits because Tylenol® capsules were not on the shelves.[15] Another consideration is the lawsuits filed on behalf of the victims which Johnson & Johnson must indemnify. Finally, as mentioned in the previous section, the losses to Johnson & Johnson were greater based on its relationship with the overall market as opposed to the over-the-counter drug industry, simply because all capsules specifically, and other drug forms generally, became associated with a higher probability of tampering. Thus, the losses suffered by Johnson & Johnson that were shared with the other drug companies must also be accounted for in order to isolate losses solely due to the depreciation of Johnson & Johnson and Tylenol's brand names.

Johnson & Johnson assigned a $100 million cost to the disposal of the capsules and the subsequent repackaging. It also claimed a $50 million business

[15] Although Johnson & Johnson attributed the decline in profits during this period to the fact that Tylenol® capsules were not on the shelves, one might argue that the capsules could have been left on the shelves, and yet the same loss may have occurred. Furthermore, consumers could have substituted tablets for capsules since tablets were never removed.

interruption expense, which included loss of profits and fixed charges during the time capsules were off the market.[16] Also, Johnson & Johnson was subject to suits filed by families of four of the victims. According to the *Wall Street Journal* ("Johnson & Johnson Unit," 1983), each suit asked for $5 million in damages.[17] Neither the *Wall Street Journal* nor *Business Insurance* revealed any information as to the filing of suits on behalf of the other three victims.[18] Based on this information, an upper-bound estimate of the out-of-pocket costs resulting from the poisonings approaches $200 million.[19]

To assess the loss in brand-name capital suffered by Johnson & Johnson due to the Tylenol® poisonings, the out-of-pocket loss ($200 million) was subtracted from the decline in value experienced by equity investors. The initial forecasted loss based on the relationship between Johnson & Johnson and the over-the-counter drug market portfolio is used as the measure of losses from which to calculate the decline in brand-name capital. The argument for this is simple. The estimates based on the relationship between Johnson & Johnson and the overall market do not distinguish between the losses suffered uniquely by Johnson & Johnson and the losses suffered generally by the marketers of over-the-counter drugs. That is, Johnson & Johnson suffered some losses simply because over-the-counter drugs, primarily the capsule form, became associated with a higher probability of tampering; hence, these losses should not be included as part of the Johnson & Johnson and Tylenol® brand-name capital loss, since they are shared by the other over-the-counter drug companies. The estimated loss to Johnson & Johnson based on its relationship with the over-the-counter drug market is $1.44 billion. Subtracting $200 million from this loss leaves $1.24 billion as a measure of the decline in the value of Johnson & Johnson and Tylenol's brand names. Thus it appears that over half of the total losses ($2.11 billion) suffered by Johnson & Johnson came in the form of the depreciation of its brand-name capital. Due to the brand-name capital loss alone, Johnson & Johnson stock suffered a 14.3% decline relative to its forecasted value.

VI. Supporting Evidence

The empirical evidence suggests that Johnson & Johnson suffered a substantial loss in the value of its brand-name capital. Additional events occurring in

[16]In a lawsuit filed in January 1983 against its nine insurers, Johnson & Johnson claimed the above mentioned costs of $150 million. Johnson & Johnson sued for only $117 million as management did not expect to recover all of the losses claimed ("Johnson & Johnson Unit," 1983). However, Johnson & Johnson was unsuccessful in recovering any of the losses sustained in recalling the capsules as the courts ruled that Johnson & Johnson's product liability insurance did not cover its recall expenses and other charges. It was also noted by the courts that Johnson & Johnson had actually cancelled its recall insurance prior to the poisonings ("Tylenol Maker," 1986).

[17]I was unable to determine the exact settlement of the suits. Based on articles in *Business Insurance, Wall Street Journal* and other sources, there was no indication that Johnson & Johnson contested the claims.

[18]No class action suits were filed on behalf of Tylenol® users.

[19]Actually, the profits lost during the period that Tylenol® capsules were unavailable for sale are not out-of-pocket costs. One may argue that they represent a brand-name loss since the capsules could have been left on the shelves and still would not have been purchased. I have included them in the estimate simply because Johnson & Johnson claimed this loss in its lawsuit. In any event, their inclusion decreases the likelihood of finding a brand-name loss.

the aftermath of the poisonings support this finding: (a) the price of Tylenol® declined relative to other over-the-counter pain relievers, (b) Johnson & Johnson made extensive attempts to downplay the connection between Johnson & Johnson and Tylenol®, (c) Tylenol® tablet sales declined although Tylenol® tablets were not poisoned, (d) Tylenol's market share never attained the level that was forecasted before the event, and (e) Johnson & Johnson delayed the introduction of several new drugs.

The Klein and Leffler (1981) theory of brand names predicts that at least until Johnson & Johnson rebuilt the Tylenol® brand name back to the pre-poisonings value, the price of Tylenol® should continue to exhibit a reduction relative to similar pain relievers. According to *Advertising Age* ("Tylenol Competitors," 1982), Johnson & Johnson was offering as much as 25% off the list price to retailers soon after the poisonings. This discount especially supports the brand name argument as it was for noncapsule Tylenol® products since capsules were unavailable for sale until late December of 1982. Apparently no other companies were offering such lucrative discounts. During this period, a 17% discount was considered a good deal according to industry analysts quoted in *Advertising Age* ("Tylenol Competitors," 1982). Also, *Value Line Investment Survey* ("Johnson and Johnson," 1983) reported that the price of Tylenol® had not returned to its premium level over the prices of other pain relievers.[20]

One would expect Johnson & Johnson to concentrate on reviving Tylenol's brand name as well as its own. They did just that, spending millions on increased advertising and promotions. Johnson & Johnson also worked diligently at insulating the company brand name. For example, during the middle of October 1982, the company ran a newspaper advertisement across the country advising consumers how to exchange their Tylenol® capsules for a refund or tablets. The advertisement made no mention of Johnson & Johnson or its Tylenol-producing subsidiary McNeil Consumer Products; instead it referred only to the "makers of Tylenol®." This suggests that Johnson & Johnson was fearful that consumers might associate the company name with the poisonings, thereby damaging its reputation across its entire product line, from band-aids to baby powder, in addition to the other products in the Tylenol® line.

The poisonings were restricted to capsules, presumably because it would have been much harder to tamper with tablets. If consumers only attached a higher probability of tampering to capsules, but still highly valued the Tylenol® product line, then some consumers would have substituted tablets for capsules during the period capsules were unavailable, thus implying a rise in tablet sales. Substitution did not occur along these lines. In fact, Tylenol® tablet sales declined 25% after the onset of the poisonings. The significance of this reduction in Tylenol® tablet purchases is magnified by the fact that Tylenol® tablets were selling at a 25% discount during this period, a lower discount than the other companies were offering. Also, in a survey reported in *Fortune* ("The Fight," 1982) one month after the poisonings, 58% of those polled said

[20]The 1982 and 1983 issues of *American Druggist Blue Book* provide additional price information. The data is limited, however, as only prices for the Anacin line are comparable to Tylenol® over these two periods. According to this source, the wholesale price of Anacin increased 50% in 1983, while the price for Tylenol® increased only 19%. Although this sample of price information is limited, it does lend support to the *Value Line Investment Surveys* and *Advertising Age* reports. Given this change in relative prices, it is not surprising that Anacin, which was Tylenol's largest competitor, was backordered following the poisonings.

they would not buy Tylenol® tablets in the future, even though they knew that it was capsules that were poisoned. Even sales of Tylenol® cold remedies declined. This consumer reaction is consonant with the notion that the entire Tylenol® product line suffered a depreciation in the value of its brand-name capital.

Actually, the empirical results demonstrate that the loss of brand-name capital for the entire Tylenol® line is more than just a notion. According to *Value Line Investment Survey* ("Johnson and Johnson," 1982), the estimated annual profits for the Tylenol® line for 1982 were $80 million. Assuming a real interest rate of 5%, the present value of a perpetual stream of $80 million a year is $1.6 billion which is roughly 33% greater than the estimated brand-name capital loss suffered by Johnson & Johnson; hence it figures that investors forecasted that most of the brand-name capital of the Tylenol® line was destroyed. Admittably, the choice of the real interest rate is arbitrary. The point to be made is that the numbers do show that such a large brand-name capital loss is possible. It is also likely that some of the brand-name loss was due to the decline in the value of Johnson & Johnson's own brand name.

Prior to the poisonings, Tylenol® controlled 37% of the over-the-counter pain reliever market and it was forecasted (in "The Fight," 1982) that Tylenol® would control 50% of the market by 1986. Tylenol® had been taking over the over-the-counter drug market rapidly ever since it began an aggressive consumer advertising campaign in the mid-1970s. However, these expectations were not realized. In early 1986, prior to the second wave of Tylenol® poisonings, Tylenol® had a market share of only 34%, a little less than it held prior to the 1982 poisonings.

According to *Value Line Investment Survey* ("Johnson and Johnson," 1982), Johnson & Johnson was also supposed to have stretched out its lead in the drug market by means of the introduction of a dozen or more new drugs by the mid-1980s. The introduction of the new drugs predicted earlier did not occur as soon as had been supposed. The reason for the delay of the new drugs is unknown. It could have been due to the actions of the Food and Drug Administration (FDA)—or to a management decision made in the face of lagging consumer acceptance of Johnson & Johnson drug products. On the other hand, many of Johnson & Johnson's competitors introduced new drug products soon after the poisonings in an effort to pick up shelf space vacated by Johnson & Johnson ("Several Drug Companies," 1983; "The Race," 1982).

VII. Evidence from Subsequent Drug Poisonings

Several drug poisoning incidents have occurred since the 1982 Tylenol® poisonings; however, most of them were minor and do not warrant mention. Five of the subsequent drug poisoning cases received national attention, though nothing of the magnitude generated by the Tylenol® poisonings. These five cases are listed in Table 12.3, along with their associated abnormal returns on the announcement day of the poisonings and cumulative abnormal returns for event windows of two, five, and ten trading days beginning the announcement day.

The *CARs* in Table 12.3 were calculated using the same methodology presented in section IV. As a comparison, Table 12.3 also contains the *CARs* associated with the 1982 Tylenol® poisonings; these are from the first column of

*CAR*s in Table 12.2, CAR^{m}_{jj}. Recall that the first column of *CAR*s in Table 12.2 is based on Johnson & Johnson's relationship with the overall market. Similarly, each *CAR* shown in Table 12.3 is based on the respective company's relationship with the overall market.

For the five subsequent poisoning cases, only the 1986 Tylenol® poisonings are associated with statistically significant negative *CAR*s, whereas none of the other poisonings appear to have any stock market impact. On the announcement day of the 1986 Tylenol® poisonings, the abnormal return was –3.8% and declined to –6.1 for the two-day cumulative abnormal return. Johnson & Johnson immediately decided to discontinue permanently the sale of Tylenol® capsules, a step which the firm had refused to undertake following the 1982 poisonings. The negative stock price reaction resulting from the 1986 poisonings and the immediate discontinuance of capsules provides additional support for the argument that Johnson & Johnson suffered a loss of brand-name capital following the 1982 poisonings.

The *CAR*s displayed in Table 12.3 suggest that Johnson & Johnson suffered significant losses when Tylenol® capsules were poisoned not only in 1982, but also in 1986, and yet the other drug companies were not affected when their capsules were poisoned. Why did Johnson & Johnson suffer enormous losses, whereas the other drug companies were hardly affected when their pain relievers were poisoned?

Several plausible reasons exist. First, Tylenol® was not only the leader of over-the-counter pain relievers, but was increasing its lead prior to both the 1982 and 1986 poisonings. Being number one conveys to consumers that your product is superior along relevant quality dimensions, including safety, for which consumers are willing to pay a price premium. Evidence presented in the previous section shows that Tylenol® lost its considerable price premium over other pain relievers after the 1982 poisonings. Second, the 1982 Tylenol®

Table 12.3. Cumulative Daily Abnormal Returns Following Six Drug Poisoning Cases

		Cumulative abnormal returns [a]					
Date	Firm	Drug	Deaths	Day 1	Day 2	Day 5	Day 10
9/30/82	Johnson Johnson	Tylenol	7	–.060 (–3.95)	–.054 (–2.42)	–.129 (–3.67)	–.229 (–4.56)
12/9/82	American Home Products	Anacin	0	–.068 (–5.98)	–.012 (–0.75)	.020 (0.79)	.021 (0.61)
2/11/86	Johnson & Johnson	Tylenol	1	–.038 (–2.85)	–.061 (–3.28)	–.132 (–4.44)	–.117 (–2.74)
3/20/86	Smith Kline Beckham	Contac	0	.005 (0.47)	–.015 (–0.96)	–.011 (–0.44)	.012 (0.34)
5/28/86	American Home Products	Anacin	1	.001 (0.09)	–.006 (–0.41)	–.017 (–0.74)	–.030 (–0.91)
6/17/86	Bristol Meyers	Excedrin	2	–.012 (–1.01)	.024 (1.47)	.040 (1.55)	.033 (0.89)

[a] *t*-values are in parentheses.

Data source: Daily stock returns tapes, Center for Research and Security Prices, University of Chicago, 1981-86.

poisonings were by far the most publicized and the 1986 Tylenol® poisonings were the next most publicized of the six cases. It is not argued here that the negative publicity caused the stock market losses, but instead that the negative publicity was a by-product of the large negative returns. Third, Tylenol® accounts for a much larger proportion of Johnson & Johnson's total profits than the other poisoned drugs account for their respective company's profits. For instance, Tylenol® accounted for approximately 17% of Johnson & Johnson's profits prior to the 1982 poisonings and 13% prior to the 1986 poisonings. The other poisoned pain relievers accounted for only 2 to 4% of their respective company's profits. Fourth, the number of deaths was by far the largest for the 1982 Tylenol® poisonings. For the 1982 Anacin III poisonings and the 1986 Contac poisonings, no deaths were reported, and the 1986 Anacin III poisoning was widely believed to be a suicide. It is also worth noting that over one-third of consumers questioned in a survey reported in *Advertising Age* ("Tylenol Rivals," 1986) believed that the same party was responsible for both Tylenol® poisonings. In the same survey, over one-half believed that the 1986 Tylenol® poisonings were an internal job. Finally, the other drug companies did suffer substantial losses contemporaneous with the 1982 Tylenol® poisonings. Thus, much of the adjustment in the probability of drug tampering had already taken place and consequently little stock market effect should be expected, especially given the relative size of the four non-Tylenol® subsequent poisonings.

VIII. Concluding Remarks

The damage to a company due to the actions of a party not associated with that company potentially may be much greater than previously appreciated. This study shows that Johnson & Johnson suffered a $1.24 billion wealth decline (14% of the forecasted value of the company) due to the depreciation of the company brand name and the Tylenol® brand name as a result of the 1982 Tylenol® poisonings. This brand-name capital loss not only reflects the loss of Tylenol® capsule sales, but also the loss of sales in the entire Tylenol® line and possibly the delayed introduction of expected new drugs. Johnson & Johnson claimed out-of-pocket costs of $150 million, an amount that seemed large at the time. Yet considering evidence developed here, this sum was small relative to the loss of brand-name capital suffered by the firm.

Prior to the poisonings, the Tylenol® brand name assured consumers of safe high-quality pain relievers. It may never be known why Johnson & Johnson was the target of the tamperer; but whatever the reason, the results imply that Johnson & Johnson was held responsible, as the assurance of product quality was severely weakened. In addition to the enormous losses suffered by Johnson & Johnson shareholders ($2.11 billion), the other over-the-counter drug companies realized a $4.06 billion wealth decline as the probability of drug poisoning increased for all drugs, especially those of the capsule variety.

Following the 1982 poisonings, Johnson & Johnson expended resources to restore the brand-name capital of the Tylenol® product line and regain its top position in the pain-reliever market. Though its stock market losses were never recovered, Johnson & Johnson was successful in recovering its brand-name capital as Tylenol® regained its position. However, the 1986 poisonings erased much of the brand-name capital that had been restored, as demonstrated by the contemporaneous stock market losses and permanent discontinuance of Tylenol® capsules.

With the exception of the 1986 Tylenol® poisonings, subsequent drug poisonings have not been associated with large losses. This can be largely explained by the fact that they were smaller cases and also did not account for nearly as high a proportion of their respective company's profits. In any event, firms must take into account possible actions by outside parties, especially when the product represents a significant proportion of the company's profits. They can be extremely painful.

Acknowledgments

I wish to thank Chuck Knoeber, Cotton Mather Lindsay, Robert McCormick, Roger Meiners, J. Harold Mulherin, two referees, and especially Daniel K. Benjamin and Michael T. Maloney for valuable comments and contributions. I am also appreciative of guidance and patience by Richard J. Sweeney in the paper's final stages. An earlier version was presented at the 1986 Western Economic Association Meetings. The U.S. Securities and Exchange Commission as a matter of policy disclaims responsibility for any private publication or statement by any of its employees. The views expressed here are those of the author and do not necessarily reflect the views of the Commission.

13 Mad Cow Disease and the Stigmatization of British Beef

Douglas Powell

On March 20, 1996, British Health Secretary Stephen Dorrell rose in the House to inform colleagues that scientists had discovered a new variant of CJD in 10 victims and that they could not rule out a link with consumption of beef from cattle with BSE, also known as mad cow disease.

Overnight, the British beef market collapsed and politicians quickly learned how to enunciate bovine spongiform encephalopathy and Creutzfeldt-Jakob disease. Within days, the European Union banned exports of British beef; consumption of beef fell throughout Europe; and the tell-tale triumvirate of uncertain science, risk, and politics was played out—and continues—in media headlines. Beef consumption across the European Union (EU) dropped 11% in 1996, and the BSE crisis cost the EU US$2.8 billion in subsidies alone to the beef industry. The U.K. cost is now estimated in excess of 4 billion pounds (Deane, 1997).

Yet the March 20, 1996 announcement, rather than the beginning of the ongoing BSE crisis, as it is now commonly called, was instead the culmination of 10 years of bureaucratic mismanagement, political bravado, and a gross underestimation of the public's capacity to deal with risk (for a full account, see Powell & Leiss, 1997). More important than any of the several lessons to be drawn from the BSE fiasco is this: the risk of no-risk messages. For 10 years the British government and leading scientific advisors insisted there was no risk—or that the risk was so infinitesimally small that it could be said there was no risk—of BSE leading to a similar malady in humans, CJD, even in the face of contradictory evidence. The no-risk message contributed to the devastating economic and social effects on Britons, a nation of beefeaters, the slaughter of 1.5 million British cattle, and a decrease in global consumption of beef at a cost of billions of dollars.

Further, even to refer to the events of 1996 as the BSE crisis is a misnomer, just as scientists are quick to point out that mad cow disease should more appropriately be called sad cow disease or unco-ordinated cow disease. Instead, the events before and after the March 20, 1996 announcement were the latest in a series of public controversies in the United Kingdom related to public perception or confidence in the food supply and the U.K. Ministry of Agriculture, Food, and Fisheries (MAFF), involving listeria, botulism, salmonella, *E. coli* O157:H7 (sometimes called hamburger disease), and BSE.

British beef became stigmatized, but was it as a health risk or a symbol of political and managerial incompetence? If stigma functions as a powerful shortform for public anxiety, then those same forces can be marshaled to protect health and catalyze change. In short, British beef deserved to be stigmatized.

Background

BSE is a slowly progressing, fatal nervous disorder of adult cattle that causes a characteristic staggering gait and is similar to a handful of rare, neurological diseases that affect humans and other animals. The most common of these diseases is scrapie in sheep. The human equivalent of scrapie is called Creutzfeldt-Jakob disease and occurs in about one-person-per-million every year throughout the world. All of these ailments have long incubation times, from two to seven years in cattle and up to 30 years in humans, but once symptoms appear, the victim rapidly degenerates. There is no known treatment.

In 1986, BSE was first discovered among Britain's cattle. In June 1988, the U.K. government made BSE a reportable disease and by July had instituted a ban on ruminant offal in cattle feed. (When animals such as cows or sheep are sent for slaughter, meat is processed for human consumption, but the remaining portions such as entrails, hooves, and whatever else cannot be used, form what is called offal. This offal is then rendered using temperatures in excess of 100° C, yielding a meat and bone meal that has many uses. One such use is as an inexpensive source of high-quality protein in animal feed. The leading theory—although still just a theory—is that diseases like scrapie and BSE are spread when an infected sheep or cow is slaughtered and its by-products fed to other animals.) In August, the government decided to slaughter and incinerate all cows suspected of having BSE and provide compensation to farmers at 50% of the animal's estimated worth. By December 1988, the milk from any suspect cows was also destroyed. Although the number of new cases of BSE has dropped dramatically since the ban was imposed (accounting for the long incubation period), by the end of March 1997, 168,382 cases of BSE at more than 32,400 British farms had been reported (IFST, 1997).

The infectious agent thought to be responsible for BSE, scrapie and CJD remains a mystery that challenges the basic tenets of biology. In 1982, Dr. Stanley Prusiner, a neurologist at the University of California, San Francisco, first provided evidence that scrapie in sheep was caused by an agent which contains no nucleic acid, something he dubbed a prion or infectious protein (for review, see Prusiner, 1995). Prusiner's suggestion was greeted with ridicule. Bacteria, viruses, and every living organism contain either deoxyribonucleic acid (DNA) or in some instances only ribonucleic acid (RNA) which encode the proteins required for life. Surely, his critics argued, there must be nucleic acid, it just hasn't been found yet. Nevertheless, after more than a decade of research, the idea of prions is holding up and winning respectability (also see Rhodes, 1997a).

The current thinking is that the normal prion protein is required for some type of cell-to-cell communication between nerve cells in the brain, but when the disease-causing form of the protein is present, it somehow interacts with the normal prion protein, making it difficult to be broken down. Experimental evidence is accumulating to support such a model (Paushkin, Kushnirov, Smirnov, & Ter-Avanesyan, 1997; Brandner et al., 1996; Patino, Liu, Glover, &

Lindquist, 1996; Telling et al., 1996). Further, whatever is causing mad cow disease and related ailments is able to survive temperatures in excess of 100º C. This mutant form of the protein accumulates and, over time, leads to diseases like CJD, scrapie, and BSE.

At the crux of the BSE crisis is the question of whether humans can develop CJD after eating beef from cattle infected with BSE. In other words, can the infectious agent jump the species barrier?

United Kingdom newspapers promptly reported the new malady in the mid-1980s, although it was not until 1990 that North American newspaper coverage began. Throughout this time, British government sources were adamant that meat and other products from cattle infected with BSE posed no risk to humans. The 1990 coverage was prompted by the discovery that the infectious prion did in fact appear to be crossing the species barrier—to cats. Several cases of English cats developing feline spongiform encephalopathy were discovered and thought to be due to the use of rendered ruminant protein in cat food. This spawned the first crisis of confidence in British beef and a torrent of denials from government ministers who insisted there was no risk of humans contracting BSE. Then agriculture minister John Gummer sought to reassure the public that all was well by urging his four-year-old daughter, Cordelia, to eat a hamburger with him for the benefit of the television cameras. *The Economist* (May 19, 1990) described events more succintly, stating that the widely reported death of the first cat "proved beyond scientific doubt that nobody trusts MAFF any more."

While consumption eventually did recover, questions continued to be raised by the scientific community, the British public, and trading partners. The United States banned British beef imports in 1989, and Canada followed in 1990. All the while the government insisted that humans were not at risk.

In the summer of 1995, questions about the causes, risks, and controls of BSE began to appear more frequently in the European press. On July 12, 1995, a U.K. consumer group called for a new inquiry into BSE, after it became clear some cattle were still being fed ruminant protein, despite the 1988 ban, demonstrating that good rules need great enforcement.

The other factor driving increased coverage of BSE was the discovery of two teenagers who had developed CJD in the United Kingdom. The disease is so rare in teenagers that there had only been three other reported cases in the world. Scientific uncertainty, parental anguish and suspicion, and an obstinate government which refused any public inquiries fed the media mill and helped accelerate the social amplification of risk (Kasperson et al., 1988).

By the middle of November 1995, media coverage began to increase. BBC Television reported that nearly a quarter of Britons had either stopped eating beef or were eating less, fearing BSE could infect people. Another television program claimed that up to 600 BSE infected cattle were being consumed in Britain every week. And by November 17, 1995, the first report of a school banning beef from its cafeterias surfaced. The chief cook wrote to parents that, "I am a farmer's daughter myself and while I don't want to see farmers put out of work I don't feel the Ministry of Agriculture is doing enough to reassure the public." In response, a Meat and Livestock Commission spokeswoman said, "There is no scientific basis for banning beef—there's a lot of misinformation going around" ("School Bans," 1995). These conflicting perspectives were increasingly cited in media coverage, with industry and government saying there was no scientific evidence that BSE could cross the species barrier into humans, and the public saying, in effect, "We don't believe you."

Scientists were eager to join the fray. For example, in a series of letters to the *British Medical Journal* that appeared on November 24, 1995, U.S. and European doctors called for extensive new research into whether people actually could catch CJD by eating beef from cattle infected with BSE. Sheila Gore of the Biostatistics Unit of the Medical Research Council said that by April 1995, BSE had been confirmed in more than 53% of dairy herds in Britain and that the new cases of CJD in humans were worrisome. In a widely cited quote, she said, "Taken together, cases of Creutzfeldt-Jakob disease in farmers and young adults are more than happenstance."

Two days later (November 26, 1995), Dr. Stephen Dealler, writing in the *British Food Journal*, claimed that most adult British meat-eaters would, by 2001, have ingested a potentially fatal dose of meat infected with BSE. He also said that the medical and dietary professions should question the present policy of "waiting passively" to see if the incidence of CJD rises in the United Kingdom.

It was in this climate of public discussion that, on November 30, 1995, Professor Sir Bernard Tomlinson, one of the UK's leading brain disease experts, told BBC Radio 4's *You and Yours* consumer program that he no longer ate meat pies or beef liver and "at the moment" would not eat a hamburger "under any circumstances." He said there was growing evidence that CJD could be caught from infected beef products (although he offered none; there was, however, growing suspicion that the U.K. government was not forthcoming with relevant information).

Dr. Tomlinson's statement once more accelerated the social amplification of risk. Even British Prime Minister John Major became embroiled in the debate (December 7, 1995), telling Parliament, "There is currently no scientific evidence that BSE can be transmitted to humans or that eating beef causes CJD." In response, more schools banned beef. Sue Dibb, co-director of independent watchdog, the Food Commission, responded that, "Nobody is trusting what is coming out of Government any more. The Government seems to be more interested in propping up the beef industry rather than admitting that there may be a risk, however small it may be. I think what is happening is that every time a minister gets up and says beef is safe, there is absolutely no danger, there is absolutely no risk, a whole lot more people stop buying beef because they don't trust the Government" (Farmer, 1995).

Another week went by, the accounts of the concerned public continued, and so on December 14, 1995, Agriculture Minister Douglas Hogg and a range of Government experts spent two hours with journalists, reviewing the scientific evidence that the latest BSE scare was unfounded. Within hours of the meeting, 200 more schools had banned beef, confirming the work of Nelkin (1995) who has suggested that efforts to convince the public about the safety and benefits of new or existing technologies—or in this case the safety of the food supply—rather than enhancing public confidence, may actually amplify anxieties and mistrust by denying the legitimacy of fundamental social concerns. The public expresses a much broader notion of risk, one concerned with, among other characteristics, accountability, economics, values, and trust.

To summarize, in the run-up to Christmas, 1995, beef consumption in the United Kingdom fell 20% and 1.4 million British households stopped buying beef. Thousands of schools took beef off the menu. And several prominent scientists predicted a medical catastrophe if a link was proven between BSE and CJD. So while the announcement of a potential link between CJD and BSE on March 20, 1996, was a news event that had all the elements of an attractive risk story—uncertainty, dread, catastrophe, and it possibly involved children (Covello, 1983)—the announcement was made in a climate of extreme mistrust of the United Kingdom.

From this retelling, it is clear, using the evaluative criteria of Gregory et al. (1996) that British beef was bound to be stigmatized:

- A standard of what is right and natural was violated or overturned, summarized in the repeated quotations of critics who said that cows are herbivores, that feeding ruminant protein to cattle is a form of industrial cannibalism, that cows shouldn't eat other cows;
- Impacts were perceived to be inequitably distributed across groups, as innocent, younger consumers suffered the dreadful consequences of CJD;
- The massive scientific uncertainty involving transmission, infectivity, even cause; and
- The perceived ineptness of the U.K. Ministry of Agriculture, Fisheries, and Food, publicly charged with obstinance, secrecy, incompetence, even criminal negligence.

The Health Risk

British beef became stigmatized, if not before then certainly after the March 20, 1996 announcement, which lead to a worldwide ban on the consumption of British beef—with some bizarre results. Newspapers around the world began to call CJD—even traditional CJD—the human form of mad cow disease; a man in the Egyptian port city of Alexandria was arrested for stabbing his wife when she refused to cook imported meat because she was scared it was infected with BSE; and vegetarian monks from a Scottish Cistercian order dined on a meal of steak and kidney pie one Sunday (J. Brown, 1996) to show their confidence in Scotch beef after the monks branded the European ban on beef products immoral and irrational (the tiny monastery, which runs a cattle farm to make money, had lost a fortune because of BSE).

During 1994 and 1995, 10 U.K. victims displaying an unrecognized and consistent pattern of CJD were identified by the U.K. CJD Surveillance Unit. The victims were much younger than in previous cases with the oldest 42 years old and an average age of 27.5; people with CJD are usually over 63. The illness also lasted longer than usual, an average of 13 months compared with the normal six months after the onset of symptoms. Some of the outward characteristics of the patients differed from traditional CJD but most importantly, the pattern of neurological damage in brain tissue was unique. Specifically, there were much larger aggregates of prions in the brains of what came to be called new variant-CJD (nvCJD) patients than had been seen in previous cases. Scientists found no evidence of genetic or medical factors common to the 10 victims and consistent with acquiring CJD. This led the U.K. Spongiform Encephalopathy Advisory Committee in March 1996, in the absence of other explanation at the time, to state that the U.K. cases were "most likely" to have been caused by exposure to BSE-infected cattle brain or spinal cord before 1989. Mr. Dorrell shared these results with the British Parliament on March 20, 1996.

Suddenly the no-risk scenario collapsed, and critics of MAFF seemed basked in credibility. Professor Richard Lacey, in a widely cited figure, predicted a rapid rise in the incidence of CJD, with between 5,000 and 500,000 stricken by 2000 (such predictions are, based on the available evidence, wildly overstated). Dr. Stephen Dealler predicted that by 2010 as many as 10 million Britons may have CJD, assuming high infectivity. Tim Lang, professor of food policy at Thames Valley University and chairman of the National School Meals Campaign, blamed intensive farming for the "tragedy" of BSE and called for modern farming techniques to be fundamentally changed.

By August 1997, 21 cases of nvCJD in the United Kingdom, including one in France, had been identified. The latest involved Donnamarie McGivern, whose mother Marie was told the day after their daughter's 15th birthday that she was probably suffering from CJD. Naylor (1997) said that all the McGiverns can do for their helpless daughter is to make her comfortable.

Dealing with Stigma

As Slovic (1997) has noted, "We live in a world in which information, acting in concert with the vagaries of human perception and cognition, has reduced our vulnerability to pandemics of disease at the cost of increasing our vulnerability to social and economic catastrophes of unprecedented scale. The challenge before us is to learn how to manage stigma and reduce the vulnerability of important products, industries, and institutions to its effects, without suppressing the proper communication of risk information to the public." How then, did North American governments and industry manage the revelation of a putatuve link between BSE and nvCJD, to reduce their vulnerability? Quite differently.

Following the March 20, 1996 announcement, questions in North America were raised about the practice of feeding herbivores ruminant protein or, as it was often described, feeding dead animals to animals that only ate plants. Immediately following the March 20, 1996 U.K. announcement, the U.S. Department of Agriculture (USDA) and the FDA announced beefed-up inspection of live cattle imported from the United Kingdom prior to 1989 and an expansion of current antemortem inspection for BSE.

The USDA statement also noted the agency's BSE surveillance program had examined over 2,660 specimens from 43 states and no BSE had ever been detected in cattle from the United States. And USDA said it was working with state and public health counterparts, scientists, and industry representatives to review current policies and regulations concerning BSE. These were all concrete actions, issued in a timely manner, and subsequently widely reported.

This did not happen in Canada. Agriculture and Agrifood Canada (AAFC) issued no public statement but chose instead to respond to individual media inquiries. The Canadian Minister of Agriculture was also quite adamant there was no risk of BSE developing in Canada. The actions of the ministry in handling the 1993 discovery of an Albertan cow with BSE—it had been imported from the United Kingdom and AAFC quickly ordered the destruction of 363 cattle still in Canada that had originated in the United Kingdom, as well as the entire herd from the infected animal—were widely lauded and used as evidence that regulators knew what they were doing.

On March 29, 1996, nine days after the original U.K. announcement, the U.S. Department of Agriculture and the U.S. Food and Drug Administration announced they were expediting regulations prohibiting ruminant protein in ruminant feeds, boosting surveillance and expanding research. The same day, several producer groups issued a statement supporting the moves and instituting a voluntary ban on ruminant protein in ruminant feed. A public statement from Agriculture and Agri-Food Canada, announcing its intention to ban mammalian protein in ruminant feed, first appeared in a *Toronto Globe and Mail* story on January 23, 1997.

Whereas American regulators—at least in the public domain—emphasized controls to prevent the occurrence of BSE and moved forward with legislation, Canadian regulators insisted there was "no risk." Appropriately, the Canadian Feed Industry Association exploited the fallacy of the no-risk argument. If, as the Canadian Minister of Agriculture stated, there was no risk of BSE developing in Canada, then why impose a ban?

U.S. regulators often talked about scientific uncertainties and the need to reduce or minimize risk, while Canadian officials openly questioned the validity of the scientific link between ruminant protein in ruminant feed and the development of transmissible spongiform encephalopathies (TSEs, as the family of diseases are known), stating instead that the science "wasn't there" and that the decisions were being driven more by public perception and trade considerations than science. Such a stance shows a flagrant disregard for the British experience with BSE.

But even good management can sometimes fail to be communicated. During the April 16, 1996 Oprah Winfrey show, the host stated she would stop eating hamburgers because of fears over BSE and that she was shocked after a guest said meat and bone meal made from cattle was routinely fed to other cattle to boost their meat and milk production. The camera showed members of the studio audience gasping in surprise as vegetarian activist Howard Lyman explained how cattle parts and downer cattle (downer is the generic term used to describe cattle who can simply no longer stand) were rendered and fed to other cattle. The chief scientist for the U.S. National Cattleman's Beef Association, rather than stressing the risk-management actions that had been taken, was left arguing that cows were not vegetarians because they drank milk. News of the popular show's content swept through the cattle futures markets, contributing to major declines in all the beef contracts as traders feared it would turn Americans away from beef. Nevertheless, when FDA announced its ban on ruminant protein in ruminant feed, the move was widely praised as prudent given the severity of the consequences should BSE be discovered in North America (Altman, 1997; "Pre-emptive Strike," 1997). Even the U.S. rendering industry, which in the early days of public attention after the March 20, 1996 announcement argued that negative public perception of the industry was simply a function of inflammatory language (one industry official, during a panel discussion in July 1996, said that part of the problem was that the word downer was a negative term; instead, industry was urging producers and others to describe such animals as nonambulatory) eventually supported the measures, with the U.S. National Renderers Association quoted as saying the ban on mammalian protein in ruminant feed put "a protective blanket around the cattle industry" ("FDA Bans," 1997).

Lessons

BSE is just one of several high-profile, stigmatizing events involving the food supply in recent years. Ever since the Jack in the Box outbreak of foodborne illness in January 1993 involving *Escherichia coli* O157:H7 (sometimes called hamburger disease) which sickened some 600 people in the Pacific Northwest and killed four children, media attention (Powell, Harris, & Griffiths, 1997), and subsequently public outrage and concern (FMI, 1997) has been focused largely on the microbial aspects of food safety. There have been dozens of other, well-publicized outbreaks since Jack in the Box, and the story helped catalyze changes to the U.S. meat inspection system that were signed into law by President Clinton in July 1996.

But still, outbreaks of foodborne illness continue. In the summer of 1996, over 9,500 Japanese, largely schoolchildren, were stricken with *E. coli* O157:H7 and 12 were killed. In November 1996, over 400 fell ill and 21—largely pensioners who had attended a church supper—were killed in Scotland from infection with the same bacterium. That same month, 65 people in four U.S. states and British Columbia fell ill after drinking juice manufactured by Odwalla Inc. of Half Moon Bay, California and found to contain *E. coli* O157:H7. A 16-month-old girl died in Denver. The common vehicle was unpasteurized cider, probably from improperly cleaned apples that had fallen to the ground and were contaminated with cattle or deer feces.

Other outbreaks have captured public headlines and consumer attention. In the spring and summer of 1996, some 1,465 people across North America were stricken with *Cyclospora cayetanensis*, a parasite initially linked to the consumption of California strawberries. However, the common vehicle was later thought to be Guatemalan raspberries (Hofmann et al., 1996). Most citizens did not hear the correction, and the California Strawberry Commission estimates it lost $20 million in sales. Yet despite increased surveillance and risk management of Guatemalan raspberries, cyclospora emerged again in 1997, associated not only with consumption of fresh fruits but with mesclun lettuce in Florida and fresh basil in Washington, D.C. Sales of fresh herbs immediately dropped (Masters, 1997).

In these cases and dozens of others, there is an enormous potential for economic damage, even damage to health as consumption of nutritious foods may decline. The potential for stigmatization of food is enormous. But there is also, in all these cases, a real risk to human health and safety. The most widely quoted figures in the United States peg the incidence of foodborne illness at 6.5 million people each year, accounting for an estimated 9,100 deaths (CAST, 1994) at a social cost of anywhere from $4.8 billion (Roberts, 1989), to $8.4 billion (Todd, 1989), to $23 billion (Garthright, Archer, & Kvenberg, 1988), just in the United States.

How then to reduce stigma? The components for managing the stigma associated with any food safety issue seem to involve all of the following factors:

- Effective and rapid surveillance systems;
- Effective communication about the nature of risk;

- A credible, open, and responsive regulatory system;
- Demonstrable efforts to reduce levels of uncertainty and risk; and
- Evidence that actions match words.

With BSE in the United Kingdom, there was a reasonably quick identification of the emergence of the new ailment, BSE, but communication about the human health risks—the no-risk message—were unfounded and ineffective, especially in the face of contradictory evidence. The regulatory system was widely perceived as closed and even contributed to the election of a Labour Government in 1997, which campaigned on a pledge to move food safety functions from the beleaguered MAFF to an independent, consumer-oriented food safety agency. While MAFF suggested reasonable measures to reduce levels of BSE-related risk, they utterly failed to enforce such measures and provided no evidence that action matched the soothing words of politicians and senior bureaucrats. Similar criticisms were raised in response to the E. coli O157:H7 outbreak in Scotland (Riddell, 1997).

In examining the North American response to BSE, the United States seems to have met all the putative criteria—a few instances like the Oprah Winfrey show notwithstanding—and helps account for the low level of concern about BSE in the United States. Canada, however, has failed on several accounts, yet public concern remains muted. This could be attributed to the more adversarial political and journalistic culture in the United States, or simply neglect coupled with good fortune.

The same criteria can be applied to other outbreaks. For example, in the Odwalla outbreak, the increased and more effective attention of the Seattle-King County Health Unit—the same one involved in the Jack in the Box outbreak—toward *E. coli* O157:H7 resulted in rapid identification of the Odwalla outbreak. The company exercised exemplary risk communication. Odwalla officials responded in a timely and compassionate fashion, cooperating with authorities after a link was first made on October 30, 1996 between their juice and an illness which was eventually linked to 65 people in four U.S. states and British Columbia. Upon learning of the child's death, company chairman Greg Steltenpohl issued a statement which said, "On behalf of myself and the people at Odwalla, I want to say how deeply saddened and sorry we are to learn of the loss of this child. Our hearts go out to the family and our primary concern at this moment is to see that we are doing everything we can to help them" (Odwalla, 1996).

Yet despite the comforting words, the company failed to acknowledge the existence of risk, let alone efforts to reduce levels of risk. Steltenpohl told reporters at the time that the company did not routinely test for *E. coli* because it was advised by industry experts that the acid level in the apple juice was sufficient to kill the bug. Because they are unpasteurized, Odwalla's drinks are shipped in cold storage and have only a two-week shelf life. Odwalla was founded 16 years ago on the premise that fresh, natural fruit juices nourish the spirit. And the bank balance: in fiscal 1996, Odwalla sales jumped 65% to $60 million (U.S.).

Odwalla insisted the experts in this case were the U.S. Food and Drug Administration. The FDA isn't sure who was warned and when. However, researchers from the U.S. Centers for Disease Control and Prevention wrote in the May 5, 1993 *Journal of the American Medical Association* that a 1991 outbreak of *E. coli* O157:H7 which struck 23 people in Fall River, Massachusetts—and was well-publicized at the time—was caused by unpasteurized, unpreserved cider. The story received national media attention and noted that researchers had found that *E. coli* could survive for 20 days in unpreserved, refrigerated cider. Further, the authors cited two previously reported outbreaks of illness associated with drinking apple cider.

In December 1994, the Columbus Salami Co. of South San Francisco recalled 10,000 pounds of salami after health officials linked the product to at least 18 cases of *E. coli* O157:H7 in California and Washington. The bacterium was not supposed to survive the acidic environment of salami, and again the story received national coverage. In this case, the industry immediately pledged to test whether *E. coli* O157:H7 could survive the process used to make dry sausages like salami, which only involves meat curing, not cooking.

And earlier in October 1996, fresh (unpasteurized) apple cider produced at the Notch Store and Cider Mill in Cheshire, Connecticut was linked to an outbreak of *E. coli* O157:H7 in at least seven people. For Odwalla to say it had no knowledge that *E. coli* O157:H7 could survive in an acid environment is simply unacceptable (and the basis of several lawsuits now working their way through the courts) in a global food manufacturing and distribution system, especially one becoming increasingly vulnerable to outbreaks of foodborne illness.

The stigmatization of British beef, rather than the exception, is becoming the norm for food and water linked to human illness or even death. That is because stigma is a warning-system—one that is often erroneous but in these cases extremely valuable—that something is wrong. If trust is the most important component of consumer confidence in the food supply, then further work is needed to better understand if risk-management strategies like HACCP are effective ways to not only ensure microbial food safety, but to bolster consumer confidence.

14 Blood, Risk, and Stigma

Penny Chan, Ph.D., M.H.Sc

Blood–Essential for Life with Magical Powers

This is a story about risk and stigma using blood as a case study. In many ways the recent events surrounding the contamination of the blood supply with the AIDS virus serve as an interesting example of the process of acquiring stigma. Recognizing that "stigma" is a mark of disgrace, and that "disgrace" is a fall from a position of honor, it is no wonder that stigma carries a negative connotation. However, stigmatization may be a very effective risk management strategy (Kunreuther & Slovic, 1997). The thesis explored through the story of blood's fall from its position of honor, is that the stigmatization may be the public's understanding of the risks associated with a product that has always been and will always be unsafe. This story has many aspects that are similar to other stigmatized products, but, as I hope this paper demonstrates, it also has many aspects that are unique.

Blood, unlike many stigmatized products, is natural and part of us. It has been known for centuries to be essential to life. It was contemplated as a magical cure for a host of "evils." When blood transfusions became possible it was known to save people from very gruesome, messy deaths. It was very visible. There has been a great deal of symbolism associated with blood. To save a fellow citizen from "bleeding to death" is a very good thing to do. Unlike most medicines or other medical treatments, this "cure" depends on the generosity of fellow human beings to donate their own blood. It has always been associated with good deeds, charity, goodness of mankind. It is an act of altruism. This positive image has been important to encourage persons to donate "the gift of life."

Although some risks were known to be associated with blood transfusions, they were dwarfed by the magical, life-saving benefits. Persons were given blood when they were very sick or at risk of "bleeding to death." It was believed that if they survived, it was thanks to the blood, and if they died, it was probably because of their illness or other causes. AIDS, however, altered the perception of risk to one of stigma. Perhaps it was not the risk as much as the way in which the risk was managed that caused the stigmatization. Because AIDS has a long latency period (the time between infection with the virus and the manifestation of the disease) there was a lengthy period between the initial contamination of the blood supply with human immunodeficiency virus (HIV), the virus that causes AIDS, and the recognition of its impact. However, there was also a lengthy period between the time that the impact should have been recognized and before steps were taken to control the risk. During this time there was inadequate risk communication, and the risks that were discussed were downplayed.

When the consequences of the contamination of the blood supply were realized, people started to question whether the risks had been managed as well as they could have been and whether bad decisions had been made. There were calls for compensation for those affected and an investigation into "what went wrong." The inquiry, at least the one in Canada, has allowed the "victims" to tell their stories. These are very sad stories. There are a lot of them. The process took a long time. The stories were interesting and popular material for the media. The investigations, scrutiny, and questioning of large organizations involved in the collection, funding, delivery, and regulation of blood and blood products led to many more stories. The public's perception of risk increased and their trust of the organizations in charge decreased. Blood fell from its position of honor, but did it become stigmatized?

History

Knowledge of blood has existed probably as long as knowledge itself. It was known to be essential to life. If we lose too much we die. What else did we know? When? It is a very long story and one which I will not go into in detail. To say that blood has always been thought of as "good" would be misleading. I am sure the term "bad blood" goes back a long time as does the association between blood and ill health. Take, for example, the practice of using leeches to suck out bad blood.

The restorative properties of blood have intrigued human beings for centuries. Blood was thought to have magical powers and was administered in hopes of restoring vigor, curing disease, and combating mental illness. The first blood transfusions took place in the 1660s in England and France. At that time, animal blood was transfused to people in the hopes of curing mental illness. The early transfusions were unsuccessful, leading to litigation, political infighting, and an eventual ban because of real and presumed dangers. It was not until the early 1800s that James Blundell revitalized the interest in transfusion and used it successfully to treat postpartum hemorrhage. He devised a device for autotransfusion, returning the person's own blood (Spence et al., 1993).

During World War I isotonic saline infusions were used to combat traumatic blood loss and hypotension. Its failure to provide long term restoration fostered the belief that transfusion of whole blood was necessary. In the second World War, rehydrated serum was used instead of saline with better results. By the end of the war banked blood treated with an anticoagulant was successfully used by the military. The first blood banks were opened in the late 1930s, and the first blood donor clinic was opened in Canada in 1940. So began the era of modern blood banking.

Development of an efficient blood banking system coupled with an understanding that blood transfusion was an innocuous procedure with the benefits far outweighing any risks led to a widespread use of blood transfusions (and some might argue—abuse). People believed that not only was the procedure relatively free of risks, but also that it was essential for a favorable surgical outcome. Blood was ordered for all surgical procedures, both to be sure that it was available if needed, and to avoid accusations of carelessness (Spence et al., 1994). It was given to "pink up" the cheeks of patients before they left hospital.

Some risks were known, but major complications such as hemolytic transfusion reactions happened so infrequently that they were not a major concern. These reactions were nearly all attributable to human errors in labeling the blood, in identifying the blood type or in giving a person the wrong blood. Human error, while not totally avoidable did not mean that the blood was inherently unsafe. Most other reactions were febrile responses. Although hepatitis was a known risk, the link to transfusion was often missed because symptoms would develop after the patients left the hospital, or because other significant complications of the patient's illness or surgery masked the effects.

BLOOD PRODUCTS

As noted earlier, even before the extensive use of blood transfusions, at least one "blood product" was used for therapeutic purposes. Serum was collected from clotted blood and dehydrated. When rehydrated, it was used to treat trauma on the battlefields. Serum was also used as a source of antibodies to protect persons from imminent infection in addition to being used as a biological medium for vaccines. It was the medium of choice because human serum does not contain proteins that would be recognized as "foreign" and lead to allergic reactions in persons being inoculated.

Technologies improved. It became possible to separate uncoagulated blood into its components: red blood cells, platelets, and plasma. This allowed patients to be treated with only those components that were needed. For example, plasma, instead of whole blood, could be used in treatment of hemophilia.

HEMOPHILIA AND ITS TREATMENT

Untreated, persons with severe hemophilia had a life expectancy of about 11 years. Unlike our popular conception of hemophilia, it is not uncontrolled bleeding from small cuts or abrasions that endangers life, but internal hemorrhages. Internal bleeding can occur spontaneously or as a result of even minor trauma. Bleeding into joints is common and leads to crippling disabilities. Sufferers from severe hemophilia. almost exclusively males, could not participate in normal activities, especially sports, without risking life-threatening injuries. As young boys they were often labeled "sickies" and "cripples." Therefore in addition to their chronic pain and disabilities, they were social outcasts—minorities.

Initially, the treatment of hemophilia was whole blood transfusions. This was later replaced by infusions of plasma. However, neither of these treatments was very effective. Because such large volumes were needed to replace the missing clotting factors, the treatment could lead to circulatory overload and heart failure. The infusions had to be administered in hospital, and this caused a delay between the onset of bleeding and treatment, which in turn led to long-term damage of areas where the blood was accumulating, such as joints.

A great advance in the treatment of classic hemophilia came with the discovery that cryoprecipitate, the part of plasma that does not dissolve when frozen plasma is thawed, contained a high concentration of many of the clotting factors. Cryoprecipitate was more easily stored than plasma and a smaller volume was required for treatment. In the 1970s, improved technology provided methods for further separation and some degree of purification of the clotting factors to produce clotting "factor concentrates." These products represented a revolution in the treatment of hemophilia. Treatment changed from reactive to prophylactic. Internal hemorrhages could be prevented by frequent administration of the factors that were lacking in the blood of hemophiliacs. Severe hemophiliacs could store the dried product in their refrigerators at home and administer it themselves when they needed it. For these people this was freedom from hospital. Prophylactic treatment freed them to live relatively normal, pain-free lives.

COMMERCIAL PRODUCTION OF BLOOD PRODUCTS

There was a problem, however. These factor concentrates were commercially produced from hundreds or even thousands of liters of plasma that had been collected from thousands of donors and pooled. One contaminated sample could contaminate the pool and all the factor concentrate derived from it. The commercial production of factor concentrates, albumin, and immunoglobulins, all derived from plasma, meant that the demand for plasma was greater than that for the other components of blood. Plasma can be collected without collecting whole blood from persons by a process called plasmapheresis. Because a person can replenish plasma more quickly than whole blood, this process allowed frequent collections of large volumes of plasma from one person. Canada, like many other countries, did not have enough plasma to manufacture its own factor concentrates and bought much of its factor concentrates and other blood products from the United States. In the United States people are given money for their plasma. Because factor concentrates became a valuable product, several companies were competing for supplies of plasma, and an international trade in plasma developed. Paying for plasma attracted less healthy people as "donors" and, as we now know, plasma from paid donors is more likely to be contaminated with infectious disease agents than the plasma collected from volunteer, non-paid donors. Also some companies routinely collected blood and/or plasma from prisons and geographical areas that had high rates of infectious disease.

Hepatitis was known to be a "side effect" of treatment with blood products and was almost an "accepted risk" and some may even say it was an "acceptable risk." Compared to the benefits of prophylactic treatment of hemophilia, hepatitis, if an unavoidable risk, was considered a small price to pay.

So there are two parallel streams to this case study, that of transfusions of blood including its components plasma, platelets, and red cells, and that of blood products derived from plasma by the process of fractionation.

DEMAND FOR BLOOD AND BLOOD PRODUCTS

In the past couple of decades many technological advances in the field of transfusion medicine, in addition to blood products separation, have occurred. Clinical practice has changed the use and need for transfusions. On the one hand, surgical techniques have improved and a variety of means of preventing blood loss and alternative treatments have decreased the need for blood transfusions. On the other hand, new techniques now make open heart surgery, liver transplants, and massive tumor resections possible. These complex surgeries could never have been contemplated were it not for a secure supply of blood. The demand for blood, even in these procedures, has dropped in the last few years. Meanwhile the market for plasma derivatives has exploded. Initially it was driven by the need for clotting factors, but an increasing number of uses have been found for other derivatives. Albumin was a major product used for the treatment of massive fluid loss such as in severe burns. Now the "market drivers" are immunoglobulins, particularly the polyclonals (antibodies with many different specificities) used for the treatment of people with immunodeficiencies or on immunosuppressive therapy.

What Changed?

This long, complicated description of the history of blood transfusions and the use of blood products sets the stage for the stigmatization of blood, a product once seen to have lots of benefits and few risks. The product is natural and viewed by many as essential for the treatment of numerous medical conditions.

This perception of blood as a product with lots of benefits and few risks changed with the discovery that the virus responsible for AIDS was transmissible by blood and blood products. In contrast to post-transfusion hepatitis from which a small percentage of patients eventually died, AIDS was uniformly fatal. Moreover AIDS caught the public eye and quickly became highly politicized. Certainly in the early days there could not be a closer association than between AIDS and stigma. As noted by Slovic elsewhere in this volume, members of minority groups, the aged, homosexuals, drug addicts, and persons afflicted with physical deformities and mental disabilities are targets of stigmatization in our society. AIDS was initially known as the "gay disease." In the early stages of the epidemic in North America, 90% of the AIDS cases were associated with homosexuality. Other "high risk" groups identified were Haitians and intravenous drug abusers, which were also identified as minorities. It is no small wonder that there was stigma associated with AIDS.

The link between AIDS and blood and blood products was made slowly. AIDS was first diagnosed in hemophiliacs and blood transfusion recipients in 1982-3, a couple of years after the first description of the disease. Why was the recognition of the link so slow? The disease was a new disease of unknown origin. It had a latent period, which meant it took time for those who were infected to be diagnosed. It also took time to confirm the causal link between the disease and the transfusion of blood or blood products. The critical times in this story are the periods between a) sufficient evidence being accumulated to make the link, and the implementation of strategies to control the risk, and b) the information becoming available and its communication to the public, and most particularly those at risk.

Unfortunately, history has illustrated that making a link between cause (in this case the use of blood and blood products) and effect (in this case, development of AIDS) is often delayed if making the link means that major changes have to be made to existing practices or existing beliefs are threatened. There is a tendency to disbelieve the information and to search for more. When more information becomes available there is a tendency to downplay the "possible" risks, especially if the link is not scientifically "proven." When there was uncertainty, in this case, the reasons for downplaying risks included: "we do not want to panic people"; "the people that need life-saving blood transfusions or blood products may refuse them"; "donors may not want to donate"; "imagine the gift of life causing the dreaded disease AIDS." There was also concern that strategies to reduce the risk and prevent contamination may require actions that would be unpopular, because they were, or could be viewed as, discriminatory. Donations are based on doing good—giving to a good cause. Charitable organizations do not want to discriminate against well-meaning people without good reason. There was a stigma associated with AIDS. This led to reluctance to communicate information on risks and single out high risk groups. Asking them not to donate was stigmatizing and viewed as discriminatory. Also the types of questions that needed to be asked to identify high risk groups were considered very personal. AIDS is an infectious disease and can be passed on to others.

The story recounted above sets the stage for the case study. For the sake of convenience, I have broken the story into two phases. Phase I is the contamination of the blood supply with the virus that causes AIDS, and the identification and management of the risk. Phase II commenced after the risk management strategies were implemented and when the impact was being realized by the public. People started asking questions: "How did this happen?" "Why did it happen?" "Should things have been handled differently?" The infected persons started to demand compensation. Many groups called for an inquiry.

Key Players

This story could have been set in any industrialized country. Some of the details differ but overall it would be remarkably similar. The players and the way they interacted to make decisions differ from country to country. There are, however, many similarities.

In Canada, all blood is collected, separated, and components processed by the Canadian Red Cross, a charitable organization of international repute. The blood program is a national program and has a global budget. The money, however, does not come from one source but from the individual provincial and territorial governments. In the 1980s there was a committee with representatives of each of these governments making financial decisions for the blood program. The federal government is the regulator. The recipients are the players who, alone, bear the risk. Of the recipients, only hemophiliacs were organized and had a society. The doctors who were responsible for administering blood and blood products were key players, particularly the hemophilia treating physicians, because their patients depended on them both for blood products and information about the risks and benefits. There were many other players, including the pharmaceutical companies that fractionated the plasma into blood products, hospitals, and community groups. Some community groups came together because some persons considered that they were being discriminated against by some of the donor screening methods that were contemplated and later implemented.

In Phase II the players were different. Certainly the institutional players that were important in Phase I—the collector and distributor of blood and blood products, the funder, the regulator, and the companies that manufactured blood products—were still key players. The role they played in Phase II was to explain or to defend their actions and decisions in Phase I.[1] The recipients who were infected were very important players. In this second phase there were many groups of infected recipients: groups of hemophiliacs, groups of persons infected by blood transfusions, and groups who contracted AIDS by secondary transmission from transfused persons. Some of these groups formed for the purpose of seeking compensation for their members, and some to commence legal action. Lawyers played an important part in this phase: those representing recipients; those defending the institutional players and doctors; and those involved in the inquiry. Finally, a key player in this phase was the media.

Key Decisions

Although there were many key decisions in this very complex case study, I will not discuss any of them in Phase I. This is still the subject of inquiry. I am proposing, however, that many of the factors that were critical in the shift from a recognition of risk to stigmatization of blood, if in fact this did occur, are those that emerged in Phase II.

The reasons for, and details of, actions and inactions in the process of managing the risk of HIV contamination of the blood supply will occupy thousands of pages in a comprehensive report by Justice Krever, the Commissioner for the Commission of Inquiry on the Blood System in Canada. I will not discuss any in this paper. In public hearings the commissioner heard that, in the early 1980s, there was a lot of misinformation including serious underestimates of the risks, the labels on products were not providing accurate risk information, and persons receiving transfusions were not told of risks. In fact, in many instances, persons were not told that they may receive a blood transfusion, let alone provided the opportunity to give "informed consent." Many persons did not even know that they had received blood or blood products.

For the sake of controversy, I propose that the critical decision was the decision to call an inquiry. I am not proposing that the inquiry itself caused the stigmatization. I do believe, however, that the stigmatization of blood, if it occurred, happened as an unfortunate "side effect" of the process that led to calling the inquiry and as a consequence of the information that became available to the public during the inquiry process. First, the calling of an inquiry leads one to assume that something "did go wrong"—perhaps someone is to blame. Second, it attracted an enormous amount of publicity. The stories of the "victims" are very sad. They are very human. It could have happened to any one of us. For hemophiliacs it had a devastating toll. Whole families were affected: brothers, uncles, and cousins. Persons who had suffered pain, disability, and ridicule, whose hopes of a normal life had been raised with the introduction of factor concentrates were now dying from a stigmatizing disease that was a consequence of the treatment. In addition the disease was infectious and could, and in many cases was, passed to their sexual partners.

---

[1]Interestingly enough, these are the same institutions that took legal action to challenge the powers of the commissioner conducting an inquiry into the events that occurred in Phase I.

The impact of the contamination of the blood supply is enormous. Perhaps as many as 80-90% of severe hemophiliacs contracted AIDS. It is estimated that 1,200 persons contracted AIDS as a result of the contamination of blood and blood products in Canada. Many have already died and many more are very sick. It is a massive public health disaster.

The impact of the stigmatization is not so easy to measure. Surveys indicate that many persons do not trust the safety of the blood supply and some would refuse a blood transfusion. There are increasing demands for, and use of, autologous blood transfusion (in which a person banks his or her own blood prior to an elective surgical procedure). There are demands for directed donations (in which a person can direct his or her blood to be used for the treatment of a friend or family member). The search for alternatives to blood and blood products has increased dramatically. Efforts to improve the safety of the blood supply have escalated. All of the above may be considered positive consequences of recognizing that blood and blood products are not and never will be completely safe. There are, however, other consequences which could be considered negative if they occur. For example, a person may be so frightened of the risk associated with blood and blood products that he or she may refuse life-saving treatment. Because of the negative publicity and lack of trust in the system, people may refuse to donate blood, which could create critical shortages of blood and blood products and endanger the lives of those who depend on them. The "gift of life" is no longer viewed in the same positive light. In fact, some have called it the "gift of death" (Picard, 1995). With such a label, donating blood could hardly be considered an act of altruism. There are economic consequences also. These include: demands for safety "at any cost"; withdrawal of millions of dollars worth of products because of a *possible* contamination with an agent that poses a *theoretical* risk; the move to new synthetic products that are more expensive and not proven to be safer. The cost of research into alternatives and means of avoiding blood cannot be considered a negative consequence. The negative consequences are only considered negative when some of us decide that they are unwarranted or unnecessary. Whether or not they are "necessary" depends on whether we consider the assessment of risk to be reasonable or unreasonable. If we, as the "experts," decide that the public demands are unreasonable for the amount of risk that we estimate to be associated with that product, we conclude that that product is stigmatized. If, however, we believe that the public, the persons who are the users of the products, have the right to make informed choices and we are honest and open about the information on the risks of using the products as well as the risks of not using them, then it is not for the experts to judge the public's decisions as unreasonable.

The Role of Science, Risk Perception, and Risk Assessment

This is a classic case study of risk management: identify the risk, estimate the magnitude, assess the impact, and make decisions on how to manage it. In Phase I, the risk identification was the realization that HIV was blood-borne and could be transmitted by blood and blood products. The estimate of the magnitude of the risk was a problem because of the length of the latency period after infection and before diagnosis. Early estimates were based on the number of diagnosed cases of AIDS per million transfusions. This may be an epidemiological measure of the prevalence of the disease at that time, however, it is not an estimate of risk. A complicating factor was that, although Canada has a contiguous border with the United States, the prevalence and incidence of AIDS were, initially, much lower in Canada. Added to this was the belief that it was a "gay disease," "It is not going to happen here," "We are self-sufficient for our blood," "We rely on volunteer donors."

My personal assessment is that, in the early 1980s, when the risk of blood transmitting AIDS was at its highest, the estimates of risk that were made at that time were at their lowest. There are probably many reasons for the low estimates of risk and the slowness for implementing control strategies including: poor scientific data and understanding of the disease, denial, lack of experience, lack of vigilance, and complacency. When there is a failure of the scientific community and the operators of the system to recognize the risk or to downplay it, the risk is not likely to get much publicity, especially when the impact has not yet been realized. The risk did not get a lot of publicity. The public's perception of risk was low. A journalist who has been covering the issues surrounding the contamination of the blood supply when asked, "Could the media have done a better job covering the tainted-blood tragedy?" commenced a presentation:

> We journalists are guilty of the same "crime" as the main players in the blood system...a failure to inform the public. Like them we have excuses...but collectively, our mistakes have cost hundreds of people their lives. There can be no excuse for that. We cannot be forgiven. But we can make amends by learning from our failures, by never again repeating them. (Picard, 1995)

The problems in the identification, estimation, management, and communication of risk that occurred in Phase I of this case study precipitated Phase II. The publication and analysis of this risk management process and its consequences is, in my opinion, what has led to the stigmatization of blood and blood products, if in fact it has occurred. With the publication of the stories of the impact; the stories of human pain and suffering, coupled with the stories of mismanagement of risk, the public's perception of the risk increased. Figure 14.1 is a graphic illustration of my thesis about the events that influenced the rise and fall in the risk as assessed by the experts and as perceived by the public. The point where the lines cross represents the time that the public's perception of the risk was greater than that estimated by the experts. Obviously this illustration should not be carried too far, but for the sake of the thesis I estimated that the cross over occurred in the early 1990s. The point that the public's perception became "unreasonable" compared to the experts' estimates, which some call the real risk, is the point that blood is stigmatized. "Unreasonable " is defined as the point when the demands for reducing the risk are out of proportion to the experts' estimates and that persons will refuse treatment and put themselves at greater risk. I do not presume to make any judgment about this point. If one believes in a strictly democratic process of government, then blood is not stigmatized. If the people want the risks reduced regardless of the costs, then the risks should be reduced. This, however, puts an enormous and perhaps unfair burden on the media because it is primarily the media that bring the risks to the attention of the public. Should the

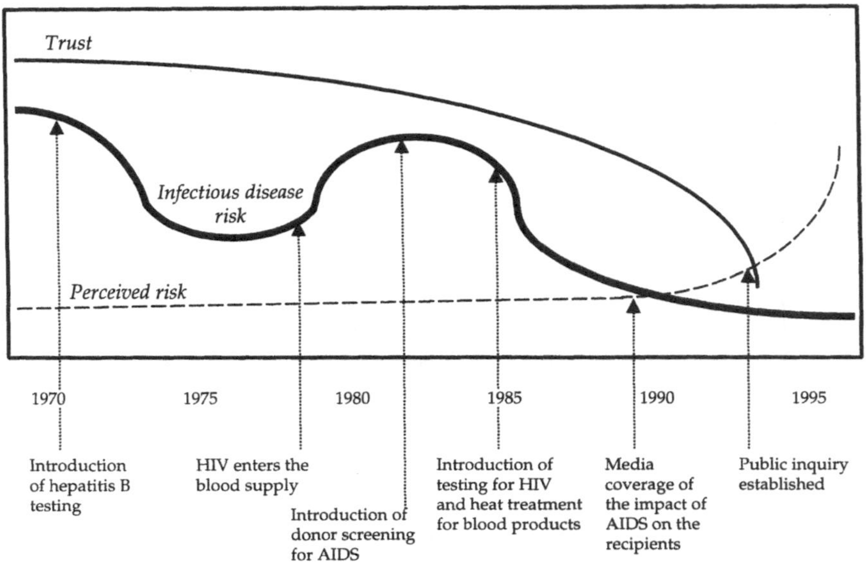

Figure 14.1 Thesis: factors affecting risk and perception of risk.

media be responsible for presenting all the information? the best information? a balance? Who should make the decisions about balance?

If, as I propose, it was the management of the risk rather than the risk itself that caused the stigmatization, one may ask whether it is the product "blood" that is stigmatized, or the institutions that we associate with the stewardship of our blood program. Therefore the blood may be stigmatized, either because of the risks, or because there is a lack of confidence in the institutions responsible for the management of the risks. There certainly is a lack of trust. If you think about this question, you realize that, this is another way that this case study differs from that of other stigmatized products. The blood and plasma, at least in Canada, are donated by fellow citizens as a "gift of life." The institutions that have stewardship over blood are not the manufacturers. In this case the institution is a charitable organization. For-profit companies were involved in the manufacture of the blood products, but they also commence with blood, or the plasma separated from it, as a raw material and, in this case, it was the risk in the raw material that contaminated the product. Interestingly enough, the stigma appears to be more visible for blood than it is for blood products. In fact, many of us do not know the extent to which blood products are used in other products.

Dealing with Stigma

In both Canada and the United States, the "industry" and the institutional players recognize that one of the primary problems for the blood system is that public confidence is at an all-time low. In both countries attempts are being made to improve public confidence. The blood safety director, Dr. Philip Lee, Assistant Secretary for Health, said to a meeting of the American Association of Blood Banks in December 1996:

> My serious message to you today is that blood product and service providers share with the Federal government the very difficult challenge of restoring public confidence both in the safety of the nation's blood supply and in our ability to respond to new threats to blood safety.

He continued:

> Without a doubt, the current environment has been defined mainly by reaction to the tragedy of HIV transmission by blood and blood products.... The public has come to fear blood to a degree out of proportion to the actual known risks...and shows unwillingness to accept even small risks when they affect blood products. Underlying these attitudes is the enduring memory that the products were promoted as safe in the early 1980s when they were not, and that the involved health professional industry and the government did not interdict the dreaded threat of HIV until after many thousands of infections had already occurred. ("Blood Safety," 1996)

What Is Likely to Happen in the Future?

Things will change. The stigma has become so great there has to be a change. Major changes are being contemplated in the organization, structure, and operations of the blood program in Canada.

Epilogue

This story was about risk and stigma using blood as a case study. It is a complicated story, and I chose to focus on the contamination of the blood supply with the virus that causes AIDS. I proposed that the critical factor in stigmatizing blood was not the risk of AIDS, but the "telling of the story" of how this risk was managed. This led to questions about whether the risk could have been managed differently and whether some of the impact could have been avoided. At the same time these questions were being asked, the impact was better understood and better publicized. Many people who felt wronged by the process came forward to tell their stories, and the stories received a great deal of media coverage.

I conclude that, like any other risk management strategy, a critical issue for minimizing risk is honest communication to the public, and particularly those at risk, of information relevant to the risk and the associated uncertainty. The success of the risk management strategy in preventing stigmatization depends on the same critical factors and the trust that the public have in the institutions providing the information and managing the strategies. I therefore finished with the question: Is it "blood" that is stigmatized or is it the whole program and institutions involved? To help answer this question I shall briefly mention another aspect of the story.

At the same time as the key players were struggling with managing the risk of HIV in the blood supply, there was another risk. There was another infectious agent already in the blood supply, a more insidious risk, less "sexy" than the cause of AIDS. This was not a new risk. Initially it was known as post-transfusion non-A, non-B hepatitis. It is now known to be mostly caused by the hepatitis C virus (although there are and may be still more hepatitis viruses in the alphabet soup). The disease is not necessarily fatal. It is, however, very commonly chronic and after many years leads to serious complications and often death. It has affected more recipients of blood and blood products than AIDS and will probably lead to more deaths.

An analysis of the means of dealing with this risk, leads one to conclude that our institutions may not have done much better with this challenge. Have we learned lessons? I sincerely think and hope we have. Have we learned the right lessons? The trouble is that new risks pose new challenges. The lessons we learn cannot be so specific that they will allow us to respond only to another AIDS virus or another hepatitis virus. What about a prion? What happens if the next risk is not an infectious agent? Blood will never be risk free. If we continue to rely on it as such an important aspect of therapy, we must be honest. We must be constantly vigilant. We must communicate the information we gather. We must act upon it, even when it is politically unpopular. We must recognize and communicate the fact that blood is not a miracle cure, that it is not risk free, but that it has and is still needed to save many thousands of lives. It should not be stigmatized, either for its own sake or for its mismanagement. If we were more honest about risks we would not have such unrealistic places of honor and the fall from these positions would not be so hard. We should not be in the position of trading "blind trust" for "blind distrust" when one risk is mismanaged.

15 Technological Danger Without Stigma: The Case of Automobile Airbags

John D. Graham, Ph.D.

Introduction

The United States is the only major country in the world that compels manufacturers to equip all new passenger vehicles with airbags. Airbag enthusiasts predicted that the airbag would be the equivalent of a vaccine for crash-related trauma. Numerous professionals (including the author of this article) devoted substantial fractions of their careers to making the case for mandatory airbags.

In recent years, however, it has become apparent that these devices, at least the designs installed in 60 million new U.S. vehicles during the 1989-1996 period, pose an unexpected risk of injury and death to both drivers and front right passengers. As of early 1999, airbags were known to be responsible for over 100 deaths of drivers and young children, and thousands of cases of nonfatal injury to the upper extremities, faces, and upper torsos of motorists (National Highway Traffic Safety Administration, NHTSA, 1996b, 1996d, 1997). Petitely statured motorists, particularly children, females, and the elderly, are believed to be at special risk of deployment-induced injury (Melvin et al., 1993; NHTSA, 1996d; 1997).

Airbag-induced injuries often occur in crashes of low to moderate severity, which means that the rapid deployment of the airbag may cause more danger to motorists in these crashes than the danger of injury or death from the crash itself. The most serious bag-induced injuries are being incurred by vehicle occupants (e.g., small children) who were not properly restrained by safety belts when the crash occurred (National Transportation Safety Board, NTSB, 1996). Yet both belted and unbelted drivers are experiencing bag-induced injuries at a rate that is disturbing to clinicians and safety experts (Dalmotas, Hurley, & German, 1995; NHTSA, 1996d).

The new airbag technology has saved far more lives than it has cost, at a ratio of about 75 to 1 on the driver side and 5 or 10 to 1 on the front-right passenger side (NHTSA, 1996d; Ferguson, Braver, Greene, & Lund, 1996). Yet risk-perception specialists are fond of reminding technocrats that lay people do not judge the perceived risk of technologies on the basis of net expected mortality (Slovic, 1986; Freudenburg, 1993; Freudenburg & Pastor, 1992;

Heimer, 1988). For example, nuclear power and industrial chemicals do not look very risky on the basis of actuarial risk statistics but have historically ranked near the top of the public's list of risky technologies (Slovic, 1987). Moreover, social scientists predict that a technology with favorable actuarial experience can become stigmatized if public expectations are not met, especially when a single adverse event triggers amplification of risk in the media (Gregory et al., 1995; Kasperson et al., 1988).

A critical question, then, is whether the unexpected dangers of airbags will heighten perceived risk of the device, provoke public outrage, and cause long-term stigmatization of the technology. Depending upon the extent of the adverse public reaction, it is possible to imagine the U.S. Congress—under pressure from the public—repealing the airbag requirement, restricting airbags to be an optional purchase, or actively discouraging manufacturers from equipping vehicles with these "killer bags" (Streisand, 1996). Alternatively, an outraged public could turn against manufacturers, demanding that highly sophisticated "smart airbags" be installed in the very near future, even though the reliability, effectiveness, and cost of such technologies are highly uncertain.

In this article, risk-perception models are used to predict whether public outrage will develop in response to the dangers of airbag technology. Airbag stakeholders are quite concerned that outrage will develop. Although psychometric models suggest that outrage should occur, the available evidence—from focus groups and national opinion surveys—does not provide much evidence that public outrage is brewing. Unlike what happened to nuclear power and pesticides, where public outrage caused permanent stigmatization and regulatory stringency, the public response to the dangers of airbags has been fairly restrained.

Even the coalition of parents whose children were killed by airbags have made only modest demands on the policy process (e.g., better use of warning labels for parents), and the government response has been exceedingly cautious. There have been none of the grass-roots advocacy efforts that characterized the public opposition and ultimate stigmatization of nuclear power and pesticides.

Three possible explanations are offered for why the predicted public outrage has not yet occurred. One explanation is that most of the public remains ignorant of important yet unsettling facts about airbags. People may yet become outraged, when they learn of the facts. They may do so particularly if the "scandal" is revealed in a dramatic or sensationalized fashion, perhaps similar to what happened to chemicals in Rachel Carson's *Silent Spring* or to nuclear energy in the movie *China Syndrome.* A second explanation is what psychologists call "optimism bias," a tendency to believe that unfortunate things "won't happen to me" and are only relevant to incompetent or irresponsible people (Weinstein, 1984). If airbags are dangerous, the rationalization might be that they are only dangerous to people who don't wear their safety belts. A third explanation is that the forces generating worry are overwhelmed by a countervailing factor: a perception of the lifesaving benefits of airbags. Public perceptions of the airbag's lifesaving benefits may diminish the importance of its risks and induce people to place confidence in "fungibility" (tradeoff) judgments made by technocrats in government and industry (Margolis, 1996).

In this chapter, I explore these hypotheses. I also examine whether public support will be adequate for the major steps that appear to be necessary to minimize the risks of airbags while retaining the substantial benefits of this promising safety device. As we shall see, this may be a case study of a technology that has not been stigmatized adequately for the larger public good.

Prediction of Public Outrage

In the classical psychological research of Slovic, Fischhoff, and Lichtenstein (1979), perceived risk is used operationally to represent the public's degree of concern or anxiety about a technological hazard. According to this view, technologies that rank high on perceived risk will become targets of collective action to reduce, prevent, or mitigate risk—or even abolish the technology.

More precisely, Slovic argues that "risk" means different things to lay people than it does to scientists and engineers. While technical people focus on actuarial evidence of danger, lay people focus on a more complex array of qualitative attributes of technologies, such as whether the associated hazards are familiar or dreadful (Slovic, 1987). Familiarity attenuates perceived risk; dread accentuates it. As Appendix 15.A indicates, familiarity is actually a surrogate for five underlying factors; dread is a surrogate for another nine underlying factors.

In a related literature, Peter Sandman, Vincent Covello, and other social scientists have expanded the number of qualitative attributes of hazards that may predict "public outrage" (Covello, 1991). These factors include whether the hazard is made by man or is natural, whether children are at risk, whether the victims of technology are identifiable before and/or after the harm occurs, whether there is a mass media craze about the risk, whether institutions responsible for the technology are trusted by the public, whether information about the risk was suppressed by the government, and whether the hazard has the potential for "social amplification" (i.e., a ripple effect of concern throughout society based on novelty, mysteriousness or other special circumstances). Appendix 15.A treats these additional factors as "aggravating circumstances" that can heighten public concern when they are present (Kasperson et al., 1988).

Based on this literature, I devised a simple scoring system to predict perceived risk for five hazardous technologies: nuclear energy, pesticides, large power lines that emit electric and magnetic fields, smoking, and airbags. I rated each of the five technologies "yes" or "no" on each of the 21 attributes listed in Appendix 15.B, based on how a lay person (or scientist!) would probably see things. (The reader may find it instructive to perform this exercise and compare his or her ratings to mine, which are documented in Appendix 15.B to this paper, with some associated commentary). A predicted outrage score was computed by subtracting the number of positive "familiarity" ratings from the sum of the number of positive ratings for "dread" factors and "aggravating circumstances." The highest possible score is 16. A more complex analysis might assign more weight to some factors than others. It would also be interesting to undertake this exercise with a representative sample of lay respondents.

As expected, the scores are high for nuclear energy (16) and pesticides (14). The scores for EMFs from power lines (11) and airbags (11) were equal. Interestingly, the score for airbags is much higher than the score for smoking (4), even though smoking ranks highest of all on perceived-risk rankings published by Slovic and colleagues. In recent research, this anomaly has been attributed to the failure of the classic risk-perception model to account for different levels of public confidence about whether a named hazard exists (i.e., is real). Smoking scores extremely high on this measure while EMF's score relatively low (Graham et al., 1999).

If risk-perception models are valid as predictive tools, then the public outrage against airbags may not be as virulent as the reaction to nuclear power or pesticides, but should be substantial. The major reasons that nuclear power and pesticides score higher than airbags is that the former are perceived to present global, catastrophic, and future-generation issues that an airbag does not. Otherwise, the airbag—or at least the U.S. mandate of airbags—seems to be a great candidate for public outrage.

Why Airbags "Should" Cause Outrage

This "safety device" is a manmade technology, installed in all new private vehicles, that has killed and injured people, particularly frail and vulnerable populations such as children and short, elderly women (NTSB, 1996). The victims are not strictly statistical; they often can be named after the fact, such as Frances Ambrose, a five-year-old girl from Nashville, Tennessee who was killed by an airbag on September 12, 1996, even though—by all accounts—she was wearing her lap belt and shoulder belt system in the passenger seat when the crash occurred (AP, 1996a, 1996b). This was a signal event in the history of the airbag that was amplified by mass media coverage both on national television and in the newspapers. Another signal event occurred during the Thanksgiving holiday of 1996, when a one-year-old girl, Alexandra Greer of Boise, Idaho, was decapitated by an airbag in a "shopping mall fender bender" (Fick, 1996). In a visual display of such tragedies, the front page of the *Washington Post* in a 1997 Sunday edition, featured pictures of the faces of numerous young children who were killed by airbags.

Airbags do save lives and prevent injuries. However, these savings are more statistical (though no less real) in nature because we typically cannot know the names and addresses of those who were saved or see their faces or those of their loved ones. Those who survive a crash have no way of knowing whether they would have died had their airbag not deployed. This particular pattern of technological risk and benefit, where risk is identifiable and benefit is statistical, has been predicted to produce public reactions that will stigmatize a technology—even if the ratio of risk to benefit is low from an actuarial perspective.

Moreover, the airbag is no longer a voluntary purchase in the United States. Consumers who are buying a new car, jeep or van are now compelled by law to purchase airbags, on both the driver side and the passenger side. No such law exists in Europe or Australia. Because the United States cherishes the notion of free choice, it should be expected that the absence of choice about purchasing airbags should provoke at least some public opposition.

Historically, it was illegal for a dealer or mechanic to disconnect an airbag system. (It was permissible for a manufacturer to install an "on-off switch" for the passenger-side airbag that the owner can operate on a trip-by-trip basis, but only on a small number of vehicles that are not designed with a back seat, such as pick-up trucks.) There were proposals to allow a disconnect option but they faced significant opposition from insurers, automakers, consumer advocates, and airbag suppliers. NHTSA decided to allow more freedom of choice in this area but the agency has made it cumbersome for consumers to exercise this choice.

The U.S. airbag regulation can also be seen as inequitable, since people who wear their safety belts must pay the same price for this technology as those who don't buckle up, even though most of the safety benefits from airbags occur to people who do not wear belts. More importantly, innocent children are being killed in order to protect unbelted adults, at the same time that belted motorists are being injured in low-speed crashes by airbags that are far too explosive relative to what a belted passenger needs. The compliance test that governs airbag design entails crashing an unbelted male's vehicle into a fixed barrier at 30 miles per hour (mph). This test induces airbag designs that are too aggressive for the safety of belted occupants, especially children, the elderly, and short drivers.

The media has publicized some of the dangers of airbags with sensationalist vigor, including repeated stories on the front page of *USA Today* as well as sizzling prime-time TV coverage on ABC's *48 Hours* and *Dateline NBC*. Particularly juicy for the mass media were indications that both government and industry knew about possible dangers to children for twenty years but did little to warn parents when airbags came into the fleet in large numbers in the 1990s. There has been an effort by segments of the media to frame a scandal, but they have not yet been successful in triggering any concerted or sustained public outrage (Brown & Ottaway, 1997).

Response of Airbags Stakeholders

The adverse publicity about automotive airbags has caused a significant degree of concern in the United States, at least among various airbag stakeholders. In particular, airbag advocates are worried that the public might turn against their favored technology (Nomani & Stern, 1996). They are taking modest steps to curb the danger while pointing fingers at who is to blame for the unexpected side effects of a highly-touted safety device (Kaplan, 1996).

NHTSA has received communications from roughly 20,000 motorists who express concern about the safety of airbags or want to have their airbag systems disconnected (Meredith, 1996). Much of the concern comes from short women drivers, but there has also been concern expressed by parents of children. A coalition of parents whose children were killed by airbags has formed and urged manufacturers to issue stronger warnings to owners of vehicles with passenger-side airbags. Automakers, under encouragement from NHTSA, mailed stronger warnings and labels to each of the 27 million owners of such vehicles.

On October 27, 1995, NHTSA issued a press release and public advisory urging care givers to protect children by making sure that all children are properly restrained in the back seat. On November 27, 1996, NHTSA issued a final rule that extended the stronger warning labels to all new motor vehicles sold in the United States.

In 1996 the motor vehicle manufacturing industry petitioned several governments to allow them to manufacture new vehicles that have "depowered" airbags. The Canadian government was relatively quick to grant manufacturers permission to move in this direction, but NHTSA responded more cautiously.

Depowering means reducing the speed and force at which the bag is deployed in crashes. The ultimate effects of depowering are not known with certainty, but modest depowering should reduce (though not eliminate) the injuries now being experienced by out-of-position children and short drivers. There is a broad technical consensus that depowering is a prudent step, and there is certainly no evidence that U.S. airbag designs are insufficiently powered (Lund, Ferguson, & Powell, 1996). Yet there are legitimate concerns that depowering will reduce the ability of airbags to protect occupants in some high-speed frontal crashes when a rapid speed of deployment is critical. After considerable consternation, NHTSA granted the petition of automakers to depower airbags as an interim measure, but only until better solutions (e.g., "smart" airbags) are found (NHTSA, 1997).

The Airbag Coalition, an ad hoc educational organization funded by automakers, insurers, and airbag suppliers, launched a $10 million public-relations campaign aimed at increasing the rate of proper restraint use among adults and children in the United States. The coalition devoted particular energy to state-level activities aimed at strengthening the provisions and police enforcement of mandatory safety belt use laws and child restraint laws. Several individual manufacturers, such as Volvo and Chrysler, have launched related campaigns aimed at changing safety-related behaviors. The Volvo television advertisements give special emphasis to the importance of placing children in the rear seat where they will not be endangered during airbag deployment.

On November 22, 1996, NHTSA also announced interest in a new rule that would require smart airbag systems by the 1999 model year. Airbag suppliers are now in a fierce competitive struggle to develop new "smart" airbag systems that would vary deployment characteristics as a function of parameters such as the weight, size, and location of the occupant (W. Brown, 1996). A February 1997 public meeting held by NHTSA concluded with a broad consensus that smart airbags are worthy of further development but are not ready to be mandated in the foreseeable future.

There were two important congressional hearings on airbags, one chaired by Congressman Frank Wolff (R-VA) of the House subcommittee on transportation appropriations in December 1996, and the other chaired by John McCain (R-AZ) of the Senate committee on commerce, science, and transportation in January 1997. Senator Dirk Kempthorne (R-ID) expressed interest in a revamped airbag regulation that would focus protection on the safety needs of the belted occupant. NHTSA, resisting this idea, claimed that Congress would have to write new legislation in order to permit NHTSA to scrap the unbelted compliance test. Congress ultimately passed legislation forcing NHTSA to require advanced airbag systems early in the 21st century.

At the encouragement of key members of Congress, NHTSA held a meeting of stakeholders in January 1997 to discuss "process issues." The NTSB held a four-day public hearing in March 1997 to air the entire issue. NTSB played a particularly important role in highlighting the dangers of airbags to children and placing constructive pressure on it's sister agency, NHTSA, to take more serious steps to address the flaws in current airbag designs and change the unsafe behavior of motorists (Wald, 1996). In the summer of 1997, NTSB issued a major recommendation calling for states to pass new laws requiring children under the age of 13 to be seated in the rear seat if a seat is available.

An undercurrent of these activities was an attempt to assign blame for the unexpected side effects of airbags. Automakers have pointed the finger at government's compliance test that forces highly aggressive airbag designs (American Automobile Manufacturers Association, 1996), while consumer advocates have pointed a finger at automakers for failing to install "smarter" airbag designs that would provide superior occupant-crash protection (Claybrook, 1996). Many simply blame parents and drivers for irresponsible behavior (Wald, 1996).

Absence of Public Outrage

Despite the flurry of precautionary activities by stakeholders, there are no indications that a widespread public protest of airbag technology is brewing. To elicit public knowledge and attitudes, the Harvard Center for Risk Analysis (HCRA) commissioned Market Facts Inc. of Chicago, Illinois to ask 1,000 Americans a series of 15 questions about airbags (Nelson, Sussman, & Graham, 1999). The sample of respondents, obtained by random digit dial procedures, with three callbacks, is broadly representative of the U.S. population, although the views of "hard-to-reach" respondents are unknown. Respondents were interviewed over the weekend of February 28–March 2, 1997.

Perhaps most revealing was the public's response to the question: "Do you favor or oppose the current law requiring all new vehicles to be equipped with dual-front airbags?" A hefty majority of respondents favored the current mandate (66.3% versus 29.4% in opposition). Those who felt *strongly* about their opinion tended to favor the airbag mandate (39.8% favor versus 14.7% oppose strongly). There was at least majority support for the airbag mandate in all major population subgroups: both sexes, all regions of the country, all education and income levels, and households with and without children in the home.

There was an indication that the recent events have caused some respondents to have a less favorable attitude toward airbags. Each respondent was asked: "Compared to three years ago, would you say your attitude towards airbags is more favorable now, less favorable now, or about the same as it was three years ago?" Despite the recent barrage of negative publicity about airbags, 54.0% of respondents reported the same attitude now, 26.0% reported a less favorable attitude, and 18.8% reported a more favorable attitude. The difference between 26.0% and 18.8% does not appear large but it is statistically significant by conventional measures. Some population subgroups are particularly likely to report less favorable attitudes toward airbags. These subgroups include women (31.5% less favorable versus 17.0% more favorable) and respondents from the western region of the United States (32.3% less favorable now versus 13.6% more favorable now).

About 29.0% of respondents report that, when they purchase their next vehicle, they are likely to request that their airbags be disconnected by their dealer or mechanic (assuming it is lawful to do so). The subgroups most likely to want to disconnect their airbags are women (34.6%), nonwhites (35.2%), the elderly (42.1%), those making less than $15,000 per year (46.9%), those living in the West (35.2%), and those with a high school education or less (35.5%). If a manual cut-off switch for airbags were installed in their next vehicle, about a third (33.0%) of respondents can imagine trips where they would be inclined to turn off the airbag system. An open-ended follow-up question found that the most common reasons stated for wanting to operate the cut-off switch were "child or baby in front seat" (42.2%) and "short trips" (15.4%).

Despite these clear indications of concern about airbags among a significant minority of respondents, it is important to recognize that the level of opposition to an airbag mandate is no greater today than it was fifteen or twenty years ago. In 1977 and again in 1984, Gallup asked national samples of Americans to respond to attitude questions about mandatory airbags that were worded similarly to the one we employed above. We can discern no downward trend in support for mandatory airbags compared to the levels of support recorded in 1977 and 1984.

Why So Little Outrage

Airbags pose a paradox of danger without outrage. Three possible explanations for the airbag's Teflon image are considered.

PUBLIC IGNORANCE

The absence of outrage about airbags might be attributed to public ignorance of the facts. In its simplistic form, this explanation is not very plausible (Healey & O'Donnell, 1996). Recent surveys by various organizations indicate that Americans are aware that the airbag has caused harm to occupants, both children and adults, in various situations (NHTSA, 1996a, 1996c; Public Opinion Strategies, 1996).

We posed the following true-false statement to respondents: "If a driver is seated too close to the steering wheel, the airbag can cause serious injury or death to a driver in a crash?" The correct answer, true, was supplied by 71.0% of respondents. A hefty majority of every population subgroup supplied the correct answer.

With regards to infants, we posed the following (false) statement with a wording designed to induce approval: "Airbags are not a danger to an infant in the front seat if the infant is restrained in an approved, rear-facing child restraint device." Twenty-six percent (26.3%) of respondents answered "true," with males getting it wrong more frequently than females (33.2% versus 19.5%). Other subgroups likely to supply the incorrect answer were nonwhites (31.0%) and those working full-time (29.3%). Those with children in the house gave the incorrect answer 24.8% of the time. Yet a majority of Americans know about the airbag's danger to infants.

On more subtle matters, there is widespread ignorance about airbags. In particular, people do not seem to be aware of a variety of unsettling facts that might make them less favorable towards current airbag designs. Perhaps this phenomenon should not be surprising, since only 6.9% of respondents have been in a crash where an airbag deployed or had a family member who has been involved in such a crash.

The following incorrect statement was cited as "true" by most respondents: "If a driver wears his or her seatbelt properly, the chance of being injured by an inflating airbag is minimal." A "true" response was supplied by 77.8% of respondents, including 79.9% of males, 90.2% of 18-24 year olds, 84.5% of Midwesterners, 82.2% of nonwhites, and 81.1% of those with postgraduate education. Yet belt users are experiencing a significant rate of airbag-related injury to their hands, wrists, arms, and faces. These injuries are often minor but the rate of moderate and serious "upper extremity" injury has been much greater than experts anticipated. NHTSA believes that upper-extremity injuries from airbags may actually be more common among belted than unbelted occupants, since the occupant's arms may flail about while the upper torso is restrained by the shoulder belt (NHTSA, 1996b).

Respondents are aware that airbags save more lives than they kill, but their understanding of the uncertain impact of airbags and the incidence of injuries is limited. True or false: "Airbags cause at least as many injuries to drivers as they prevent." The percentage of respondents who answered "false" was 67.3%, again with every population subgroup exhibiting the same modal response. Here the actuarial data is not entirely clear, but it appears that the most frequent type of injury in car crashes, the minor bruise or abrasion (scored as a 1 on the 0-6 point Abbreviated Injury Scale), is at least as likely to be caused by airbags as to be prevented by airbags. For injuries of moderate severity or greater (AIS 2+), airbags appear to prevent more injuries to belted and unbelted occupants than they cause, but the differences are not statistically significant. When injuries are subdivided by body region, it appears that airbags are reducing the incidence of moderate and serious head injuries to occupants while increasing the frequency of moderate and serious upper-extremity injuries (NHTSA, 1996d).

Perhaps the most interesting public misperception is suggested by responses to the following statement: "The lives of more children have been saved by airbags than have been killed by airbags." The percentage of respondents who said "true" was 59.9%, with this answer more common among males (66.4%), those aged 35-45 (65.5%), those with incomes over $50,000 (69.8%), and those with a post-graduate education (68.4%).

If "children" are defined as anybody under the age of 18 or 21, then the statement is true and there is no misperception. However, if (as we intended and suspect) respondents were thinking of younger children (under age 10 or 12), the statement is false, based on the available actuarial evidence. The children killed by airbags have been primarily kids who were either not restrained or who were restrained improperly when the crash occurred. Whether airbags have resulted in a net lifesaving effect among young children who were properly restrained is not known. Future surveys of the public should include more precise questioning on the perceived effects of airbags on children of various ages.

We also replicated a finding from a previous survey that most Americans do not know what impact speed is sufficient to trigger airbag deployment. U.S. manufacturers have set sensors to trigger airbags in crashes with forces equivalent to a 12-mph impact into a fixed barrier. Early sensors were designed with a "guarantee no deploy" around 9 mph and a "must deploy" around 15 mph. In contrast to these design values, less than a quarter of respondents supplied a threshold value between 9 and 15 mph. About 37.2% of respondents supplied a number greater than 20 mph, with 11.4% of respondents volunteering a "don't know" response. Public ignorance about this matter means that people are in a poor position to articulate one of the most interesting technical objections to current airbag designs: that the sensors are allowing too many deployments, particularly those involving belted occupants in low-speed impacts.

People are aware that lives are being saved by airbags but they tend to disbelieve the following actuarial statement: "A majority of the lives that have been saved by airbags have been among people who were not wearing seatbelts." Overall, 51.3% say this statement is false, 38.7% true, and 10.0% don't know. The incorrect answer is supplied by a disproportionate share of respondents with incomes greater than $50,000 (61.4%) or a post-graduate education (61.9%).

Current airbag designs are less effective for unbelted occupants than experts predicted years ago (NHTSA, 1984). Yet the best available estimates of the airbag's lifesaving effectiveness remains larger for unbelted than belted occupants: 13% reduction in fatality risk for the unbelted compared to 9% for the belted (NHTSA, 1996b). In serious crashes, the ratio of unbelted to belted occupants is still about 1 to 1 (in the United States), and thus the majority of lives that are being saved by airbags are in the unbelted population.

In summary, the public is aware that airbags can be dangerous, and thus the absence of outrage cannot be attributed to a simple "ignorance of danger" hypothesis. However, the public is ignorant of a variety of unsettling facts that, if widely known, might contribute to public concern or even outrage. Future surveys should test public reaction to the following plausible hypotheses now being examined by the technical community: airbags kill more young children than they save; airbags cause a significant rate of injury to belted occupants; the risk-benefit ratio for airbags is unfavorable—at least for belted occupants—at low and moderate impact speeds; a majority of the lives saved by airbags have been among unbelted motorists; and the net effectiveness of current airbag designs in reducing moderate and serious injuries may be close to zero. Unsettling information about the effects of airbags on short people, the elderly, and pregnant women also needs to be tested for public reaction.

PERSONAL OPTIMISM AND IRRESPONSIBLE BEHAVIORS

Another possible explanation for continued public support of airbags is that people believe that "it won't happen to me," an optimism bias that has been noted in research into public attitudes towards floods and earthquakes (Kunreuther et al., 1978). Early psychological research in the field of traffic safety found that most drivers believe that they drive more safely than the typical driver. The failure of even educated adults to wear safety belts was seen as a tendency of people to treat low-probability events as if they were zero-probability events (Fischhoff, 1977; Slovic, Fischhoff, & Lichtenstein, 1978).

If people don't perceive crash-injury risks as significant, then it is difficult to understand why they would have a favorable view of a regulation that compels them to purchase an injury mitigation device. Yet people may support an airbag mandate because they believe other people need to be protected (family or friends), even though the respondent considers himself or herself as one who is unlikely to need airbag protection (e.g., because of his or her superior driving performance). Those who support airbags may also believe that teenagers and immature adults need to be protected. Alternatively, people may see airbags as a source of protection for themselves against reckless and drunken drivers who too frequently travel the same roads as they do.

Prior to conducting a national opinion survey, HCRA sponsored several focus groups on airbags with "soccer moms" and other middle-class adults living on the West Coast. Several participants were quite vocal about the fact that the dangers of airbags were attributable to irresponsible behaviors: placing infants and toddlers in the front seat; failing to restrain children properly; and the failure of adult drivers to wear safety belts. The basic rationalization is that the technology is good but some of the human behaviors are bad. If this sentiment is widespread, it certainly provides a plausible explanation for the airbag's Teflon image.

Consistent with this theory is our finding that people don't believe that the majority of lives saved by airbags have been among the unbelted population. People are also unaware of (or perhaps resisting the notion) that airbags can injure drivers who are properly belted. Even if people recognize that airbags can injure properly belted occupants, they may rationalize this unfavorable prospect with the idea that the dangers from the crash itself would normally or always be greater than the dangers from the airbag. Several of our focus-group participants expressed confidence that any injuries caused by the airbags would be dwarfed in severity by the crash-related injuries that are prevented by the airbag. The available technical data on nonfatal injuries do not warrant such high levels of confidence in the airbag's injury-mitigation performance (NHTSA, 1996d).

It is difficult for this theory to explain the persistence of favorable attitudes towards airbags in the face of widespread publicity of cases where children were killed or seriously injured by airbags, even though—at least reportedly—they were properly belted (AP, 1996b). For example, one article in *USA Today* cites the research of Martin Eichelberger of Washington's Children's Medical Center as follows: "His studies found that several children 6 and younger received serious head injuries from passenger bags, even though the kids were buckled up or in child seats" (O'Donnell & Healey, 1996).

The notion that airbags are only dangerous to irresponsible people will continue to be challenged in the near future as new data is released on the potential adverse effects of airbags on belted people, short drivers, pregnant women, the elderly, and people involved in low-speed crashes (Ottaway, 1996). Particularly unsettling may be information that airbag designs in Australia and parts of Europe have features that are better tailored to the protection of belted occupants than are the current airbag designs found in the United States. These features include a higher deployment threshold (near 20 mph into a fixed barrier), a smaller and less aggressive bag, and different folding patterns to mitigate bag-induced injury.

ON-SCREEN BENEFITS JUSTIFY PERCEIVED RISKS

An alternative and more complex explanation for the high degree of public support for airbags would be a perception that any risks that they create are acceptable in light of their substantial lifesaving benefits. In other words, people allow perceptions of benefit as well as risk to shape their viewpoint of the acceptability of a technology. Such a rationalist perspective seems implausible to some social scientists but is beginning to demand a new hearing.

Howard Margolis of the University of Chicago has argued that public reaction to many technological dangers can best be understood as following a danger-opportunity calculus. If a technology's danger is "on-screen" but its benefits are "off-screen," there will be public demands for a "better safe than sorry" approach to government policy. This is in fact the public response to nuclear power and agricultural pesticides, technologies that have imaginable dangers yet benefits that are difficult for ordinary people to fathom. If a technology's benefits and dangers are both "on-screen," then Margolis' prediction is that people will approach the issue with an attitude of "fungibility" (Margolis, 1996). This attitude is far more tolerant of danger than the "better safe than sorry" attitude. Fungibility is characterized by a tendency to trust qualified experts in the proper balancing of risk and benefit.

The continued trust in the medical establishment may have its roots in this fungibility attitude (i.e., doctors will help us make sound risk-benefit decisions about complex medical technologies). There is also compatible literature in the psychology of risk perception indicating that benefit perception attenuates perceived risk, such as in the case of the ionizing radiation from medical X-rays (Gregory & Mendelsohn, 1993; Alhakami & Slovic, 1994). Even the original psychometric paper by Slovic, Fischhoff, et al. (1979) on perceived risk documented the importance of perceived benefit in public reaction to technological hazards.

The benefits of airbags are "on-screen" for ordinary Americans, both literally and in the sense suggested by Margolis. Television portrayals of airbag deployment create an image of a soft, cuddly, billowing bag that insulates the motorist from harsh and sinister crash forces. Even the disturbing deaths of children are frequently misperceived by people as a consequence of inadvertent suffocation rather than severe brain or spinal cord injury.

In response to the statement, "The lives of more female drivers have been saved by airbags than have been killed by airbags," 68.4% supplied the correct answer that the statement is true. The correct answer was supplied by a hefty majority of all population subgroups, though somewhat fewer correct responses were found among women (62.7%), drivers over age 65 (62.2%), those in lower-income households (53.6%), nonwhites (59.3%), and those with a high school education or less (62.9%). As noted above, people also believe (incorrectly) that children are better off due to airbags.

If Margolis' theory is right, it may be difficult to mount sufficient public concern about the dangers of airbags to take the major steps that are needed to prevent them. The widespread public perception of airbag benefits may dampen public perception of danger. The real risks of the technology, while preventable, may draw too little concern to motivate effective action by government, industry, and ordinary citizens.

Insufficient Stigma?

In order to minimize the risks of airbags while retaining their considerable benefits, major changes in human behavior and technology may be required. It is difficult to imagine that these changes will occur unless the public becomes more concerned about the risks of airbags.

Proper safety belt use among adults is even more important now that vehicles are equipped with airbags. Current airbag systems are not as effective as originally anticipated in mitigating the crash-related injuries experienced by unbelted occupants, and the unbelted occupant faces elevated risk of deployment-induced injury due to being out of position when the bag deploys. Increasing the rate of adult safety belt use is not easy; it requires substantial police enforcement in a social context where the public will support the activities of police. The American people are not yet committed to this approach.

For children the challenge is to reduce the rate at which children ride in the front seat while also increasing their rates of proper restraint use (whether it be a child safety seat for small children or lap and shoulder belt use for older children). Again, changing the behaviors of parents and children is not easy and will require considerable public resources and support from opinion leaders, clinicians, educators, and elected officials.

Modifying technology to suppress airbag deployment when children are detected in the front seat is a promising idea that will require some technological refinements. Yet the technological fix is not realistic for the 27 million vehicles with airbags already on the road that have a lifetime of 10 to 20 years. It is also not feasible for at least the next two or three model years, when another 30-40 million vehicles will be sold to the American public with passenger-side airbags. The behavioral strategies are thus critical but will require substantial public support and courage by elected officials.

Tailoring the design of airbags to the safety needs of the belted occupant will also require substantial public interest in the issue. The optimal airbag for a belted occupant may be smaller and less aggressive (perhaps using variable inflation rates according to crash circumstances), deployed at a higher threshold, and folded differently than the airbag system now being sold to the U.S. public under government regulation. Some manufacturers serving Australia and Northern Europe (where belt-use rates exceed 80%), have shown sensitivity to the needs of the belted driver in airbag design. Interestingly, airbags are not mandatory in most developed countries. In fact, consumers in many of these countries have not yet made an unequivocal commitment to passenger-side airbags. In Sweden, for example, rear-facing child restraints are promoted for use in the front seat and passenger-side airbags are discouraged or recommended to be disabled during the child's early years (Weber, 1995). Serious questions have also been raised in Europe and Australia about whether a passenger-side airbag is worth its cost, since belt-use rates are high and there is no steering wheel to harm the belted occupant's head and face.

Conclusion

Quite unexpectedly, automobile airbags have emerged as a new technological hazard in the United States. Since 1989, airbags have been responsible for over 100 deaths and thousands of cases of nonfatal injury to the hands, wrists, arms, chests, and faces of motorists. Airbags have saved many more lives than they have cost: a ratio of roughly 75 to 1 on the driver side and 5 or 10 to 1 on the passenger side. Yet risk-perception specialists have argued that quantitative risk information, particularly in this actuarial form, is relatively unimportant in public reactions to hazards.

Allegedly hazardous technologies often carry a stigma that is disproportionate to the actuarial danger they present. Nuclear energy and agricultural chemicals come to mind as salient examples to many risk analysts. Airbags, popular as they are today, do not fit this profile. Favorability attitudes could change as people learn more of the subtle facts about airbags. Yet airbags seem to have a Teflon image, even though psychometric models of risk perception predict that people should be outraged about the unexpected dangers of this highly-touted "vaccine."

The proper explanation for the airbag's Teflon image requires further examination. The perceived benefits of the device may induce a high degree of tolerance of the technology's well-publicized risks. Alternatively, it may be that people don't have enough information about the risks of airbags, conveyed in a sufficiently compelling fashion, to engender significant concern. More likely perhaps is the possibility that people know about the risks of airbags but believe that these risks only happen to others (the "personal optimism" bias). The relative explanatory power of these hypotheses will require much more in-depth research into what people do and do not know about airbags and how they sense their own degree of benefit and risk from this technology.

Without heightened public concern about airbags, it is difficult to imagine how the side effects of airbags will be minimized while the benefits of the technology are retained and enhanced. More lives may be lost unnecessarily if this technology does not become somewhat more stigmatized in the minds of the public.

Acknowledgments

The author thanks Howard Kunreuther, Michael Finkelstein, Tobien Nelson, Maria Segui-Gomez, and Kimberly Thompson for helpful comments on an earlier draft of this paper.

Appendix 15.A
Qualitative Attributes Associated with Perceived Risk

DREAD
1. Fatal versus unfatal
2. Global versus local impact
3. Involuntary versus voluntary
4. Uncontrollable versus controllable
5. Unfair versus fair
6. Catastrophic versus unclustered victims
7. Future versus current generations
8. Increasing versus decreasing
9. Not easily reduced versus easily reduced

FAMILIARITY
1. Observable versus not observable
2. Known to exposed versus unknown to exposed
3. Immediate effect versus delayed effect
4. Old risk versus new risk
5. Known to science versus unknown to science

AGGRAVATING CIRCUMSTANCES
1. Manmade versus natural
2. Children at risk versus adults at risk
3. Untrustworthy institutions versus trustworthy institutions
4. Media focus versus media neglect
5. Identifiable victims versus statistical victims
6. Suppressed data versus accessible data
7. Amplifies well versus doesn't amplify well

Sources: Slovic (1987); Covello (1991).

Appendix 15.B Author's Scoring of Five Technological Hazards on the 21 Factors in Appendix 15.A

DREAD	Nuclear	Pesticides	Smoking	Powerline EMF's	Airbags
1.	Y	Y	Y	Y	Y
2.	Y	Y	N	N	N
3.	Y	Y	N	Y	Y
4.	Y	Y	N	Y	Y
5.	Y	Y	N	Y	Y
6.	Y	N	N	N	N
7.	Y	Y	N	N(?)	Y
8.	Y	Y	N	Y	Y
9.	Y	Y	N	Y	N(?)
FAMILIARITY					
1.	N	N	N	N	N(?)
2.	N	N	Y	N	N(?)
3.	N	N	N	N	Y(?)
4.	N	N	Y	N	N
5.	N	N	Y	N	Y(?)
AGGRAVATORS					
1.	Y	Y	Y	Y	Y
2.	Y	Y	Y	Y	Y
3.	Y	Y	Y	Y	Y
4.	Y	Y	Y	Y	Y
5.	Y	N	N	N	Y
6.	Y	Y	Y	N(?)	Y
7.	Y	Y	Y	Y	Y
TOTAL SCORE (S = D + A – F)	16	14	4	11	11 (9 – 14)

Note: See Appendix 15.A for words associated with each of the factors.

Commentary

Evaluating airbags against these factors is not entirely straightforward. Dread-9 refers to whether the risk is not easily reduced. The risks to children can be reduced but parental behavior is hard to change. The risks to short drivers are much more difficult to reduce, particularly in the short run. Familiarity-1 refers to whether the hazard is observable, which it certainly is for people who experience an airbag-induced injury, though most people have never witnessed such an event. Regarding familiarity-2, airbag hazards are not really known to respondents in any meaningful way, though they have certainly heard about them in the same way that people have heard about the adverse effects of pesticides. Familiarity-3 refers to the immediacy of the effect. For airbags, the effect of the hazard occurs right after the deployment but well after the vehicle (and airbag) is purchased. The adverse effects can persist well after the injury but there is pain from the trauma at time of the event. Familiarity-5 refers to whether the hazard is known to science. The physical mechanisms of an airbag's adverse effects are fairly well characterized, although the dimensions of the problem for short drivers are not well characterized by science. Given these uncertainties, a person might score airbags anywhere from 9 to 14 depending upon how these interpretative issues are resolved. The author recognizes that Slovic et al. originally asked respondents to rate each hazard on a 7-point scale of riskiness, and then to rank each hazard on each qualitative dimension using a 7-point scale. It would be interesting to see such an exercise carried out for airbag hazards.

16 Dioxins, or Chemical Stigmata[1]

William Leiss

"[Dioxin is] by far the most toxic compound known to mankind."
–Dianne Courtney, U. S. EPA official, 1974

"Dioxins are the deadliest chemicals known to guinea pigs."
–Clyde Hertzman, Professor of Epidemiology, University of British Columbia

"We've always believed dioxin is a killer."
–Sheila Copps, Minister of the Environment (Canada), 1994

"We have targeted the mega-uglies—this includes dioxin and PCBs—for virtual elimination."
–Sergio Marchi, Minister of the Environment (Canada), 1996[2]

Introduction: Stigma and Cachet

Stigma (pl. stigmata) is the Latin form of a word originating in ancient Greek which means a mark or tattoo, especially a mark of opprobrium. The stigma is the external mark which enables a community to recognize instantly the presence of individuals whose situation differs from what is normal and socially desirable. Equally as important as the element of instantaneous recognition is the notion that the mark itself is only the external manifestation of an "inner" state of being (criminality, social deviance, slavehood, etc.). The strongly negative connotations, which predominate in usage, are challenged only by a part of the Christian tradition in which the wounds of Christ, suffered on the Cross, are signs of an inner state of holiness when they manifest themselves on a person.[3]

In fact, the meaning of stigma is best understood through semiology: A stigma is a sign, and signs have a double aspect, as signifier and signified. The

---

[1]The details of the dioxin case study in this article are excerpted from the much longer version found in chapter 3 of Powell and Leiss (1997).

[2]David Burnham (1974); Clyde Hertzman, personal communication, July 1995; *Kitchener-Waterloo Record* (1994); and Environment Canada (1996).

[3]See, for example, the reproduction of Tiepolo's painting, "St. Francis Receiving the Stigmata," in *The New Yorker* (1997, p. 57).

signifier is the physical aspect which refers to and releases—when it is comprehended—the intended meaning that otherwise lies hidden. Finally, in its dominant (negative) set of connotations stigma is the opposite of *cachet*, which is a mark of approval or prestige. One turns the notion of stigma on its head, so to speak, with that of cachet: For example, whereas smoking now carries a stigma in many social circles (a fairly recent and still-developing phenomenon), among many young persons who decide to take up cigarettes, smoking has a cachet that may be a factor in a person's willingness to tempt fate in the face of some well-described risk factors. A similar process may be at work in automobile driving or, indeed, in many activities where familiarity and positive social connotations lead individuals to underestimate risks or offset risk information with other considerations.

Stigma as an Indicator of Risk Communication Failure

A risk-information vacuum originates in a systematic failure of effective risk communication. This represents a failure on the part of a risk manager to make a sufficiently determined effort to assist interested parties, with respect to a risk issue, to cross the divide separating expert assessment from publicly-perceived risk.[4]

A risk-information vacuum for dioxins was allowed to develop, starting in the early 1970s, and then—beginning in 1987—to be filled by the perspectives of a single interested party (Greenpeace), whose views went virtually unchallenged in the public domain itself (in so far as *effective* communication is concerned). As a result, dioxin was "put into play" as a risk issue.[5] The quoted phrase comes from the financial sector and refers to the situation in which a corporation's publicly-traded stock is made the subject of competing bids in order to secure control of the company. The analogy here is control over the positioning of a risk issue in the public's mind within a hierarchy of priorities, or "agenda-setting." The desired outcome is to evoke a demand in the name of the public to "do something" about the issue, in such a way that politicians and their civil servants will find it impossible to ignore, or at least will find that lack of resistance to this demand is the easiest course of action.[6]

It is possible to claim that dioxin became a stigmatized substance at the very first moment when a broad public audience learned of its existence—in the reporting of Dianne Courtney's characterization of it as "by far the most toxic compound known to mankind." Courtney's phrase was transmuted into

[4]Powell and Leiss (1997), chapter 2, "A Diagnostic for Risk Communication Failures."

[5]Usage varies between the singular and the plural form, with the latter preferred because dioxins are a "family" of 75 related chemical compounds, which differ markedly in their toxicity profiles. Collectively these 75 are referred to as polychlorinated dibenzodioxins (PCDDs). They are closely related to another family of 135 compounds, the polychlorinated dibenzofurans (PCDFs), and so one will often see the phrase "dioxins and furans," or the collective abbreviation PCDD/Fs. The singular form "dioxin" usually refers to the most toxic member of the PCDD/F family, 2,3,7,8-TCDD (tetrachlorodibenzo-*p*-dioxin), sometimes just TCDD.

[6]In September 1994 representatives of a number of Canadian environmental groups asked the federal Minister of the Environment, Sheila Copps, to "declare a dioxin emergency in Canada" (*Kitchener-Waterloo Record*, 1994). It is impossible to say what such a declaration might have meant in terms of public policy or environmental regulation. There has been no further mention of this request since that time.

many others in the ensuing years, notably "the deadliest poison ever made by man," "more poisonous than cyanide or strychnine," "so deadly that 1/200 of a drop will kill a human," phrases that conjure up images from detective fiction of a substance deliberately made to cause harm in minuscule doses being administered to innocent victims by cunning criminals.

This overly dramatic phraseology shows an essential part of the stigmatizing process, namely the effect it has of creating a short-cut for human understanding: We have at least a vague idea of what strychnine is and what it does to people, and now we know that dioxin is "like" strychnine. But the short-cut is of course most detrimental to our appreciation of what dioxins are, how and why they are generated in the first place, and then end up in our environment, what effects they may have on different species, and what it is prudent to do about all this. But such an appreciation in the public mind could arise—in view of the enormous complexity in the scientific research on dioxins—only if a reasonable effort had been made to convey that complexity from scientists to the public in understandable terms. On the whole this has not happened, not for the entire period exceeding twenty years in which dioxin has been a risk issue in North America, and *it is still not happening now*, despite the fact that controversy over dioxin risk was re-ignited six years ago, when "hormone mimicking" was first reported in the press.[7]

The stigmata that dioxins bear are signs representing the sum total of all the potentially hazardous and health-threatening effects attributable in public understanding to the products of a fecund industrial chemistry. In other words, dioxin "stands for" the many-layered *and perfectly understandable* confusions, contradictions, ambiguous responses, and misapprehensions in the public's attitudes towards chemicals generally (Gregory, Slovic, & Flynn, 1996, p. 217). For the process of stigmatizing something like a chemical takes root and flourishes in the vacuum created by the absence of other information that is sufficiently clear and compelling to make a difference in the public mind. My working hypothesis is that this stigmatizing process itself, as well as most (but not all) of the confusions in the public's understanding of chemicals, is rooted in the pervasive risk-information vacuum.

A Concise History of the Dioxin Risk Controversy.

The general public first heard the word "dioxins" used in the context of the war in Vietnam, in 1972, in connection with Agent Orange and with the beginning of allegations from veterans that they were experiencing a range of severe adverse health effects as a result of coming into contact with both the herbicides and the dioxin contaminants in Agent Orange. About two years later the story of Times Beach, Missouri, received widespread media coverage, with the death of both domestic and wild animals and serious injury to children added to the elements previously mentioned. Thus in a very short period of two years dioxins had accumulated more intensely negative associations than many other chemical substances do over the course of a century.[8]

---

[7]Luoma (1990) discusses Ah receptor binding.

[8]A rich background discussion of the Agent Orange and Times Beach episodes will be found in Wildavsky (1995), chapter 3.

The initial *New York Times* story in 1974 about Times Beach offered the first mention in the press of the acute lethality of dioxins to guinea pigs (0.6pg/kg/bw)—or, as it was expressed more clearly two years later in the context of reporting on Seveso: "A dose of less than a billionth of a gram is fatal to guinea pigs." But the very first story at all—the report from Vietnam in September 1972—had initiated the blending of references between animal study results and fears of human health effects from exposure to dioxins, as well as the parallel conflation of items associated with 2,4,5-T, on the one hand, and with dioxins on the other. Thus that first story referred to animal experiments as showing effects ranging from "liver injury, to chromosomal changes, to embryonic tumors called teratoma, and to cancer," and then immediately moved on to allegations of unusual numbers of cases of primary hepatoma, or liver cancer, in human patients.

The first story on Times Beach (August 1974) dealt with human injury in Missouri, suspicions of birth defects among the Vietnamese, and the animal study results (birth defects, acute lethality). The first story on the Seveso explosion also mentioned human illnesses, animal deaths, birth defects in Vietnam, and the guinea pig acute lethality number. This theme continued in *New York Times* coverage throughout the decade; the June 1980 article of the important National Cancer Institute dioxin studies (animal cancer bioassays) discussed the lab animal findings in connection with birth defects in Vietnam as well as alleged cases of "cancer, loss of sex drive, personality changes and inexplicable weaknesses the limbs" among U. S. Vietnam war vets (*The New York Times*, 1972, 1974, 1976, 1980).

Although there are not a lot of press stories about dioxin in this period, by 1976 the phrases "one of the most toxic chemicals known" and "one of the deadliest chemicals known" were firmly entrenched in journalistic usage. What was missing, however, was a way of making these expressions more meaningful to the ordinary understanding of citizens; this lack was repaired by the common device of analogy or comparison. The first one had been supplied early on, by the Harvard University biologist Matthew Meselson, who had gone to Vietnam in 1973 to test for dioxin residues in fish. Meselson remarked then that dioxin was "a nasty poison that is 100 times more toxic than the deadliest nerve gas" (quoted in Lyons, 1973). The graphic notion of dioxin as a "poison" resurfaced five years later and then stuck fast, again prompted by specific events, in this case the discovery of large quantities of buried toxic waste in Love Canal and Hyde Park in northern New York State. Referring to the trichlorophenols in the waste, a long article in the *Times* published in late 1978 noted:

> [O]ne of the byproducts is dioxin, a poison as virulent as botulism or shellfish toxin. A millionth of a gram is said to be potent enough to kill a rabbit, and 3 ounces, minced small enough into New York City's water supply, could wipe out the city. Experts calculate that the 200 tons of chlorophenol in the Love Canal may contain about 130 pounds of dioxin. (McNeil, 1978)

In addition, both of the initial context-setting themes returned again and again: The Agent Orange story kept resurfacing as Vietnam veterans pressed their case for recognition and compensation for what they believed to be their herbicide-caused health problems, and the U. S. EPA kept trying to restrict more and more the uses of 2,4,5-T for herbicide applications.

In a lengthy 1979 *Times* article on allegations of serious adverse health effects attributed to both the wartime and domestic agricultural uses of phenoxy herbicides, Matthew Meselson was quoted again: Dioxin is "the most powerful small molecule known and it is now beginning to appear that it is the most powerful carcinogen known. Nobody argues about the toxicity of this poison" (Severo, 1979). Over the next three years the following characterizations appear in articles in the *Times*: "a poison 100,000 times more deadly than cyanide," "dioxin, a highly poisonous chemical," the "virulent poison and carcinogen dioxin," the "deadly poison dioxin" (AP, 1980; Cerra, 1980; "Inmates," 1981; Biddle, 1982).

The powerful and emotive "poison" characterization was picked up by the environmental reporters for Canada's *The Globe and Mail* in late 1980, in the very first story ever published on dioxin in this newspaper, and was recycled faithfully by them over the next two years. In addition to finding the rubric "the deadliest man-made chemical" in that first article, readers were told that dioxin is "so deadly 1/200 of a drop will kill a human." The reference to possible or probable human deaths was of course always latent in the poison analogy, but the expression "1/200 of a drop" made it much more vivid. The reporters (Jock Ferguson and Michael Keating) were very fond of it, and it is found in every one of the five articles that they wrote for the *Globe* between November 29 and December 10, 1980 and in a number of their other pieces published in the first half of 1981. By the time an editorial on dioxin, entitled "Slack rules for a poison," appeared in August 1981, it is clear that the phrase had attained the status of conventional wisdom: "Lest we forget, about 1/200 of a drop will kill a human being" (Ferguson & Keating, 1980a, 1980b, 1980c, 1980d, 1980e; "Slack rules," 1981).

The poison analogy predominated. Dioxin is "the most poisonous substance made by man" or "the deadliest poison ever made by man. It is more toxic than cyanide or strychnine." Another article expanded the analogy in an informative way—TCDD is "far more toxic than curare, strychnine or cyanide though less toxic than botulism, tetanus or diphtheria"—although one cannot know whether readers derived any reassurance from dioxin's relative placement on *this* list of horrors (*The Globe and Mail*, 1980, 1981a, 1981b, 1981c, 1981d, 1982a, 1982b, 1982c, 1982d)! Coverage of dioxins in *The Globe and Mail* faded thereafter, but it would be hard to imagine that this barrage of consistent and dramatic terminology, dominating a dozen articles appearing over a period of about 18 months (representing the initial spate of coverage in this newspaper), would not have made some lasting impression on the Canadian public.

Dioxin as a Stigmatized Substance

Dioxin risk as a public issue has escalated in direct proportion to the falling level of both directly evident human effects from dioxin exposure and falling levels of total dioxins burden in the environment. Is there a good explanation for this paradox? I think that it may be found in the multi-dimensional character of dioxin risk perception, including certain unique aspects of the way in which the issue originated and developed. Although these unique aspects play a role in the place of dioxins on the risk issue agenda, other aspects are generic: In those other aspects dioxin became the chief bearer of the stigmata of industrial chemistry as such, eliciting in many persons a set of ambiguous reactions to the interplay and inseparability of risks and benefits and the dependence of our lifestyle on modern technologies.

The dimensions of the dioxins risk issue may be summarized as follows:

1. *Agent Orange:* Its attachment to this controversy, which among other things played out over an extended period of time (about 1965-1985) because of the long-drawn-out legal battles, kept dioxins in the public eye.

2. *Pesticides*: Its connection with endemic controversies over pesticides use generally, and a very contentious substance (2,4,5-T) in particular, had the effect in newspaper accounts of weaving together the reports of alleged adverse human health affects of 2,4,5-T with those of dioxins.

3. *Dramatic Incidents:* This is the series of high-profile accidents or unexpected discoveries at Times Beach, Seveso, Love Canal, and Hyde Park.

4. *Dramatic Language:* A unique stereotyping in language ("most toxic," etc.) did not discriminate between the intrinsic hazard of the most toxic dioxins and the relative human exposures to them.

5. *The Risk-Information Vacuum:* An enormous scientific research effort, yielding results (such as the variations in observed effects among species) the meaning of which are inherently difficult for non-specialists to comprehend, was not matched by any reasonable effort to communicate credibly, in a publicly understandable language, the ever-changing and sometimes inconsistent interpretations of that research.

6. *A Dramatic Turn in the Science:* After twenty years of being framed primarily in terms of cancer risk (and where some later analysis had begun to lower estimated human cancer risk), the emphasis in public awareness shifted in 1990 to an entirely new dimension, first called "hormone mimicking," and to molecular-level biological events, which virtually started the dioxin risk controversy all over again.

7. *A Capable New Actor:* Greenpeace's leap into the dioxins issues in 1987 brought to the ongoing controversy an organization capable of reaching broad audiences, and attracting media attention, with both high-profile "actions" and effective technical analysis in its publications.

Only the sum total of all these dimensions, and not any one or any subset of them, can explain the paradox represented by the persistence of our society's dioxin risk controversy.

Moreover, the specific temporal sequencing of these dimensions contributes in no small measure to the intractable nature of the controversy. The initial framing of the issue in the context of chemical warfare, an unpopular war, and aggrieved war veterans cast it in exceptionally bitter terms which in turn inevitably seeped into all subsequent aspects. Second, the extreme toxicity of 2,3,7,8-TCDD in animal tests was itself newsworthy, but in the coverage of the series of dramatic incidents in the 1970s the newspaper accounts often conflated the discussions of experimental animal effects with both actual and alleged human health effects, blurring the otherwise important distinctions between the two types of findings. The animal results had shown some remarkable effects at comparatively low doses, but by the mid- to late-1980s the relevance of those effects for human impacts was widely questioned by scientists; it was at that point that the new turn in the science, focusing on the molecular-level events in the receptors within cells, became publicly known. In these and other respects, therefore, not only the multiple dimensions themselves but also the sequence of particular events within them had a part to play in the outcome.

What exactly is the nature of the risk-information vacuum for dioxins? So far as governments are concerned, they appear to believe that, in cases (like dioxins) where they throw huge amounts of resources at scientific research and risk assessment programs, the *meaning* of the results from these efforts will somehow be diffused serendipitously throughout the public mind. In Canada, so far as one can tell, those in government who are in charge of environmental and health protection programs simply do not believe that constructing an effective risk communication dialogue with the public is a part of their responsibilities. It may be only a slight exaggeration to consider the one-liner on dioxin attributed to a senior official in a 1982 newspaper article—"It's a nasty material to have around"—as the only statement on dioxin ever issued by Health Canada to the Canadian public in a clearly-understandable language throughout the entire history of the risk controversy over dioxins.

In the late 1980s the Canadian Council of Resource and Environment Ministers issued a ten-page pamphlet summarizing the scientific database on dioxins and furans, including sources and exposures, and listing some government policies for the same (Canadian Council of Ministers of the Environment, 1993). While this document is perhaps better than nothing at all, it is extremely tentative in its conclusions and has the usual emphasis for government publications on the need for more research. It even complains about how expensive a proposition it is to test for dioxins! For further information the reader is referred to a list of technical scientific publications issued by various Canadian government agencies since 1981.

In the United States, EPA devotes so much energy to battling with both industry and environmentalists over regulatory agendas, both in and out of courts, that there does not seem to be much left over for public communication. Given the evident complexity and changing character of the scientific research program results for dioxins, on the one hand, and the corresponding lack of appropriate public communication on the other, just how an informed understanding of the risk assessment and management issues is supposed to take root and to influence attitudes among the citizenry is one of the great mysteries of our time.

Industry's share in creating and maintaining the risk-information vacuum for dioxins was different in nature but not less weighty. Here again the issue framing was critical, because the decade of high-profile legal battles with the Vietnam vets over Agent Orange set the tone for much of the subsequent period. This meant that industry would not say publicly what was said privately and that the public messages would be largely ones of denial and stonewalling. Moreover, it was an era when industries relied heavily on public-relations firms for "issue management" and the crafting of elaborate strategies for dealing with the public, politicians, and the press. In retrospect, had industry taken but a small fraction of the moneys it paid out to public relations (PR) firms over the years, and spent it instead on finding a way to create a credible risk communication dialogue with the public, the dioxin issue might not be today as intractable as it still is.

The fact that this issue remains in play after so long is, therefore, the outcome of both a specific chronology of events and the persistence of a substantial information vacuum. This means, among other things, that the issue was driven by a number of accidental features, including the skilled interventions by Greenpeace. But another one of those accidental features is that, even as environmental dioxin levels have been dropping, our analytical capability for measuring these levels has been increasing in sensitivity by leaps and bounds. To some extent, therefore, so long as it remains in play the dioxin issue will be carried along simply by our ability to measure dioxin at progressively lower levels in the environment, regardless of whether or not toxic effects are found at those levels.

All of which makes the most recent Canadian federal government actions pointless and indeed self-defeating. As noted at the outset, the Minister of the Environment, upon tabling the new *Canadian Environmental Protection Act* in the House of Commons in December 1996, labeled dioxin and PCBs "mega-uglies" and said that they and similar substances were destined for "virtual elimination" through resolute action by his government. The labeling was bad enough, but what is worse, the draft legislation actually defines virtual elimination in terms of no detectable level.[9] Since it is certain that dioxins *will* be found in the environment at progressively lower levels, what the Minister and his officials have done is to guarantee that the dioxin risk issue will remain in play in Canada for many years to come.

Discussion

In a number of publications Gregory, Flynn, and Slovic have applied the concept of stigma to environmental risk issues, and especially to the aspects of those issues that pertain to the public perception of risk.[10]As used there stigma appears to mean the attribution of a higher risk to a technology, place, or event than is otherwise warranted by evaluations based on well-founded expert risk assessments. However, it is not clear what Gregory, Slovic and Flynn think that the concept of stigma adds to what they already know about the public perception of risk: In other words, is it just a generic designation for the set of characteristics that have been described on the basis of the survey work? Second, it is unclear whether or not the stigmatizing process is (sometimes or always) an indication that "the public" is *making mistakes* in the evaluation of various risk factors, when perceptions are arrayed against expert opinion. Third, the illustrative application of the concept of stigma to specific risk controversies has been to date done in a rather loose way, which leaves the meaning of the interpretive framework unclear. In my opinion this is certainly

[9]The actual wording is as follows: Virtual elimination "means, in respect of a substance released into the environment as a result of human activity, the ultimate reduction of the quantity or concentration of the substance in the release below any measurable quantity or concentration that is at or approaching the level of quantification,..." *Canadian Environmental Protection Act*, tabled in the House of Commons on December 10, 1996, section 64. Since dioxins are already classified as "toxic" under this Act, the virtual elimination requirement pertains to these compounds. See generally Leiss (1996).

[10]See especially Gregory et al. (1995, 1996).

the case for the references to dioxins,[11] Alar,[12] genetic engineering,[13] and EMF.[14]

For anyone who affirms the truth of the doctrine "there is no such thing as real risk," as I do, the state of perceived risk for any issue is something which demands of the analyst an honest effort at understanding. Second, it *can*—depending on the circumstances—demand a risk communication effort, because individuals *may* inadvertently act in ways that are not in their own best interests.[15] In the limited set of cases to which this stricture applies, the concept of stigma may be a useful addition to our thinking about social controversies over risk management. In this setting stigma represents an "excess" or "surplus" of negative factors associated with perceived risk, representing a qualitative rupture in a continuum of concern among various groups of citizens.

Let us look at the dioxin case briefly from this angle. Dioxin (i.e., TCDD) is chemical substance with some remarkably hazardous properties (and no known useful ones, so tradeoffs are not at issue). There are undoubtedly some dioxins that arise from natural causes, and they may be presumed to have been ubiquitous in the environment at low levels for as long as humans have existed. It is also highly likely (although not a certainty) that the more toxic dioxins have been created largely as a result of modern industrial processes. Almost as soon

---

[11]"Dioxin is another product that is a candidate for stigmatization; west-coast pulp producers have been criticized for many years regarding low-level releases of dioxins to coastal waters,..." (Gregory et al., 1996, p. 218) What exactly is the point being made here? The British Columbia shellfish grounds around coastal pulp mills were closed by Health Canada in 1990—as a direct result of a fuss created by Greenpeace—after Health Canada determined that its acceptable daily intake guidelines for dioxins—which are 1700 times *higher* than EPA's—could be exceeded by those, especially members of native bands in the vicinity, who made a steady diet of the shellfish. They were reopened in 1996 after the dioxins levels in pulp mill effluent had been reduced dramatically. It is also arguably the case that elevated dioxins levels in the late 1980s were the cause of the dramatic reproductive failures in the blue heron colonies in those areas.

[12]"A well-known case of risk-based product stigmatization occurred in the spring of 1989, when millions of consumers stopped buying apples and apple products due to their fears that the chemical Alar (used then as a growth regulator by apple growers) could cause cancer" (Gregory et al., 1996, p. 217). But EPA's risk analyses had shown that it *could* (possibly) cause cancer; those analyses were flawed, but they were not inherently implausible. The public panic was based on a brilliantly conceived publicity plan to accompany the release of the NRDC's study, *Intolerable Risk*, which followed a major segment on CBS's "60 Minutes." The NRDC had commissioned its own (very conservative) risk estimate from a reputable academic scientist. A full account of the Alar episode shows that the public response was not unreasonable, given the information base in the public domain: see chapter 6 in Leiss and Chociolko (1994).

[13]"Even 'natural' technologies, such as bioremediation..., have been limited in their use by public perceptions of linkages to genetic engineering and impending environmental catastrophe" (Leiss & Chociolko, 1994, p. 217). In Canada, at any rate, the public has been given almost no useful information about genetic engineering technologies, and a massive information vacuum is developing (which Greenpeace has started to fill); this is fully documented, for the case of environmental risks associated with applications of plant biotechnology, in Powell and Leiss (1997), chapter 7.

[14]"Power-frequency electric transmission and distribution lines are stigmatized in many places because of health fears associated with exposure to electromagnetic fields" (Leiss & Chociolko, 1994, p. 217). The huge scientific uncertainties are a reasonable explanation for this phenomenon: see Leiss and Chociolko (1994), chapters 4 and 5.

[15]This is not true in all cases where perceived risk and expertly-assessed risk diverge. And I concede the point that, at least so far, risk communication has not been shown to make very much of a difference in addressing this divergence.

as this chemistry became known, efforts got under way to control dioxin creation in industry as well as human exposures to it. Even the relatively high human exposures, due to industrial accidents, that are well-described cannot be said definitively to be linked with potentially fatal health effects for humans (the same is not true for some other animals). But in any case there is a solid consensus that dioxins exposures should be kept to very low levels, as they are in many places today. When exposures are regulated to very low levels, it is very unlikely that dioxin is or will be a significant risk factor for human health.[16]

None of what is related just above justifies the excessive language about dioxins that has been illustrated earlier in this essay. Framed in the public mind as "mega-uglies," dioxins (and PCBs) carry a special burden as risk issues—in other words, a stigma—that is not justified by *any* reasonable reading of the risk assessment record.[17] This special burden is the "excess" perceived risk. How it arose is not a mystery: It is the outcome of a long process of the "framing" of dioxins in the public mind through the episodic statements of *all* those in the following list: regulators, scientists, politicians, journalists, and environmental groups. The first fault here is in a twenty-five-year tradition of rhetorical excess, that is, the type of language used by some representatives of all those strata in seeking to explain the nature of dioxins and dioxin risk to the public. But there is a second fault as well: Those responsible for the societal management of dioxin risk allowed a massive risk-information vacuum to develop, and it is in the "empty space" created by this vacuum that the terminology of rhetorical excess becomes firmly rooted.[18] In biological terms, there was nothing else provided that could compete with that terminology. In an otherwise empty space filled with this rhetorical excess there is a social amplification of risk, in the sense that dioxin risk attains a level of priority among known human health risks that is out of all proportion to the plain evidence before us.

As Kunreuther and Slovic have suggested, chemical risks in general carry a well-described burden of stigma in our society. And in a very interesting observation, they suggest that the rise of scientific risk assessment approaches themselves in the last quarter-century accentuate this stigma (Kunreuther & Slovic, 2000). As indicated earlier here in the case study materials, dioxin risk provides an excellent example of this phenomenon: Many accounts reaching the public thoroughly mixed up the laboratory animal studies with alleged human health effects, so much so that it would have taken a most dedicated reader to sort out and evaluate the different kinds of references. In addition, the quite exceptional results of the animal studies for acute lethality and carcinogenicity of dioxin fixed this substance in the public mind as something to be feared. Because of this perception of extreme hazard, dioxin has come to represent the hazardous properties of chemicals in general. It is in this sense that dioxin has borne the stigmata of chemical risk generally during the past quarter-century.

---

[16]Dioxin is an endocrine modulating substance, and it is thought that actual adverse health outcomes related to such effects *could* result from very low levels of exposure. We will not know for some time yet whether or not this is the case.

[17]For PCBs see the case study in Michael Edelstein's chapter in this volume.

[18]See the more complete dioxin case study in chapter 3 of Powell and Leiss (1997) for an account of the risk-information vacuum.

Conclusion: Stigma as a Concept in the Analysis of Risk Controversies

There are two different kinds of stigmata, and the difference between them may be fundamental to the evaluation of the usefulness of this concept in the analysis of risk controversies. Stigmata may be indicators of (a) a "natural" or essential character, that is, the very being of an entity, or (b) an accidental feature which is the result of special medical, behavioral, or "spiritual" circumstances.[19] Racial stereotyping is the best example of the first; the stigmas attached at various times to physical deformities and certain diseases, but also to homosexuality, divorce, or unwed motherhood, of the second. If the stigmatizing process is an essential characteristic of the public understanding of risk, which is likely to remain unaffected by education or preaching, then there is little one can do, if one is in a risky business that falls into this domain, except try to offload some of the costs through insurance. On the other hand, if this process is a response to specific behavioral influences—which is more likely to be the case—it may be possible to "manage" the type of stigma that arises in risk issues.

Managing stigma in this context means seeking to counteract only the "excess" or "surplus" of perceived risk, and not the underlying continuity of public opinion itself. (Risk perception studies reveal a highly consistent form of risk judgment over time.) The rationale and justification for seeking to manage stigma is that this excess is not *necessarily* in the best interest of the public.[20] Better risk communication could be a factor in managing stigma, but only if the risk communication processes themselves are conducted fairly by social actors who have earned the public's trust.

---

[19]This distinction does not appear in Goffman (1963).

[20]This is partly a matter of definition, and the general application of the concept of stigma to the domain of risk issues is yet too new for settled definitions to arise. For example, if stigma is just the "mark" or sign of perceived excess risk, it may be entirely justified by events, even when it has real consequences such as devaluation of property, as it seems to be in the cases of Rocky Flats and the proposed Nevada high-level nuclear waste repository, discussed elsewhere in this volume. On the other hand, the excess may be entirely or partly unjustified, as it is in the case of dioxin. Where the stigma appears to be unjustified, it is appropriate for various parties to seek to moderate or eliminate it, so long as they do so through communications and management processes that are otherwise perceived to be fair.

17 An Industry Perspective on Risk and Stigma

Richard K. Long

This is a story about one large, successful multinational company and its 20-year struggle with public controversy. It is also about a huge global industry that has had to deal with public fears of technology and the health effects those technologies might (or might not) cause.

Especially for the company, it's a story of distress, confusion, and, sometimes, arrogance as segments of the American public became more concerned about the environment and new laws were enacted to force changes in industry practices that had been in place for many years.

The story has a relatively positive ending, though each day brings new challenges to both the company and its industry as they grapple with public expectations and potential crises. That the company finally understood the consequences of endless controversy and the importance of addressing public values is, in itself, evidence of management resolve to change and rank-and-file employee commitment to end two decades of turmoil. Those conflicts combined to make the organization a unique case of corporate stigmatization.

One hundred years ago, Herbert Henry Dow founded the company bearing his name on the banks of the Tittabawassee River in Midland, Michigan. For the next 70 years, The Dow Chemical Company was best-known by its customers, competitors, shareholders, and neighbors of its manufacturing operations.

In 1966 and 1967, the company saw signs that its traditional low public profile had ended. The reason was a small government contract involving only a dozen Dow employees who blended polystyrene, benzene, and gasoline into a military weapon known as *napalm* for use in Southeast Asia. Napalm had been invented and used in World War II, but through the turbulent Sixties and early Seventies, Dow became the focus of hundreds of protests on university campuses as company recruiters came seeking chemists and engineers. Hundreds of students were arrested for their efforts, and violence marred Dow visits to many of the nation's leading institutions. One placement center was set afire with a Dow recruiter barricaded *inside*. At another school, a Dow recruiter was "rescued" by chain-swinging members of Hell's Angels.

At the time, Dow's senior executives were, typically, World War II veterans who felt morally obligated to fulfill a government request for a weapon designed to protect American troops in Vietnam. Dow's response was well-intended and patriotic but also prickly at times. After several years, Dow and napalm became seen by many people as synonymous with an unpopular and immoral war, failed U.S. policy in Southeast Asia and indiscriminate killing of civilians.

The napalm experience alone could fill books about public controversy. Had this been the last of Dow's collisions with the public, it would have been quite enough for one company. Unfortunately it was far from the final skirmish.

While the early 1970s brought an end to U.S. involvement in the Vietnam war, social and political changes on the homefront created new challenges for Dow and its industry. During the Seventies, Congress created the EPA and the Occupational Safety and Health Administration and passed six significant environmental laws that directly impacted the chemical industry. Those six bills passed with an average of 89% "yea" votes in the House and 92% in the Senate, which suggests that Congress wasn't going to oppose pro-environment legislation. Unfortunately, Dow and its industry were often portrayed—correctly in most cases—as resisting these bills every step of the way.

Throughout the 1970s and the first half of the 1980s, Dow took on all comers in debates that sometimes had substance and other times could have been avoided with better results. Some of the clashes involved *perceived* risks and caused Dow to be seen as insensitive to public values. For example:

- In 1977, Jane Fonda, already well-known for opposing the Vietnam war, spoke at Central Michigan University, located 25 miles west of Midland. After her prepared remarks, Fonda took questions from the audience. When a student posed a largely rhetorical question about "imperialist companies like Dow Chemical," Fonda agreed with the questioner. The next day's newspapers reported that Fonda had "blasted" Dow. Paul Oreffice, later to become Dow's chief executive officer, promptly cut off all company support to the university on the grounds that Dow money would not be used to subsidize programs that bit the hands that fed them. Predictably the national news media had a good time with this one, portraying Dow and Oreffice as stifling free speech.

- Several months later, Dow learned that the EPA had contracted with an aerial photography service to take pictures of the manufacturing complex at Midland, ostensibly to look for violations of environmental rules. When some of these photos were seen posted on the wall of an EPA conference room in Chicago, Dow sued to protect its technology. In Dow's view, this was the same as industrial espionage, because competitors could learn much about Dow's manufacturing processes by examining the highly detailed photographs—especially when EPA did not protect their confidentiality. Dow claimed constitutional protections against unreasonable search and seizure, in that the photos were taken without a warrant. The case eventually went all the way to the U.S. Supreme Court, where Dow lost, but EPA also agreed to some changes in its tactics. Alas, the early stages of the debate were a bonanza for reporters and, even more so, for editorial cartoonists. One cartoon showed an anti-aircraft battery in a Dow plant, just waiting for an EPA photographer to fly over.

- Through an acquisition, Dow inherited highly visible litigation involving Bendectin, an anti-nausea drug that several million women had taken to alleviate the effects of morning sickness. A relative handful of these women had difficulties with their pregnancies or delivered children with birth defects. Though these experiences occurred no more frequently than in the population at large, soon there were numerous lawsuits, often with media coverage showing pictures of children with deformities. Some women who had taken Bendectin were worried about delayed effects, even if their children were born normally. Dow and its pharmaceutical subsidiary cited mountains of scientific data showing no unusual link between Bendectin and birth defects, but what the public saw was a big corporation pitted against small children and pregnant women.
- Beginning in 1979, another Vietnam-era government contract came back to haunt Dow. For several years, Dow and six other chemical companies had produced herbicides for military use in strategic defoliation of Vietnamese jungle. Agent Orange contained equal parts of two herbicides—2,4-D and 2,4,5-T—and the latter contained varying amounts of a complicated molecule known generically as "dioxin" and often called "the most toxic material known to man." Several years after the war ended, Vietnam veterans in the United States, Australia, and New Zealand began claiming medical ailments they thought resulted from exposure to Agent Orange. There was little debate about whether there were veterans who were ill or, in some cases, dying. Rather, the debate focused on three key issues. First, the chemical producers felt the levels of dioxin present in Agent Orange weren't harmful to humans. Second, were Vietnam veterans any sicker than comparable groups of men who had never been exposed or had never been to Vietnam? Third, even if there were delayed effects from Agent Orange exposure, who was the responsible party? While the chemical companies had produced the herbicides, it was the U.S. government that developed the specifications and also had total control over the formulation and application of the compounds. After five years of legal maneuvering and countless news stories—many of them highly anecdotal—a settlement was reached. Among the seven defendants, only Dow had been willing to debate the science of dioxin, which gave Dow the higher profile. Once again, Dow became synonymous with a possibly lethal chemical.
- The international activist group Greenpeace had several disputes with Dow. When Greenpeace's ship *Rainbow Warrior* was sunk in 1985 in Auckland, New Zealand, harbor, much finger-pointing followed. Ultimately a French intelligence agency was found responsible for the bombing, which killed a photographer aboard. Before blame was affixed, however, Dow had to endure the following questions from a wire service reporter in Los Angeles: Is Dow aware of the sinking of the *Rainbow Warrior*? (Yes) Is it true that Dow and Greenpeace have had numerous disagreements over dioxin? (Yes) Does Dow have any operations in New Zealand? (Yes) Did Dow have anything to do with the bombing of the Greenpeace ship? (Good Lord!).
- The same year, Dow had a much more embarrassing encounter with Greenpeace. Activists picketed the Midland complex for two weeks and,

eventually, several demonstrators were arrested for trespassing and other misdemeanors. Upon being booked, each was asked to submit to a blood test to be sure no communicable diseases were brought into the jail. One of the blood tests on a young woman came back with a positive reading for venereal disease. Someone at the jail told someone from Dow who told someone else, and soon Dow was back on the griddle for violating the woman's privacy. And someone had broken Michigan privacy laws by divulging the test results in the first place. A lawsuit followed, along with a barrage of caustic news stories and editorial comments. In the end, Dow made a highly visible public apology, settled the lawsuit and made some management changes.

- During the Reagan presidencies, there were many charges of sweetheart relationships between industry and federal regulatory agencies. This became a major thorn in Dow's side when it was revealed that the acting administrator of EPA had asked a Dow lobbyist to review a draft copy of an EPA report on dioxin sources in the Great Lakes region. As requested, Dow scientists reviewed the document and changes were suggested. When the matter leaked to the press, it was used as further evidence of the polluters controlling the regulators. Dow was soon summoned to a House subcommittee in Washington, D.C., to explain its involvement. The hearing, which had the tenor of an old-fashioned trip to the woodshed, gave House Democrats a new chance to lambaste the Reagan administration, question Dow's tactics, scold EPA officials for rolling over to industry and, in a particularly rhetorical moment, remind EPA that its job was to represent the American people, not to leave regulation to "the felons."
- A final example followed EPA's decision to *buy* the town of Times Beach, Missouri, after tests showed high level of pollutants, including various dioxins, in the soil. Within a few days, the issue shifted to Midland, where Dow was engulfed by reporters from all over North America. This feeding frenzy was supposedly driven by long-known soil tests showing trace amounts of dioxins in a portion of Dow's Michigan Division complex. Before this chapter ran its course, Dow had committed to a six-point, $3 million effort to seek independent review of analytical data, conduct new tests and study health statistics for Midland County to see if there were any abnormalities that might be linked to dioxin. Once again, Dow employees and management were annoyed, but now members of the community expressed concerns about the risks of living near the complex.

There are other examples, but these illustrate the kinds of predicaments that Dow kept encountering. However, it's also true that Dow was not the only company being afflicted with controversies over environmental or health issues. During the timeframe mentioned, the chemical industry had other high-profile issues that added to public concerns over reports of elevated risk of cancer, birth defects, and more. Love Canal, near Buffalo, New York, achieved international notoriety. Union Carbide's unfortunate experience at Bhopal, India, where some 4,000 people were killed by a methyl isocyanate release in December 1984, created a new perspective on industrial safety incidents, though there was ample reason to suspect the gas release resulted from sabotage. Whatever the cause, Union Carbide had a chemical release much closer to home, at Institute, West Virginia, just eight months later. Though there were no fatalities this time, the two incidents (and other industry accidents) prompted Congress to legislate much tougher reporting requirements for air emissions. If one includes some of the Superfund sites, such as Devil's Swamp in Louisiana and Stringfellow Acid Pits in California, the industry had more than a decade of highly public horror stories about chemical brews dumped carelessly for later generations to deal with.

A student of corporate behavior must wonder, "Why did these events keep landing at Dow's doorstep?" I was an active part of these cases, and I believe there were several causes:

1. Dow was led mostly by engineers and chemists, typically from midwestern and southwestern land-grant universities where conservative political views were more likely to hold sway. And their generation, which went to college in the 1940s and 1950s, wasn't well-prepared for the social and political changes that hit the United States—Vietnam, Kent State, assassinations of Martin Luther King and Robert Kennedy, Watergate, and other historic events. Most of Dow's leaders were pioneering sorts who had spent parts of their careers in isolated surroundings, building major chemical complexes out of Brazilian swamp or breaking into tough markets in Europe or Asia. They were proud to have put Dow on the global map in a highly competitive industry. And most of them felt they had succeeded without much help from governments.

2. Being engineers and chemists, they tended to rely on "objective scientific data" in decisions about the safety of a product, siting of a new plant or the value of a new technology. By itself, this is a commendable approach, but it tends to ignore the importance of evolving social agendas and the role of emotion (and politics) in the formulation of public policy. Further, within the Dow culture, reliance on "hard data" was often carried to the point of hubris. Too often, Dow tended to say "our science is right and your science is flawed." This disdain for non-Dow research caused the company to be portrayed as arrogant and, at times, in denial.

3. Much of Dow's discomfort was caused by its dealings with the news media. At a time when the company was getting 15,000 media calls each year, virtually *every* call was answered, often by a line manager or technical resource, rather than a PR person. Unfortunately, the company expected its candor to win the day, even if its officials were sometimes crusty or condescending toward reporters. It was a constant frustration that the opinions of Barry Commoner or Sam Epstein, viewed as quasi-scientific opportunists by many at Dow, carried equal or greater weight than those of industry experts. And Dow quickly tired of anecdotal stories about a woman who lived near a chemical plant and had a miscarriage. Or the 40-year-old man who once worked in a chemical plant and was now dying of cancer. In Dow's view, reporters never challenged self-diagnoses or addressed the reality that miscarriages or cancer occurred frequently in the population at large.

4. It took Dow a long time to understand how public opinions form. "Facts" were more important than "perceptions." It was difficult to believe that Dow's contributions to its employees and communities didn't outweigh the small risks involved in working or living near a production facility.

Given the focus of this conference, it's reasonable to ask if Dow was stigmatized by these encounters with public opinion. My opinion is an unequivocal "yes" up to the point where senior management recognized the need to change its stripes, in 1984 and beyond. With selected audiences, that stigma has endured, though not necessarily to Dow's material detriment.

By my definition of "stigma," Dow and its industry became pariahs to several important audiences—the liberal arts community in higher education, the news media and others. Here are some indicators:

- Certain words found their way into everyday use, often as pejoratives—cancerphobia, chemophobia, carcinogens. The very term "chemical" became, to many, synonymous with risk and death, rather than with innovation or life-extending research.
- In many cases, editorial cartoonists used "Dow" to represent the evil money-grubbing corporations harming the public. When "Dow" became a generic for the chemical industry, we knew more bad news was near.
- Institutional memories of "Times Beach," "Bhopal," "Agent Orange" or "*Exxon Valdez*" became new descriptive terms, as in "we may have another 'Bhopal' on our hands."
- Employees had their own sad stories to tell about Dow's frequent brushes with controversy. Some quit wearing their service pin in a lapel because they tired of having a fellow traveler see the Dow Diamond and begin asking questions about the latest fight. One of Dow's bright young researchers told of going to his high school reunion near Philadelphia. When he returned to Michigan, he was still upset by the number of questions he got from old friends about Agent Orange and dioxin.
- The industry as a whole saw regular evidence of its standing with key audiences. In poll after poll, the chemical industry was next-to-last in favorability and credibility. Small wonder that we were often grateful there was a *tobacco* industry in the surveys.

If one accepts the premise that for at least 20 years, Dow became stigmatized in issues surrounding risk and public policy, it's fair to ask how this condition impacted the corporation. The answer, predictably, is mixed.

Here are some measures of a company's success and my personal assessment of how the frequent turmoil impacted Dow:

- **Financial Performance**—As a multinational company heavily invested in commodity chemicals and plastics, Dow was much more vulnerable to cyclical swings in the world's economies than to public policy spats. While there may have been cases of a customer refusing to do business with Dow on "moral grounds," I never heard about them. There was discussion in the mid-1980s about some customers getting satisfaction from watching Dow eat humble pie, but that didn't equate to lost business. And, the company's sales and profit performance over the two decades were consistent with other companies in the cyclical industry.

- **Stock Price**—There is no evidence that Wall Street paid much attention to the controversies. Except for the Agent Orange litigation, which could have produced a huge jury award in a worst-case scenario, the financial community stayed riveted on Dow's business results.
- **Stock Ownership**—Many of Dow's controversies took place before the growth of the so-called "social investing" movement. Still, there were no significant sales of Dow stock to protest a particular issue. Dow was, and is, owned primarily by large institutions such as pension funds or mutual funds. If anything, Dow was criticized at times by its large shareholders for making some socially-conscious decisions, such as its withdrawal from South Africa at the height of the anti-apartheid movement. Individual shareholders, including employees and retirees, may have felt anguish about the battering Dow was receiving, but this seemed to have no impact on their decision to hold or sell shares.
- **New Employee Recruiting**—Given the turmoil Dow faced on college campuses for nearly a decade, one might think the company suffered in its search for new, young talent. If so, it cannot be proven. Dow recruited primarily chemists and engineers, and these proved unlikely participants in sit-ins or susceptible to an environmental group's rhetoric. If anything, Dow enjoyed *increased* success in its technical recruiting.
- **Employee Retention**—Dow, as with other corporations, tracked its employee turnover year-to-year, and there was never an indication that people were leaving to protest napalm production or dioxin issues. Some left for greener pastures or other reasons, but Dow's public persona seemingly had no influence in these decisions.
- **Employee Morale**—Employees, especially in Midland, experienced a full range of reactions. There was denial, indignation about attacks on Dow, pride in the company for taking tough stands on unpopular issues, embarrassment about the way Dow was portrayed in media coverage and, eventually, disgust over the endless controversy, which was so difficult to explain to neighbors over the backyard fence or with friends at social gatherings. Some employees were quietly satisfied as a few of Dow's more combative senior executives were buffeted publicly, but the middle management ranks were anxious to stop the distractions.
- **Community Attitudes**—Many of Dow's larger manufacturing complexes are located in what might reasonably be called "company towns." In each, Dow is a major employer, and the company's economic impact is abundantly clear. Dow has always tried to maintain positive relationships with its neighbors, and the philanthropic activities of Dow and its founding families have built libraries, performing arts centers, school facilities and other civic improvements. Dow employees served on school boards, city councils, county commissions, and other community ventures. Some critics claimed this was just a subtle way of reminding the locals they had better not cross Dow, but in my experience, the company bent over backwards to avoid even the appearance of dictating to

local elected officials. Opinion polls conducted regularly among key audiences showed that neighbors in plant communities gave Dow high marks for its contributions to local quality of life, environmental performance, and other measures. The one common negative among all audiences surveyed was Dow's immersion in controversy, which nobody saw as enhancing the company's reputation. Midland was described in a *Forbes* magazine article as "a hotbed of social rest," and residents there grew especially tired of Dow's frequent battles, as well as "outsiders" coming around to stir things up.

- One issue where Dow's steady conflict had a potentially significant impact was Agent Orange. When the many individual lawsuits were certified as a federal class action, Dow and the other defendants prepared for a lengthy jury trial. Any suggestion of a settlement, at least within Dow, was quickly dismissed. As the first trial date approached, Dow's outside attorneys began research to determine an "ideal" jury. This is a common practice in major litigation, and in this case there was a clear answer. Dow needed a jury of men who had served in World War II. (By coincidence, Dow's lead external attorney had won a Silver Star for heroism in Italy during that war.) In 1983, many World War II veterans felt something like this: "I fought in my war, and when I came home, I didn't whine about how tough it was." Any notion of delayed stress syndrome and other effects of Vietnam service fell on deaf ears. Over the next year, though, a significant shift occurred. There was continual publicity about how Vietnam vets were being ignored by the Veterans Administration, Congress, and the Reagan administration. The first Agent Orange trial was postponed and reassigned to a new federal court judge, so jury research began anew. This time, Dow found that World War II veterans had become much more sympathetic to the men and women who claimed they were harmed by exposure to Agent Orange. Without an "ideal" jury to seek, among other considerations, Dow and the other defendants settled out-of-court for $180 million. While the amount of the settlement may seem huge, Dow's relatively small share was an acceptable price to get Agent Orange behind them.

There is a final area of impact that I observed during 24 years with Dow. It had to do with the management distraction caused by the many disputes, especially those having to do with dioxin. At one point, Dow's Chief Executive Officer (CEO) lamented that he was spending more than one-fourth of his time on dioxin, Agent Orange and other conflicts. If accurate, and I suspect it was, this was time he could *not* devote to strategic planning, acquisitions, and the other complex tasks of a company leader.

Aside from distraction, there were management decisions driven more by stubbornness and denial than by good business sense. A case in point was Dow's long-running defense of 2,4,5-T herbicide. For several years, "T" gradually lost market share, mostly because of newer formulations being developed by the industry. If one agrees that business has a "law of diminishing returns," it's fair to ask why Dow didn't abandon 2,4,5-T much earlier than it finally did. The reason expressed frequently was that "if we quit making 2,4,5-T, the 'enviros' will come after the rest of our herbicides." Ironically, this rationale for defense of "T" was costing Dow money at the very time another "domino theory" was failing in Southeast Asia. At any rate, Dow stayed in the 2,4,5-T business long after the product ceased benefiting shareholders.

There were many key players in the Dow story. Highly visible environmental activists were thorns in the side of Dow executives. Usually they were discounted as having questionable credentials or being more interested in book contracts and expert witness fees than in finding scientific truth. In some cases, Dow was right, but each time this corporate giant attacked an opponent's credentials or motives, Dow came out seeming the bully.

Elected officials were important factors in many of the debates. Those who defended Dow and the chemical industry had the good sense to look at opinion polls and keep a low profile. Critics of Dow got most of the headlines and came across being "for" things, while the company always seemed to be "against" things the public valued. Dow's political sophistication eventually grew, but it had a long way to go. For years, the company paid little attention to liberals of any stripe; except for a few conservative Southern Democrats, Dow had basically a GOP-oriented lobbying effort at a time when the Democrats controlled both houses of Congress.

Oreffice was a central figure in many of the debates. Born in Venice, Italy, Oreffice and his parents emigrated to South America to escape the Mussolini regime. He arrived in the United States as a freshman engineering student at Purdue University and went on to have a tremendously successful career with Dow. Fluent in at least four languages, Oreffice was a truly multinational executive whose grasp of marketing and finance helped him progress steadily to the CEO's job. If Oreffice had a shortcoming, it was his distrust of almost anything governmental. To his credit, he later accepted the need for Dow to change.

Another important player was Bob Lundeen, who served as board chairman when Oreffice became CEO. Lundeen recognized that Dow's public fights were often counterproductive, and he took some early steps in reaching out to audiences that had been highly critical of Dow.

Keith McKennon's progression to the Dow board paralleled Oreffice's, and when Oreffice had a difficult problem to solve, it was often McKennon who got the assignment. To McKennon goes the credit for persuading Oreffice to commission an extensive study of Dow's stakeholder relationships and then convincing the CEO to endorse the recommendations that emerged.

This "turnaround"—at least with some of Dow's key audiences—resulted from substantial introspection. With McKennon's help, Oreffice sponsored an internal study group of a dozen Dow employees from varied backgrounds. While there was ample quantitative research to tell how Dow was viewed by some audiences, the group really wanted to engage the middle and upper-middle managers who could be catalysts for change.

The team, which I led, came to be known as the "Futures Group." We conducted 213 interviews, including all 18 members of Dow's board of directors, managers at various levels, a sampling of new employees and 38 non-Dow people whose insights were useful.

Several themes were repeated over and over by those we interviewed:

- Employees take criticism of Dow personally and expect top management to do something about it.
- Dow's reputation should be the responsibility of *line* managers and part of everyone's job description.

- Dow needs to be *for* things, not always *against*.
- We live in a world where perceptions are critical, and the news media are the key to forming (and changing) perceptions.
- Dow says, "There's no problem," when the better message is, "We'll fix it."
- We want to be viewed as progressive but don't want to act that way.

Each of the respondents was asked to describe the "ideal" Dow reputation, and these thoughts were expressed over and over in some form:

- Dow should be honest and tough—but extremely fair.
- The company should be prepared to fight as a last resort, but should be much more selective in choosing the fights.
- Dow should be known as a "good works" company.
- We must approach critics and patiently try to understand *their* agendas.

Six months after the first team meeting, I presented our findings and recommendations to Dow's board of directors. It occurred to me that morning that, with eight children under the age of 16, this was not a good time to stumble. After very little discussion, Oreffice said he didn't like the messages but realized that Dow needed to go in a different direction.

What followed was a multi-pronged effort driven by the need for *outreach*—to employees, neighbors, journalists, traditional adversaries, *both* parties in Congress, academia, and shareholders. The results were gratifying over the next several years, but the most rewarding change occurred when line managers embraced the concept of a more cooperative and thoughtful company. All they needed was a signal from the top that it was OK to seek dialogue with Dow's critics. This response by production, research and staff people made Dow's efforts much more than a "PR program." It became part of a true culture adjustment.

Though it was met with some initial skepticism by Dow's technical ranks, one of the most impactful elements in Dow's change was a national advertising program carrying the theme, "Dow lets you do great things." In the early stages, technical people wanted the advertising to explain the relevance of highly complex subjects such as "a part-per-billion of substance X in a plant's wastewater." Instead, the advertising used an oblique approach that was disarming to critics while setting a new, softer tone for the company.

Dow learned many useful lessons from its two decades of conflict. And, by most accounts, the lessons seem to have stuck, because Dow has become much more interested in finding solutions and treating problems, not just diagnosing them. Among the key lessons learned are:

- Arrogance attracts new antagonists via lawsuits, onerous legislation, and regulation, aggressive media scrutiny and disdain from the intellectual community. (It still amazes me when I meet a tenured university professor, now well into his or her fifties, who says, "Oh, yes, I took part in a Dow protest at ...") Just as the school bully eventually finds someone to knock the chip off his shoulder, so do corporations learn that they can be better focused by avoiding needless fights.

- Discounting public fears of technology or specific products is insulting to the worried parties. Where risks are involved, the corporation must help provide thoughtful perspective, not a blanket plea to "trust us."
- When *perceptions* and *facts* do not mesh, it is easy for a company to think that with a better communications effort, the affected audiences will come to "understand." The important distinction for a company is to know whether it has a *communications* problem or a *behavior* problem that no amount of wordsmithing will change.
- The news media love to bring down the proud and haughty. Only by demonstrating a good-faith effort to address public concerns can a company hope to avoid prolonged negative news coverage.
- One formula for repairing a corporate reputation starts with evidence of contrition, followed by visible and sincere good works and demonstrating bipartisan sensitivity to diverse opinions.
- If I were to challenge any one of you to a game of one-on-one basketball, but you never got to touch the ball, I would probably do pretty well, bad knees and all. Dow found that always being on defense was uncomfortable, unproductive and unlikely to change momentum. Worse, it led to management behavior that was described—even internally—as the "Bunker" or "Fortress" Mentality. Predictably, this sort of management isolation and paranoia does not serve the interests of a company or its leadership.

The Dow Chemical Company of the 1990s is different in many ways from its former contentious self. The company has won numerous awards for environmental excellence. There is greater willingness to discuss diverse points of view with external parties. Clearly the company has quit being "the heavy" in each new public policy dispute.

Dow is not without challenges, notably the silicone breast implant issue which has already forced its 50%-owned subsidiary, Dow Corning, to seek protection under federal bankruptcy laws. Pending judicial decisions will determine whether or not the assets of the parent companies (Dow and Corning Inc.) will be vulnerable to breast implant lawsuits or potential settlements. Though faced with an issue of great magnitude, it seems significant that the debate continues without Dow exhibiting shrillness or arrogance. Perhaps this demonstrates that the company learned its lessons well.

Risk, Media, and Stigma

This section begins with a study of images in the media presentation of risk events. Ferreira, Boholm, and Lofstedt observe that the images of hazards are pervasive in television and the news media, often presenting a different emphasis than the text or audio commentary. The authors argue that stigma is primarily created and represented by a visual mark, that photographs and visual presentations have a primary capability of creating negative imagery and thereby forming stigmatization. Flynn, Slovic, and Mertz describe the failed risk-communication program intended to obtain support for a radioactive waste facility in an area where the project was highly controversial. Flynn, Peters, Mertz, and Slovic describe how stigma results from newspaper accounts of the operations and problems at a nuclear weapons facility and subsequently effects residential property values.

18 From Vision to Catastrophe: A Risk Event in Search of Images

Celio Ferreira, Åsa Boholm,
and Ragnar Löfstedt

Much of people's knowledge about the world is gathered from the reading of visual images in television and newsprint media. This is especially so in the case of "risk events," from epidemics to ecological disasters, creating a heightened awareness of the fragility of life-systems in face of different kinds of hazards. Photographs of smokestacks, oil smeared seabirds, fish floating belly up, traffic jams, and dying forests have become integral parts of environmental imagery. The search for blame and responsibility is always present in news media coverage, sometimes leading to complex chains of causation which are visually recognizable.

Despite the increasing emphasis placed on images as a privileged medium of communication, the nature of the view which people form from this overflow of risk images and the uses they put to it is not a well researched area. This neglect is especially noticeable when it comes to the role of news photography in the print media in conveying risk messages (see Boholm, 1998). The reasons for researchers' lack of interest in the visual aspects of communication of risks may be several, from difficulties associated with reproducing pictures in research journals to copyright restrictions. The neglect of risk related images may also stem from a firmly rooted tradition of regarding pictures as mere illustrations of the surrounding text, an assumption implicitly based on the picture's ability to "authenticate" the written account by its iconic properties.

Risk images, however, do not always attune with the intentions of the surrounding text. In certain cases, images may even consistently induce interpretations of their own and should therefore be regarded as relatively autonomous of the text. This does not imply that the study of risk images belongs to the province of the visual per se, but that risk images are about the possible range of relationships enmeshed and encoded in the visual. This article suggests that most risk images operate at a metaphorical level with the accompanying propensity to create encompassing, self-contained messages from visual fragments. The construction of generality from specific instances of risk events gives images a stigmatizing potential which, it is argued, can hardly be matched by words. A brief case study of an environmental pollution event in the south of Sweden exemplifies the potential of images in the construction of stigma—the marking of people, communities, or artifacts with some undesirable qualities. The study focuses on the risk event as it was visually constructed by a evening tabloid newspaper and discusses some of the implications of visual imagery on the understanding of risk events.

Imagery and Understanding

Knowledge of any domain of human activity is not primarily organized in the sentential logical form characteristic of language. Information has to be processed and organized before it becomes knowledge, and strong evidence from cognitive psychology and communication research suggests that such knowledge, in order to be effective, has to be non-language like. Research that has been conducted into how well people understand and remember events indicates that there is generally a low level of recall of any given item of news and that people are more likely to remember the setting and people involved than what was actually said about the event. Highly visual and dramatic events, on the contrary, tend to prompt high levels of recall, but people tend instead to exaggerate the content of what is remembered. "Public exaggeration considerably exceeds anything for which the balance of media reporting can be held directly responsible" (Bell, 1991, p. 243). Most notably is that issues with high levels of recall connect particularly well with deeply embedded cultural imagery (Hansen, 1991; Patterson, 1989; Wilkins, 1987). While images must in some sense be familiar, they must also contain an element of distinctiveness and be dramatized in a symbolic way if they are to make a major impact (Hannigan, 1995). Though recall studies do not necessarily tell how well people understand a news item—understanding occurs to a greater or less depth—it seems that understanding is greatly enhanced by news stories that are visually strong (McQuail & Windahl, 1993). These research findings are congruent with the ideas which professional producers of news stories hold about the necessity of good pictures to make a good piece of news. Clearly, issues that attract media attention tend to be "mediagenic" and possess a powerful symbolic resonance (Anderson, 1997).

The study of stigma has recently come to the forefront in studies of risk perception (see, for instance Flynn et al., 1998; Slovic, Layman, Kraus, et al., 1991). The sociological notion of stigma, which began with the seminal work of Goffman (1963), has recently been extended to areas outside the social arena and might nowadays be applied to places, communities, artifacts, and technologies (Edelstein, 1987; Gregory et al., 1995; Slovic, Layman, Kraus, et al., 1991; Vyner, 1988). Research into stigma and "socially amplified hazards" (Kasperson et al., 1988; Kasperson, Jhaveri, & Kasperson, 2000) provides evidence of risk understanding as intuitive and non-probabilistic, focusing on discrete events rather than comparative risk assessments. These findings suggest that people create meaning structures of risk events, which can then be associated with, or carried metaphorically to, other risk events. The relationship between stigma and risk images is both intriguing and largely unexplored. It seems plausible, however, to suggest that stigma is a predominantly symbolic construct which carries much of its effectiveness through the use of images (Kunreuther & Slovic, 2000).

The power of images stems partly from their capacity to create fluid meaning structures that, under certain circumstances, can become "autonomous" from the event which generated them. When images are disengaged from the specific event they depict, they also dissociate themselves from the particular facts or controversies specific to the original event. In this way, images can turn into symbols, condensing the structural outlines of the event in a generalized form. Such images will become "portable" in the sense that they can become symbols for a certain class of events—e.g. close ups of agonizing oil smeared sea birds are symbols of sea pollution generally rather than a representation of a specific instance of pollution. Such images are often called stereotypes and wear out with use, but as long as they are effective they make the point well—the image itself, as an object, has become independent of the eventual textual environment. The structural similarities with stigma lie in the process of recognition and classification. If a certain class of photographs has generally been constructed as symbols of, for instance, sea pollution by means of their emotional force and wide media exposure, these photographs literally become the only visible "mark" of stigma. Stigmatizing pictures will, more or less intentionally, be used as templates in subsequent pictorial descriptions of other, similar events, thus not only identifying but also implicitly "marking" new events as instances of older classifications. We might expect an on-going process of photographic renewal—new pictures will be taken—but it will remain "stereotypical" in intention. The number of scenarios that may be construed is not infinite, and the themes themselves are easily recognizable in the symbolic imagery of life forms like birth, illness, suffering, decay, and death. It is this power of recognition, which gives pictures the ability to recall events long passed and to communicate beyond words.

From Vision to Catastrophe

Tunnel construction sites are not usually associated with environmental hazards and seldom, if ever, do they alter the esthetical characteristics of the landscape they traverse. From the point of view of the news media, a tunnel is neither controversial nor conspicuous and often lacks the mediagenic features that characterize a good story, if there is to be one. When the construction of a huge tunnel in the south of Sweden went wrong and adverse effects on the environment became known,[1] the media coverage of the event was intense,

[1]The proposed railway tunnel through the horst at Hallandsas was a key part of Banverket's plan to modernize the west coast trunk line. The two 8.6 km long tunnels were seen as vital to reduce the problems faced by freight and passenger trains as they reached the ridge. As the ridge is rather steep (200 meters tall), the train engines could drag no more than 17-18 freight wagons at a time causing delays for the freight transport in the region as well as delays for passenger traffic. These delays were exacerbated during the fall and winters when the trains were struggling up the ridge sliding and slipping on the fallen leaves, snow and ice. It was envisaged that the tunnels would save the passenger trains a minimum of 10 minutes during normal operations and much more during adverse conditions. Hallandsas is an extremely water rich ridge, however. As the tunnel was being built large amounts of water began to leak from the tunnel. The water leakage also caused severe problems for people living in the area. In some places groundwater levels fell up to 100 meters, causing many wells in the area to dry up. Skanska, the entrepreneur, was also naturally concerned about the water leakage and during 1996 it had a laboratory test 80

despite the lack of good photographic opportunities. Media interest in the tunnel construction as a risk project was triggered by a local newspaper's report that several cows, which had been grazing in the vicinity of the projecting area, fell ill and fish in a nearby stream died. Suspicion was raised that some type of toxic substance was leaking from the construction site, a suspicion soon confirmed by the discovery of acrylamide in the ground and surface waters outside the tunnel. Acrylamide is a component of Rhoca-Gil, the material used to seal the tunnel from leaking water. The region was classified as a high-risk area and produce such as vegetables, milk, and meat were banned. Cattle who had drunk water from the contaminated wells and streams were slaughtered and burned. The environment "scandal" became even greater when it became clear that construction workers were suffering from acrylamide poisoning, as they did not have enough protection from the toxic substance. In the local community, 6,000 people walked from the city center to the opening of the tunnel expressing their outrage that something could go so wrong, and farmers were worried about their future (Boholm, Löfstedt, & Strandberg, 1988; Löfstedt & Boholm, 1999).

A daily tabloid newspaper, *Expressen,* ran a 12-page supplement (October 23, 1997) explicitly positioned as the search for responsibilities. Headlined "They Kept Silent," the space devoted to pictures was roughly half of the supplement, the other half being texts organized chronologically—from vision to catastrophe." The supplement included photographs of persons quoted in the text, perhaps in order to personalize the different arguments brought forward, thus juxtaposing the text with the iconicity of the portrait. Documentary photographs highlighted the main events in the history of the tunnel construction and visualized the story told in the text. Though the event falls into the province of environmental risks where photographic templates are widely used, both the discrete nature of the building site itself and the absence of visually recognizable health hazards poses a challenge to pictorial description. The event lacks most of the dramatic visual power that other well-known risk events have been able to create and distribute to an international audience (for example, the Chernobyl nuclear disaster, see Boholm, 1998). Still, news stories need images even in the case of less spectacular events and the choice of pictures becomes more apparent when the focal event's mediagenic qualities are subdued and ready-made templates for photographic action are not easily available. It is some of these photographs that are described here at some length. A more detailed hyper-linked analysis of the pictures is to be found on the Internet at www.anthropologist.com.

---

different concrete/chemical sealant blends to see which one was the most suitable in halting the water flow. Of these 80 blends, Rhoca-Gil manufactured by the French company Rhone-Poulenc was seen as the most suitable.

By early October 1997, 1,400 tons of Rhoca-Gil was injected into the tunnel walls to halt the water leakage. In late September the acrylamide in the sealant began leaking out into the groundwater which caused considerable environmental damage. 310 wells were analyzed and traces of acrylamide were found in 29. A total of 196 people from 75 households underwent medical examination; 370 animals were slaughtered and the meat destroyed. Nine dairy farmers were not allowed to sell their milk and 330,000 liters were deemed unfit for human consumption and was dumped (Boholm et al., 1988; Löfstedt & Boholm, 1999; Tunnelkommissionen, 1998a, 1998b).

Risk Object: The Mixing of Categories

Basically, the pictorial description of a risk event involves three closely related components, namely, the culturally construed risk object or "source" of danger (Hilgartner, 1992), the victims and the environment in which the event takes place. We might also assume that a "successful" picture would simultaneously visualize all three components:

- The victims, people (or living things generally), whose existence, in one way or another, is disrupted.
- The environmental setting, natural or man-made, in which the event takes place.
- The source of danger, whether a substance or activity.

The distinction between these components of a risk event is heuristically useful for the purposes of sorting out and classifying photographs. Most important, however, is that such a classification offers the opportunity to see how these components are visually put together—how they are syntactically linked, contrasted or juxtaposed in specific pictures or sets of pictures. In many cases, the risk object is either immaterial or difficult to visualize. Its existence can only be inferred from the combination of other categories such as "victims" and "environments." The visual inference, naturally, does not follow the linguistic objectivity of cause and effect. Rather, it is achieved by the juxtaposition of imageries of environment and victims, imparting meaning to the risk object through more or less unconscious networks of associations.

In the case of the tunnel construction site, for instance, the written account unequivocally places the toxic substance acrylamide in the tightening material Rhoca-Gil as the risk object. The pictures, however, embody the visually immaterial substance in the imagery of poisoned water, initially inferred from the victims and environment categories. Meaning saturated and visually concrete, "poisoned water" is susceptible to prevail over written, factual information—the likely common-sense understanding is that poison is poison and water flows, outweighing any assessment of real hazards or the geographical boundaries of the "risk area." Although a host of other factors are at work here, visually retrieved messages operate on cultural values which are mostly taken for granted and are therefore both more generalizing and less open to argumentation than the texts which accompany them. In the tunnel project, a set of four pictures "introduces" the event (see Figures 18.1, 18.2, 18.3, and 18.4).

Of this set of photographs, two visualize the disrupting event (the photographs of the lying cow and the act of dumping of milk, Figures 18.1 and 18.3), a disruption accomplished by means of implicit contrast with the expected conventional imagery of dairy farm. The two other pictures (the family outside their house and the man squatting at the water stream, Figures 18.4 and 18.2) acquire information value by the positioning of the portrayed persons as both victims and storytellers, the snapshot character of the pictures accentuating the authenticity of the event. If the "breakdown" of the dairy farm imagery is accomplished by the photographs of the lying cow and the dumping of milk into the urine reservoir, the squatting man provides the necessary explanatory link between them. This picture, meaningless when viewed by itself,

Figure 18.1 Close up of a deformed, dead cow. The caption reads that the cow is poisoned and paralyzed (not dead).

Figure 18.2 A man squatting down at a water stream in the landscape. The caption reads that the (named) farmer had to slaughter his cows because they drank from the water.

is crucial for the creation of the visual message formed by the causal connections between the three photographs. The squatting storyteller pointing at the water stream "makes sense" of the other two pictures, explaining the apparent breakdown of the expected dairy farm imagery. The cow is "dead" and the milk is thrown away *because* the cow drank from the stream. The message is retrieved from contrast to the expected pasturing cows, water in the landscape, and the storing of milk for human consumption.

These humble photographs are, in the vocabulary of Roland Barthes, unary—devoid of "punctum," the detail that attracts or distresses. Investigating the nature of photographers and photographs in *Camera Lucida,* Barthes (1984) writes that the most widespread type of photograph is the unary, the banal, to which category most news photographs belong:

> News photographs are very often unary (without disturbance). In these images, no punctum: a certain shock—the literal can traumatize—but no disturbance; the photograph can "shout," not wound. These journal-

Figure 18.3 Milk in a tractor's scoop is dumped in a pond like, fenced reservoir. The caption reads that a (named) farmer dumps the milk in the urine reservoir because no dairy wants to buy it.

Figure 18.4 A (named) family posing by the drinking well outside their house. The caption reads that they do not dare to drink from their well. Although they live outside the contaminated area, they believe the water to be poisoned.

> istic photographs are received (all at once), perceived. I glance through them, I don't recall them; no detail (in some corner) ever interrupts my reading: I am interested in them (as I am interested in the world), I do not love them. (p. 41)

The set of photographs introducing the risk event, as most news photographs, certainly lacks "punctum." This is also one of the reasons why the joint reading of pictures and captions in this set seems to corroborate the usual notion that pictures are mere illustrations of texts, with an added dimension of authenticity, the "this is how it is." The photographs follow the text closely and the scope of interpretative possibilities left to the viewer is rather limited. This is expectable because the photographer seemingly records what he or she is told: "Here is the cow/ Here is the water/ Here I dump the milk/ Here is my family." The captions fill in the verbal narrative about cause and effect, damages, complaints, fear, and blame.

The pictures, like the texts, rely on the viewer's existing knowledge and imagery of the dairy farm as an integrated whole including people (the farmers), animals (the cows), the landscape (pasture), and produce (the milk). Herein lies the hidden dimension of news photography; the "proof," at a glance, that something is not how it should be. As Barthes puts it, "I glance through them, I don't recall them." The "glance," however, reveals the fundamental deviation from any supposed steady state—in this case the imagery of a dairy farm—and the "recall" should not be expected to refer to any particular photograph but to the simultaneous (all at once) reading of the photographs as a testimony of disorder. It is in the glance, where observation and interpretation are performed more or less simultaneously, that contradictions of order and disorder are revealed. As the anthropologist Marshall Sahlins (1976) observes in the analysis of clothing as a system of communication, "mere appearance" must be one of the most important forms of symbolic statement in Western civilizations:

> For it is by appearances that civilization turns the basic contradiction of its construction into a miracle of existence: a cohesive society of perfect strangers. But in the event, its cohesion depends on a coherence of specific kind: on the possibility of apprehending others, their social condition, and thereby their relation to oneself "on first glance." This dependence on seeing helps to explain, on one hand, why the symbolic dimensions have nevertheless not been obvious. The code works on an unconscious level, the conception built into perception itself.... On the other hand, this dependence on the glance suggests the presence in the economic and social life of a logic completely foreign to the conventional "rationality." (p. 203)

The kind of photographs that "witness" the collapse of a dairy farm imagery do not rely on notions of rationality—the weighing of alternatives derived from a "second glance"—but in the relation between how the imagery of order is culturally constructed and how its disruption is visualized. The immediacy of this relationship is in-built in the photographer's role as eyewitness recording the "truth" of order in its own contradiction, a story of disorder. The components of people, animals, landscape, and produce are cast in orderly relationships that together form the building blocks of dairy farm imagery. These relationships could be summarized, in a structuralistic manner, as: Cows are to Pasture as Farmers are to Milk, positing a dual imagery in which cows and landscape share the same context and infuse each other with meaning. Photographs of pasturing cows in "open" landscapes are long-standing stereotypes of Swedish countryside imagery. People (the farmers) and produce (the milk) pertain to another context colored by commercial associations—producing milk is the farmer's "livelihood" and source of income. These relationships are confirmed by their negation; the "farmer's family" and the "dumping of milk" pictures tell, *at a glance*, of the disruption of the orderly imagery of the farmer's livelihood (without which imagery the pictures would be meaningless).

Because we are dealing with a "risk event," our previous heuristic distinction between "victims," "environments," and "risk objects" would place the farmers as victims and the processing of milk for consumption (that is, its negation in the dumping photograph) as the (man-made) environment. In this context of "milk production," the risk object is firmly placed in the milk itself. In the context of "pasturing cows," the breakdown of the imagery was able to release messages of poisoned water as the risk object. We notice that it is in the interplay between these three heuristic categories that we are liable to find the distinctiveness of a pictorial description of risk events (in contrast to, for instance, a story run on a dairy farm improving its milk processing facilities). The basic building blocks of the dairy farm imagery and its breakdown points could be summarized as follows:

Victims	Environment	Risk object
Farmers	Milk processing	Milk
Cows	Landscape	Water

Rather than being reinforced by the pictorial description, the factual and intended risk object (acrylamide/Rhoca-Gil) has been metaphorically "transformed" into the much more powerful and encompassing imagery of poison, concretized in polluted drinking water. Without delving in the symbolic imagery of "poison" and "water," we observe that they belong to two incompatible categories in human thought and that their mixing raises issues of purposeful human agency (or issues of negligence). Like the mixing of water and poison, the mixing of milk and urine (generalized to food and excrements), plays upon one of the strongest "taboos" in every society. These pictures are in line with the claim that the crucial test of news worthiness is disorder and its most important metaphor is sight (Hartley, 1992). The eyewitness perspective and the implicit messages of order/disorder will always remain the bread and butter of news photography generally and risk events in particular. At the same time, the professional photographer will always attempt to use his or her cultural intuitions, aesthetic talent, and technical know-how to capture *the picture*—that photograph which is likely to trigger emotions and be remembered.

Metaphors and Stereotypes

News photographs depicting risk events are often visually arranged in order to form "pre-packaged" messages construed by means of available photographic devices and symbolic materials. These photographs are not aimed at a pictorial description of the imagery's breakdown in the first place. As prepackaged messages, they constitute a certain kind of "visual explanations," making creative use of metaphor as well as aesthetic and technical devices. These pictures are often recognizable in the combination of a powerful "human angle" achieved by means of deliberate composition and the imaginative use of light and, perhaps more infrequently, by the felicitous instant. Although, in a sense, all news pictures are subject to planning, manipulation and editorial selection, it is useful to distinguish between the "snapshot" whose primary ambition is to tell a story (of breakdown) and the pictures that we might call "metaphor." In a nutshell, the distinction does not refer to how the picture is read—a snapshot can also be read as metaphor and vice-versa—but to the photographer's "intentional" search for symbolic material.

Some of these photographs are subject to media "careers" in which they become trivialized by means of wide media exposure, and labeled as stereo-

Figure 18.5 The caption reads: "Celestially (above ground) beautiful."

types. Trivial and stereotype are often negatively charged terms, conjecturing the desensitization of the reader-viewer and the concomitant failure to stimulate critical engagement in issues. There is, however, the possibility to conceive the reader (the audience) as less "active" than it is commonly expected in audience research. If so, the force of the stereotype lies precisely in its aptness to evoke simplistic mental reactions derived from oversimplification and generalization of issues. Unlike, for instance, written accounts and works of art, the photographic stereotype does not need a conscious effort of interpretation. According to one scholar (Gardner, 1992),

> When visual messages produce stereotypes, they encourage the recipient of the message to judge the image through predetermined values instead of asking the viewer to judge the information accompanying the image in the context in which it is presented. In other words, the viewer uses the image as confirmation of internal beliefs rather than as information to refine or change beliefs. (p. 124)

Herein lies the stereotype's importance in the context of risk events—the Sahlins *glance* referred to previously. In the context of stigma formation and sustenance, the power of stereotypes resides in their propensity to create autonomous and fluid meaning structures which are widely used to identify (and possibly stigmatize) other structurally similar events. Stereotypes are pervasive in media mediated risk events, not least due to the difficulties inherent in visualizing concepts and immaterial sources of danger—how does one visualize, for instance, radiation? Or the depletion of the ozone layer? How is the toxic substance acrylamide to be pictured? Though stereotypes may simply "happen," they usually start as intentional acts of symbolic "bricolage" before they become widely disseminated and therefore recognizable by audiences. While the metaphor picture refers to the creative efforts of photographers and editors, the stereotype is seen as the outcome of media visibility. Figures 18.5 and 18.6, photographs of the tunnel site, illustrate the concept of stereotype and metaphor respectively.

Figure 18.6 The caption reads, "Underground Hell...."

These two photographs are large and prominently displayed in the centerfold of the newspaper's supplement. Placed side by side, they are explicitly intended as visual contrasts. Figure 18.5, displayed on the left of the centerfold, shows a man walking along a stone fence in a bright, green landscape. The caption, playing with words, reads: celestially (above ground) beautiful. The walking man is a known author said to have found inspiration in the scenery. Figure 18.6, on the right of the centerfold, is intended to be an exact contrast of the "above ground" landscape. The caption reads: "Underground Hell...." Shot inside the tunnel, the photograph is dark and barely illuminated by a yellow/orange ball of light somewhere along the tunnel. In the foreground a man with a miner's lamp and a thick book in his hand looks up at the tunnel's ceiling. It is not necessary to elaborate upon the visual semantics of these pictures, as they were judiciously chosen to conjure up and contrast the Swedish cultural imagery of the beautiful landscape with the darkness, fear, and mystique of underground imagery. (Swedish dairy farmers have long played with the cultural imagery of "open landscapes" and "stone fences" to build positive images of their products). The before/after dichotomy also present in the visual contrast suggests that the intended visual syntax is one of generalization, that is, the synthesizing and reinforcing of messages already communicated by other photographs. Clearly, the intention is to associate the "Underground Hell" with the risk object—the caption also informs us of "poisoned water dripping from 1,400 tons of Rhoca-Gil." Visually, however, the intended juxtaposition of the two photographs places the "victim" as the landscape and the risk object as the tunnel itself.

In a system where causality is always present, the imagery of a "beautiful landscape" introduced in contrast to the tunnel photograph is an explicit statement of order/disorder in terms of before/after. While the landscape photograph is undoubtedly a stereotype, effortlessly conjuring up an aesthetically charged imagery of the "Beautiful Swedish Landscape," the tunnel photograph is still a typical metaphor photograph (because of its widespread media exposure and "transportability" to other but similar contexts, this picture may eventually become a stereotype itself). As a metaphor photograph, the caption remains important in directing interpretation, but the dichotomy text/image should not, at least methodologically, be taken too seriously. This binary opposition often entails notions positing the written account as more authoritative than pictures, drawing upon other binary distinctions like knowledge/common sense or experts/laymen. Whereas "experts" tend to use a discourse addressed to "knowledge" and to action upon the events analyzed, photographs are addressed to the "common sense" of their general readership.

In their study of newsroom culture, Ericson, Baranek, and Chan suggest that "fake words" (like political terrorism to describe an unexplained bombing incident) are "required to invest these matters with significance, to make it apparent to common sense that what was being visualized was worth attending to." Fake words are "important in filling the gap between a random event in the world and what most people find meaningful" (Ericson, Baranek, & Chan, 1987, p. 339). A closer reading of picture captions reveals that "fake words" are common occurrences in the reporting of risk events, not aimed at deceit, but as help to understand the risk event by rendering it meaningful at a structural level, as a whole. This, of course, is also the main task of metaphor and stereotype pictures; to make sense of the risk event as a whole, "at a glance."

The "Underground Hell" picture, like other "metaphor" pictures, can not be translated into the medium of verbal language without losing most or all of its culturally induced symbolic and emotive power. Paradoxically, part of the "stickiness" of these photographs lies precisely in the room for interpretation allowed to the individual reader. The "range" or scope of interpretative possibilities depends, at one level, on the seemingly varied ideas that individuals may hold about any depicted event (we expect, for instance, the "underground hell" metaphor to be more pervasive to the local population, reportedly adverse to the tunnel construction project).

At another level, however, visual messages are created according to the conventions and cultural templates of the society to which the individual belongs. The force, or meaning, of a metaphor picture depends to some extent on how well these two levels are brought together in the construction of the visual message. This is not to say that a picture must be uncontroversial to gain a wide impact—agreement and understanding are not the same thing. A famous example is the 1989 picture of the student protest in Beijing where a frail man in civilian clothes stands right in front of a column of tanks. As Van Ginneken (1998) notes, the image was presented in the West as symbolizing the extraordinary courage of unarmed civilians, while the same image was presented in China as symbolizing the extraordinary restraint of soldiers in face of provocations. The logical conclusion is that "the picture does not carry either interpretation in itself: it is the reader who decides on the meaning of the image, by referring to certain notions which are taken for granted in his or her political (sub)culture" (p. 175).

The presentation of the picture—the caption, headlines, and the discursive text around it does, of course, reflect the ideology of its makers and favors a certain reading of the picture. This is also obviously the case with the pictures of the tunnel construction site; their captions and surrounding text reflect the ideological position of the newspaper on the issue. In terms of readership, we might also envisage the (hypothetical) congruence of a population strongly adverse to the tunnel construction with a population sustaining a deeply rooted and negatively charged underground imagery. In such a hypothetical case, the "underground" picture could be expected to strike a strong emotive chord and perhaps even acquire a symbol status in the current terminology of "something that stands for something else."

Fake Words and Metaphor Pictures

Metaphor pictures, like "fake words," are not intended to deceive, but to render events meaningful at a common-sense level. The joint reading of the photographs and their fake captions do not raise issues of objectivity or truthfulness. The aim is to render the risk event meaningful by making full use of available cultural templates not immediately open for reasoning. Whatever opinion any individual reader may have on the reproduced landscape, it still remains "beautiful," because the picture is a stereotype in the sense discussed above. Once it has become an integral part of deeply held structures of meaning, a visual stereotype, like a trademark, does not "stand for" something; it simply *is* that something. Like trademarks, visual stereotypes are both difficult to eradicate and resistant to wear and tear.

In contrast to the stereotype, the "metaphor" picture is nearly always dependent on "fake words" to guide interpretation or, rather, to implicate already existing networks of meaning in the reader's construction of the visual message. Obviously, this is a risky enterprise, which may have the opposite effect of that intended—the "meaning" of the picture is imposed, rather than suggested, to the reader. The attempts to circumscribe the reader's own interpretation may define the photograph as noniconic and therefore fake, thus defeating the purpose. Which strands of symbolic association the "Underground" picture might exactly evoke is difficult to pin down because retrieving visual messages explicitly from a reader's understanding of the picture is a nearly impossible task. A conceptual distinction between ideology and culture is, in this regard, heuristically useful. It helps to distinguish between the opinions that individuals may hold about any depicted risk event (or the discursive context in which the pictures are embedded, for instance, "experts" accounts and arguments) and the conventions and cultural templates that make the metaphor work (or fail).

Once a conceptual distinction between ideology and culture is achieved, we observe that there is a certain correlation between ideology and language on one hand (the argumentative component) and between culture and visual communication (the nonverbal and implicit component) on the other. While it is relatively easy to "discuss" the intentions of the photographer or even the information content of the picture at a level of individual variations, it is in the nature of the visual to be mute. Any effort to retrieve a visual message verbally necessarily transforms it to another medium - much of the force that the visual message may have simply vanishes. If the focus is primarily on the cultural/visual rather than on the ideological/verbal dimension, questions about the "meaning" of a picture shift. Questions regarding the intentions that the photographer or editor may have had in creating the picture give way to questions about how the picture implicates basic cultural values in order to evoke meaning to its readers. While photographers and editors certainly use their cultural intuitions to create and select metaphor pictures, this is mostly a process of trial and error; some pictures have profound emotional appeal and are widely disseminated, others do not.

Because the concept of meaning tends to obfuscate rather than illuminate, and "the meaning of a photograph" has become more ambiguous than the purported ambiguity of photographs themselves, it might be useful to elaborate somewhat on the concept. A very common meaning of "meaning" when used with reference to human phenomena is intentionality (Weber, 1947). The meaning of a human act is the agent's intention, purpose, motive or reason for doing it. Seen from this perspective it is obvious that most news photographs are meaningless because no one intended them. Certainly, the selection of a segment of "reality" and how this segment is pictorially arranged is intentional, but this is not the issue. These are the photographers and editor's acts and tell us about the purposes and motives in the selection of the picture.

We can ask the editor why just this or that picture was selected, and we can ask the photographer why the picture was taken in a certain way and not in another. But we cannot explain the meaning of "a man in a tunnel building with a thick book in his hands" by an account of the photographer's motives—even when the photographer draws upon his or her cultural intuitions to intentionally produce a metaphor picture. This kind of "explanations" has to focus on cultural symbolism in its own right and not on people, so explanations in terms of intentions are not fitted to them. Instead, we should focus on consequences and ask questions about context, that is, the way those things are linked by logical implication to other things and ideas. The "meaning" of a photograph operates on cultural values which are mostly taken for granted by its readers (and by the author of the photograph) whom, most probably, are unaware of the complex interrelationships of ideas which might given rise to a "powerful" photograph. They might therefore be unable to volunteer an explanation of how and why any given photograph is intrinsically meaningful to them. A photograph may condense several possible layers of implicated ideas, of which only some will be relevant at any given moment, a relevance which is given by the context in which the photograph appears and the reader's degree of attention and understanding.

Stigma and Imagery

The term "stigma" is of Greek derivation which originally denoted a mark carved or burnt into the skin to identify undesirables. Erving Goffman applied the concept to modern society, arguing that persons with traits that differ from the normal or normative in society are treated as if they bear a mark of disgrace. The person is stigmatized as others withdraw their acceptance and distort the person's real identity to fit stereotypical expectations. Environmental stigma is thought as analogous to social stigma involving a parallel process for discrediting settings, places, objects, nonhuman life forms and surroundings as well as people associated with these environments (Edelstein, 1991). Accordingly, when applied to risk events, stigma may be evident in practically everything that is feared for some reason or another. Necessary preconditions are the "visibility" of the stigma mark and the expectations of the "public" to which the mark is visible. In contrast to Goffman's work in which the object of study is people with rather well delineated social expectations, the expansion of the concept of stigma into the field of risk is less clear-cut. At the tunnel construction site, the milk in the neighboring dairy farms was judged unfit for consumption and thrown away, as were the vegetables produced in the area. The Swedish Board of Food and Health mandated the destruction of all produce from the area in order to protect consumers, who have the right to expect that the milk and vegetables they buy are free from unhealthy substances. Facing a choice, consumers were expected to have avoided produce stemming from the risk area.

The issue of self-preservation is important, not least because it is nearly always present, in different degrees and guises, in the individual's construction of the risk object. From the perspective of risk management bodies, it is obvious that self-preservation does not account for the exaggerated responses of stigma induced behavior—the "objective" threats posed by the risk object seldom bear a direct relationship to people's understanding of those threats. Accordingly, stigma may be seen as the outcome of ignorance, prejudice, inconclusive scientific knowledge or lack of trust in experts managing the risk event, and appropriate stigma reducing strategies carried out in educational terms. Didactic strategies indirectly reaffirm self-preservation as the main drive behind stigma related behavior and reintroduce the "rational" individual in the process of making enlightened choices. This is in accordance with a highly developed tradition of exegesis in Western cultures—we are taught from an early age to explain things and to check the logical exposition of our explanations. This tradition is deeply rooted—it is easy to think that we think sequentially and logically—and combines well with the idea of the "risk averse" individual.

In other words, any individual would regard fear of contagion as a more plausible and therefore appropriate explanation for, for example, beef avoidance at the time of the BSE epidemic than to introspectively elaborate upon negative meaning structures—even if it were intellectually possible to do so. What is beef, symbolically speaking, and how are ready-made universal stigmas of "madness" or "cannibalism" (cadavers in the cows' food supply) associated with it? What kinds of expectations are, in fact, violated? People fearing "contaminated" beef may not even be fully aware of how they organize their knowledge of the event or, if they are intellectually aware of some of its components, may find it inappropriate to dwell on them. The "rational" individual, when questioned, will nearly always advance self-preservation (fear of contagion, leading to illness or death) as the main reason for shunning the stigma marked beef. The theory of rational choice, however, overlooks the fact that choices are made according to available alternatives that are themselves symbolic constructs. Somewhat simplified, this is also the distinction between verbal and visual communication; the rationality of the verbal seldom accounts for how negative meaning structures are formed and reformed. Their fundamental characteristic is precisely that they occur as images necessarily different from the way in which they are verbally communicated.

In the case of the tunnel construction site we noted that the risk object "water," for instance, instead of remaining within the realm of landscape breakdown (posing a danger to the cows) "invades" the realm of people by means of the photograph of the "family" posing in front of their drinking well. The breaking of contextual boundaries allowing a putative hazard to "flow" into contexts others than those where it was generated is, of course, a main feature of risk events. Once the associative link between the two contexts of landscape and community is visually established (water in landscape/water in well), the "fact" that the family lives outside the risk area is only relevant at the level of argumentation and to notions of trust or distrust in the risk management authorities. The statement that the family "did not dare to drink the water," stems from a disrupted imagery to which factual knowledge is only secondary. The flow of meaning that imparts from the breakdown of the "normal" imagery against which the risk event is necessarily contrasted contains in itself the building blocks of the "metaphor" pictures category. Metaphor photographs will expectedly attempt to conjoin the heuristic categories of environment, victim, and risk object in both aesthetically and emotionally charged ways, artifices which are not necessary in the pictorial description of the imagery's breakdown. It is in the metaphor picture that stigma's genesis can be found.

The argument forwarded here is that stigma (its mark and visibility) is not formed, generated or created in a vacuum. Rather, stigma is *implicated* from already existing ideas about how the world is socially constructed. Further, these ideas are embodied in concrete and recognizable risk objects that may, or may not be the "real" causes of the hazards or perceived threat. The photograph shot inside the tunnel (photograph 7.), previously identified as a visual metaphor, is what might be properly called a "risk event in search of an image" as implied in the title of this article. The success of the picture as metaphor depends to a great extent on how well it is able to "implicate" relevant imagery in its viewers. The overt contrast with the "beautiful landscape" picture is one of the means by which the photograph was intentionally "implicated" in a deeply held system of ideas and values from which its meaning would be widely understood. From the perspective of the picture creator or editor the gauge of success lies in its "marketability" that is, how often the picture is reproduced in the media. Another important criterion of its economic success is just the picture's ability to visually "condense" the risk event. If these criteria are met, the picture will most likely become what we call a stereotype.

Conclusion

The argument put forward here is that the mark and visibility of stigma is generated, disseminated and sustained (as well as eradicated) by predominantly visual means. Stigma thrives on stereotypes and the visual media, in this case photographs, is a prime motor in the creation of stereotypes. Stereotypes facilitate communication, and the patterns of avoidance intrinsic to stigma formation are triggered by symbolic representations of risk objects which embody these objects in terms of common-sense understandings rather than objective, factual, and linguistic assertions of putative hazards and rational thinking about causes and probabilities. As symbols (or rather, clusters of symbols if we take into account the different elements that form the picture as a whole) stereotypes both generalize and make use of taken-for-granted assumptions about the world which are not always retrievable by linguistic means. But the important feature of the photograph as a stereotype/symbol is that under certain conditions it will stand for itself, that is, the photograph becomes a representation not only of the event of which it was a part, but other similar events as well.

In other words, a certain photograph will "not stand for" a certain risk event but becomes in itself the embodiment of the risk event. As a symbol, the photograph becomes autonomous of the context in which it was first formed and can therefore trigger an immediate recognition and understanding of other, perhaps substantially different, risk events. It goes without saying that few photographs meet the criteria of stereotypes and become symbols in the sense that they are "portable" between events, but some photographs do display an extraordinary resilience in memory—they are often capable of "tagging" to themselves new events and interpretations of these events. Indeed, once a picture has been given a central place in the construction of our imagery of any given risk event, the same image will structure and condition the interpretation of new events. Stereotypes of seawater pollution, for instance, will "recall" an imagery of marine life in danger and endow new instances of pollution with emotional power. Issues of scale and objectivity do not really matter; the context is, so to speak, already in-built in the stereotype itself, disengaged from the events which gave rise to its formation and therefore free of the facts or controversies which would associate them with that original event.

The tunnel photograph meets most of the criteria to become a stereotype. It contains enough semantic elements to develop its own meaning structure, allowing for the concomitant messages to be "retrieved" independently of other photographs. The flow of meaning from water to tunnel tends to conjure a composite imagery, where the notion of tunnel, by itself complex, semantically ambiguous and visually concrete, tends to absorb the risk object "water" into its own underground imagery, thus acquiring the potential of becoming itself a "representation" of the risk object. The "religious" imagery which the caption intends to induce by means of contrast to the stereotype picture, (beautiful/above ground//Hell/below ground) probably does more to reveal the sensation seeking biases of the tabloid newspaper than it does to reveal the associative capacities of the picture. As noted previously, the associative capacity of a successful metaphor photograph will necessarily have interpretative supremacy over the written caption, and eventually be able to entirely dispense with it. It is then interesting to note that a very similar picture was reproduced in full-page color in the Sunday supplement of *Dagens Nyheter* (November 22, 1998), Sweden's largest morning newspaper. The picture, without caption, has the text "Light in the tunnel?" printed over it. In the context of risk events, the photographic challenge to the accompanying text is most interesting when it is focused on the "risk object" itself. The reading of the metaphor picture suggests that it is the tunnel itself, and not its builders or the toxic substance, that is to be blamed. If a further analysis of other visual media sources dealing with the same risk event were to confirm the interpretations advanced here, we should expect an imagery of "tunnel" to emerge embedded in negative symbolism. In such a case, it could be expected that future tunnel projects in Sweden would be, if not stigmatized, at least much more debatable and problematic than they have been hitherto. This suggests the supremacy of visual media in the creation of negative imagery and hence, in the formation of stigma.

Acknowledgments

We want to thank Per Binde and Peter Lutz for comments on an earlier draft of this article and Betty Kahn Ferreira for language editing. The Swedish Research and Communications Board (FRN) have funded the research on which this article is based.

19 The Nevada Initiative: A Risk Communication Fiasco[1]

James Flynn, Paul Slovic, and C. K. Mertz

Introduction

The only site currently being studied as a potential location for the nation's first high-level nuclear waste (HLNW) repository is Yucca Mountain, Nevada, a barren ridge of compressed volcanic ash located about 100 miles north of Las Vegas. According to most scientists, nuclear wastes can be deposited underground in places like Yucca Mountain with minimal risks to public health or the environment (National Research Council, NRC, 1990; Nuclear Waste Technical Review Board, 1992). The public does not believe these claims (Flynn et al., 1990; Flynn, Mertz, & Slovic, 1991). Numerous surveys have indicated that Nevada residents oppose the proposed repository at Yucca Mountain by about a 4 to 1 margin. Nevadans, like the American public generally, believe that a HLNW repository program poses high risks of accidents and radiation exposure (Flynn et al., 1990).

The combination of Nevada's opposition and management problems within DOE have resulted in an extremely slow-moving program—on average, the anticipated completion date recedes two years, every year. Combined with funding shortfalls, questions about the ability to meet regulatory requirements, and a history of inadequate management of the scientific work, the program often appears on the verge of collapse. The commercial nuclear power industry, which collects fees to fund the federal effort, has a significant stake in the program because it plans to use the repository as permanent storage for spent fuel from its reactors. Occasionally, the industry expresses great frustration and concern with the lack of progress (Rogers, 1992; "Nuke Exec Mad," 1992).

The Advertising Campaign

In 1991, an industry group, the American Nuclear Energy Council (ANEC), designed and implemented a large-scale advertising campaign in Nevada to gain support for the federal repository program. ANEC hired a top Las Vegas

---

[1]Reprinted from *Risk Analysis, 13*(5), 1993, pp. 497–502. Copyright 1993 by Society for Risk Analysis. Reprinted with permission.

advertising and political consulting firm to develop the ads and employed a popular former sportscaster, Ron Vitto, to narrate them. The ads emphasized safety and attempted to change the public's perception that a repository and transportation of HLNW wastes to it would be highly dangerous. According to an ANEC vice-president the purpose of the campaign was to "inform and educate the public."

One ad demonstrated the safety of transporting HLNW by showing a truck and trailer bearing a cask being rammed at high speed by a locomotive. In this spectacular accident the cask and various parts of the transport machinery fly in all directions after which the announcer declares that the cask had safely survived the collision (Flynn et al., 1990).[2]

Other presentations featured scientists contracted to work for DOE. In one ad the narrator asked if high-level nuclear waste can explode. "Absolutely not!" the scientist replies. Another ad confronted the issue of radiation risk from nuclear facilities. "Does living near a nuclear power plant cause cancer?" Yet another scientist claimed that scientific studies say no. The campaign spokesman conducted televised interviews at a Las Vegas mall. After questioning about 50 persons and learning that a majority thought high-level nuclear wastes could be liquid or gaseous, the narrator triumphantly shows how wrong this idea is. Holding up a pellet of simulated spent fuel, he demonstrates that it is solid and, presumably, safer than liquids or gases.[3]

Effects of the Campaign

How well did this ad campaign work? To learn the answer, we conducted a representative statewide telephone survey of Nevada residents between October 25 and November 5, 1991 (Flynn et al., 1991). Five hundred and four interviews were completed with a response rate of 52.3%. The survey showed that a few weeks after the start of the ad campaign, about three-quarters of Nevada residents (72.4%) had seen or heard the advertisements.

However, despite this impressive coverage, the survey results were not encouraging for the ANEC campaign managers. Fewer than 15% of the respondents who had seen the ads said the ads made them more supportive of the repository, while 32% said the messages made them less supportive. More than half those surveyed (52%) said the messages did not change their opinion. Despite the barrage of pro-repository messages, almost three-quarters of the respondents (73.8%) said they would oppose the repository if they were to vote on whether it should be built—almost exactly the same proportion as before the ad campaign (Flynn et al., 1991).

Respondents who reported having seen the advertisements were asked: "In thinking again about the advertising campaign in support of the repository at Yucca Mountain, what is your single most important opinion or feeling about those advertisements?" These comments were examined and assigned to categories, as shown in Table 19.1.

[2]The claims about the integrity of the cask were later contested by state officials. See Manning (1992).

[3]No mention is made of wastes from nuclear weapons facilities which are not in the form of spent fuel but which are also expected to go to the repository.

Table 19.1 Responses to the ANEC Advertising Campaign by Category and Referendum Vote (in percent)[a]

Category	% Response	Referendum vote[b] % Yes	% No
1. Disbelief (Lies, gimmick, dishonest, scam, con, false, misleading, propaganda, hype, tainted, snake oil, misleading, deceiving, one-sided, etc.)	48.5	13.5	42.7
2. Insulting (Insulting, stupid, dumb, not informative, resented)	3.3	1.0	2.4
3. Disagree (Disagree, not telling the whole story, not fair, not safe, too risky, hazardous, accidents will happen, etc.)	11.8	2.1	11.0
4. Positive (Good, educational, true, effective, necessary, helpful, interesting, etc.)	15.1	39.6	3.5
5. Positive . . . "but" (Helpful but biased, good but not enough, if safe, then OK, helpful but still has reservations, etc.)	4.1	3.1	2.7
6. Don't know, no position, indifferent (Don't care, don't know, doesn't make any difference, can't answer, no opinion, neutral, etc.)	8.2	8.3	3.8
7. Fatalistic (It's up to the federal government, what people think doesn't matter, they'll do it no matter, etc.)	0.8	0.0	0.5
8. Analytic comments on the strategies, source, or purpose of the ads (Suggestions for ads, repeats of what the ads say, need for more detail, ads are a waste of money, ads are not effective, they (ANEC) have freedom of speech, who pays for the ads?, etc.)	6.8	8.3	3.5
9. No answer	1.4	24.0	29.8
Number of respondents	365	96	372

[a]May not equal 100% due to rounding.
[b]Excludes those who said they would not vote or did not know how they would vote ($N = 36$).

Almost half (48.5%) the respondents who had seen the advertisements said that they did not believe the ads (Category 1), while 3.3% felt insulted by the ads (Category 2) and 11.8% disagreed with the ads for a variety of reasons (Category 3). These three categories of negative comments make up 63.6% of the recorded responses.

About 15% (15.1%) of the respondents provided positive comments (Category 4), while a few people (4.1%) had positive comments but with reservations (Category 5). Fatalistic opinions (Category 7) were expressed by less than 1%, while 8.2% were indifferent (Category 6) to the messages. About 6.8% made comments about the advertisements that did not reveal how they evaluated the ad campaign (Category 8).

Prior to asking about the advertisements, respondents were asked to provide a hypothetical vote: "For the purpose of this question, let's suppose that the Department of Energy selected the Yucca Mountain site for the nation's first high-level radioactive waste repository, but it wouldn't be located there unless state residents voted in favor of it. If this were the case, would you vote for it or against it?"

The right-hand column of Table 19.1 shows that respondents who would vote against the repository tended to view the advertisements in negative terms; the Disbelief, Insulting, and Disagree categories accounted for 56.1% of this vote, only 13 people who would vote against the repository thought the ads were good, necessary, informative, or helpful. About three-quarters (74.5%) of the positive comments were from those who would vote for the repository.

Table 19.2 shows how respondents who would vote for or against the repository rated the influence of the advertisements on their support. Those who would vote in favor of the repository were 10 times more likely to see the ads as positive and to report that the ads made them more supportive. On the contrary, only 2.3% of those who would vote "no" thought the ads increased their support, and 29.3% of those "no" voters said the ads made them *less* supportive. At this point the advertising campaign appeared to have had, at most, very minor impacts on people's existing positions. In terms of convincing opponents to support the repository program, it may have been counterproductive.

The Advertising Campaign Becomes an Issue

Following the first six weeks of the advertising campaign and after this survey had been completed (mid-November 1991), the campaign itself became a controversial public issue. An anti-nuclear group released confidential ANEC documents providing details of the advertising campaign's strategies. The package included "The Nevada Initiative" a key document drafted by a political consulting firm in Las Vegas (Oram & Allison, 1991; A.J. Kessler, personal communication, October 25, 1991; P. Schoen, personal communication, September 6, 1991). These documents were highly embarrassing to ANEC. Whereas the vice president of ANEC, in televised testimony before the Nevada Commission on Nuclear Projects, asserted that the advertising campaign only intended to correct misinformation and educate the public, the documents revealed that the purpose was to sway public opinion toward acceptance of the repository or at least toward a belief that the repository was inevitable (Oram & Allison, 1991).

Table 19.2 Referendum Vote and Reported Influence of Advertisements

	Referendum vote[a]	
Influence of advertisements	% Yes	% No
More support	36.5	3.2
Less support	4.2	29.3
About the same	35.4	38.4
Had not seen ads	24.0	29.0
Number of respondents	96	372

[a]Excludes those who said they would not vote or did not know how they would vote ($N = 36$).

The ANEC documents recommended spending $9 million over three years to: (a) increase public support for the repository program, (b) bring pressure on the State of Nevada to cooperate with the program, especially in issuing permits for work at the Yucca Mountain site, (c) recruit and support allies in Nevada who would support the industry program and the repository work, and (d) obtain support for the repository program from key media (newspaper and television) sources (Oram & Allison, 1991; A.J. Kessler, personal communication, October 25, 1991; D. Schoen, personal communication, September 6, 1991; Schneider, 1991). These documents did not mention informing and educating the public.

"The Nevada Initiative" spoke in military jargon of an effort to provide "air cover" so elected officials could negotiate benefits in exchange for the repository. The ads were to be directed particularly at women aged 25 to 49 because they are the people "with the highest statistical potential for favorably affecting polls if they can be informed, reassured and moved." Local reporters were to be hired to present the "industry's side of the stories" to their peers. Department of Energy scientists were to be trained by the ad campaign managers, in an attempt to "convince the public that nuclear energy is safe."

The response in Nevada to the leaked strategy documents was outrage (O'Callaghan, 1991; Rogers, 1991; Eldreidge, 1991; Flynn, 1992). Newspapers and television coverage featured scathing attacks by state officials that continued for weeks. Senator Richard Byran (D., Nev.) demanded an explanation from Energy Secretary James Watkins regarding the role of his department and its scientists in the industry campaign (Riley, 1992; Hickox, 1992).

Las Vegas newspapers published defensive letters from scientists employed by the DOE who claimed that their professional objectivity was not being corrupted.[4] Local TV newscasts contrasted film clips of ANEC vice president Ed Davis ("we only want to inform and educate the public") with counter quotes from the strategy document. Governor Bob Miller wrote asking the governors of states with nuclear power plants to investigate the propriety of using utility ratepayer funds to persuade Nevadans that they ought to accept nuclear wastes that no other state wants.

[4]For example, see letters in the December 1, 1991 issue of the *Las Vegas Review Journal.*

Perhaps the most devastating rejoinders to the ANEC campaign came from a pair of Las Vegas disk jockeys who began to parody each of the new TV ads. The main character in these satiric skits bore the mock name "Ron Ditto," whose rather simple-minded pronouncements were heaped with ridicule. "Hi! This is Ron Ditto, your formerly respected sportscaster, trading in your respect for much needed dollars."

Local businesses joined in. A TV advertisement showed the disk jockeys in a huge pair of overalls as a two-headed mutant, "Yucca Mountain Man," in a commercial for a Las Vegas auto dealership. A restaurant extolled the quality of the tomatoes in its salad bar by putting one through the same tests that nuclear waste casks were subjected to in the ANEC ads: After the tomato survives being run into a cement wall, hit by a speeding train and dropped from a high tower, "You can be sure that it's one high-quality tomato."

The ANEC campaign, faced with disbelief, ridicule, and little measurable influence on public opinion, was discontinued in the spring of 1992. By that time, the campaign's credibility had been damaged considerably. A survey conducted in June 1992 by researchers from Arizona State University and the University of Nevada, Las Vegas showed that after seeing the ads only 3.3% of respondents reported an increased level of trust in the repository program while almost 41% were less trusting and the remainder were unchanged. The levels of opposition to the repository project remained at about 75% (Mushkatel & Pijawka, 1992).

After a hiatus of several months, a redesigned and much reduced effort was introduced in the fall of 1992. In comparison, this new series of ads is low-key, featuring a woman's voice-over to the TV spots and less assertive claims to scientific certainty. The effects of this new effort remain to be seen.

Reflections on the ANEC Campaign

Information and advertising campaigns designed to address people's risk perceptions are common. The federal government and numerous advocacy groups frequently communicate about risks from cigarette smoking, AIDS, alcohol consumption, drug use, and numerous other activities that involve threats to health and safety.

In the face of concerns about chemical and radioactive hazards, the chemical, oil, and nuclear industries have also developed information programs. The recent ads by the U. S. Council on Energy Awareness in support of nuclear power are an example.[5] There have been a number of attempts to demonstrate that industry campaigns such as these can change irrational, unscientific, or misinformed public perceptions. One experiment found that a pro-industry brochure increased people's confidence in the ability of managers to dispose safely of nuclear wastes (Bisconti, 1991). Another study measured public opinion on nuclear power before and after advertising programs and found that people who remembered the pro-nuclear ads were more likely to have an improved attitude toward nuclear power (Bisconti, 1989). Department of Energy officials in Las Vegas report that people who take the free tours to Yucca Mountain (approximately a six- to seven-hour trip from Las Vegas) in the company

---

[5]These advertisements appear almost monthly in the large circulation magazines. For a critical commentary on the nuclear industry ads, see Grinspoon (1992).

of personnel who describe and advocate the repository program are more likely to support the project after the trip.[6]

Most often these risk communication efforts have been conducted as experiments in non-adversarial settings. In the adversarial world where public opposition to a specific nuclear facility is at issue, different conditions apply. The public response to the ANEC advertising campaign was dominated by disbelief and other negative evaluations. This skepticism seemed justified with the subsequent revelations of the Nevada Initiative, which triggered another, even more negative, round of public opinion. Nevada residents responded to ANEC's covert strategy with satire, derision, and political attack. The response to the ANEC campaign verifies the results of experimental research that has found the influence of advocacy positions in risk communication to be fragile and easily reversed by counter arguments (Slovic, Kraus, & Covello, 1990).

Attempts to convince the American public that nuclear facilities are acceptably safe have recorded few successes in the past two decades. No new nuclear power plant has been sited during this time, nor has there been much success in efforts to develop radioactive waste disposal facilities. Public response to such projects has been strong opposition derived from perceptions of risk, a profound lack of trust, and a concern with issues of equity. This legacy comes in part from the Three Mile Island accident, the revelations about the mismanagement of the nation's nuclear weapons programs and facilities, and a deep social and cultural skepticism about things nuclear (Nealey & Hebert, 1983; Slovic, Flynn, et al., 1991; Weart, 1988). Countering such attitudes and opinions will require extraordinary honesty, patience, and skill in communicating the appropriate information to the public, information that will have to go well beyond claims of safety and repeated messages of assurance that everything will turn out well. Little of the requisite qualities were evident in ANEC's "Nevada Initiative."

Conclusions

The ANEC campaign to support the Yucca Mountain repository did not merely fall short of its goals; it was a spectacular failure. A well-financed, professionally developed and implemented advertising campaign was forced into premature closure due to strong public opposition—indeed a truly vicious reaction —that aggressively attacked the campaign and its sponsors. The advertising campaign was based upon a belief that the public did not understand the issues and that what was called for was a dose of scientific assurances. This attempt was misguided. The public opposition to Yucca Mountain is not driven entirely by the risk of radiation exposure. Distrust of the nuclear industry and the Department of Energy, concerns about equity and fairness in site selection and evaluation, and the desire for local and state government to have an effective role in project decisions also play key roles (Flynn et al., 1990, 1991).

The adversarial context in this situation obviously made risk communication difficult. Important public officials, including the governor and a U.S. Senator, spoke out strongly against the ad campaign. The public viewed the ANEC and DOE scientists not as disinterested experts but as "hired guns,"

[6]DOE claims a 25% increase in support for conducting studies of Yucca Mountain after taking the tour (Richert, 1992). For another view of this program, see Ward (1991).

and ridiculed them as quickly as other aspects of the pro-repository presentations.[7] The strong existing distrust of DOE and the industry was activated and focused on the scientific testimonials presented on TV. In its turn, the ANEC campaign made no attempt to deal with the issue of distrust beyond choosing scientists for key roles as spokespersons, and this was clearly inadequate for providing a convincing message. Finally, the revelations about the campaign strategy presented the Nevada Initiative as a cynical move to manipulate the public attitudes, which led, as might have been expected, to further rejection of the messages (Morgan & Lave, 1990; NRC, 1989).

Public response to the Nevada Initiative provides some important lessons about risk communication. It demonstrates that risk communicators must confront the psychological, social, cultural, and moral values that inform and shape public perceptions. Scientific rationality by itself will not substitute; it is neither the final answer to addressing public concerns nor the only consideration for guiding public policy and the management of hazardous technologies.

Acknowledgments

This material is based upon work supported in part by a contract between Decision Research and the Nevada NWPO with federal funds pursuant to the provisions of Public Law 97-425. The government has certain rights in this material. Any opinions, findings, and conclusions or recommendations expressed in this material are those of the authors and do not necessarily reflect the views of the federal government or the Nevada NWPO. The authors are grateful for the suggestions and comments made by Steve Johnson and Robin Gregory.

[7]Cartoonist Ray Collins in the *Boulder City News* created "Dr. Klone, " an ANEC-sponsored scientist with a smiling "happy face." In one cartoon a frumpy citizen asks, "Do high level nuclear waste transportation casks emit harmful radiation, Dr. Klone?" The happy-faced scientist, standing in front of a decrepit and obviously leaking cask loaded on a truck replies, "ABSOLUTELY NOT! You get more radiation from door knobs and your cat than the teeny-weeny amounts emitted from THIS cask!" (December 12, 1991).

20 Risk, Media, and Stigma at Rocky Flats[1]

James Flynn, Ellen Peters,
C.K. Mertz, and Paul Slovic

Introduction

This article examines one of the more dramatic yet poorly understood claims about the impacts of nuclear science and technology, the stigmatization of places associated with a nuclear facility. In this case the facility is a nuclear weapons site and the associated places are residential properties. The studies reported here focus on a dramatic news story about conditions at the Rocky Flats, Colorado nuclear weapons facility. The studies are conceptualized within the social amplification of risk framework (Kasperson et al., 1988; Kasperson, 1992) and provide an analysis of both newspaper coverage and a survey of actual and potential home buyers in the adjacent metropolitan area.

The development of public responses to the dangers of radioactivity has a long history. Nuclear science and its technologies are among the great achievements of the 20th century—at once magnificently impressive and greatly feared. The fear emanates from the massive destructive power of nuclear weapons and from the dangers of other sources of man-made radiation. Short- and/or long-term exposure to radiation results from nuclear medicine (e.g., x-rays, radiation therapy, etc.), nuclear power, nuclear weapons, and the disposal of nuclear wastes. Although some advocates of nuclear technology argue that small doses of radiation are good for human health, and others think small levels are acceptable, a number of scientists believe that almost any exposure presents some health risks.[2]

Over the past few decades, the public opinion reversed from support of nuclear technologies. For example, public opinion that once supported nuclear power by more than 60%, now opposes it by more than 60% (Rosa & Dunlap, 1994). This reversal of public opinion has played a role in halting the develop-

---

[1]Reprinted from *Risk Analysis, 18*(6), 1998, pp. 715–727. Copyright 1998 Society for Risk Analysis. Reprinted with permission.

[2]Several authors have argued that low doses of radiation may be beneficial to health. See, for example, recent comments by Wolfe (1997). A recent issue of the *Belle Newsletter* presented a paper by Heitzmann and Wilson (1997) that argues for a policy of assuming low-dose linearity of response to radiation and chemicals. This was critiqued by Pollycove

ment of nuclear power, slowed and stopped the search for nuclear waste sites, and complicated management of the nuclear weapons complex.

Slovic (1987) observed that laypersons evaluate health and environmental risks on a number of characteristics, which he and his colleagues analyzed with factor analysis and combined into two dimensions of (a) dread risk and (b) unknown risk. Dread risks are catastrophic, deadly, uncontrollable, etc. Unknown risks are poorly understood, unknown to those exposed, with delayed effects, etc. Nuclear power and nuclear waste were rated extreme on the dread and unknown dimensions. Validation of these psychometric studies occurred when survey respondents were asked for word associations to a high-level radioactive waste repository. The resulting images were overwhelmingly negative, dominated by thoughts of death, destruction, pain and suffering, and environmental damage (Slovic, Flynn, et al., 1991).

Technological hazards with such affective imagery can shape individual and societal behaviors. A conceptual model of social responses to hazards developed by Kasperson and his associates was termed the "social amplification of risk" (Kasperson et al., 1988; Kasperson, 1992). It incorporated the earlier work by Slovic and his colleagues on the perception of risks. The social processes described in this model are complex and interactive with a wide range of outcomes, of which stigmatization is one possible result (see Figure 20.1). Basically the model states that once a risk-relevant event enters the processes of social communication, awareness and concern about the risk often increase and initiate individual and group behaviors. Information about risk, broadcast through the mass media and producing widespread concern, can result in important social, economic, and political impacts. This social process underlies "technological stigma," an aversion or shunning of places, products, or technologies with impacts that often include substantial economic losses (Gregory et al., 1995).

Public acceptance of technological hazards will depend in large part on confidence in the ability of hazards managers, on how well they understand and control hazards and how trustworthy they are in fulfilling their protective duties. The history of the nation's nuclear weapons complex, including Rocky Flats, have produced serious concerns about the motives and performance of DOE and its contractors resulting in loss of trust and confidence in the operations of these facilities (Secretary of Energy Advisory Board, 1993).

A stigmatized place, the subject of this article, exists when people view an area as undesirable because of its association with a technological danger, in this case radiation. Such dangers, it is believed, should not exist and would not exist with proper procedures, controls, or management. Underlying technological stigma is the sense that those responsible for public safety have failed their fiduciary responsibilities and acted in a way that is antisocial, illegal, or immoral—they are to blame for the danger. The public avoids property in areas

---

(1997, see especially p. 18), who argued for health benefits from low levels of radiation exposure. For a number of years, the U.S. Nuclear Regulatory Commission has considered adopting regulations for radiation "below regulatory concern," on the basis that human physiology is capable of repairing minor damages to the body that might occur from very small exposures to radiation. Other scientists believe that even very small exposures to radiation can result in damage to DNA and possibly cancer. For example, the International Commission on Radiological Protection (1991) lowered radiation exposure limits for workers based on a conservative interpretation of a linear low-dose (no threshold) theory.

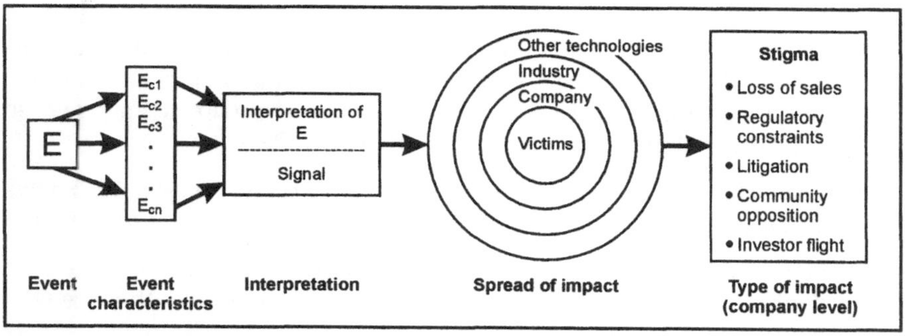

Figure 20.1 Model of social amplification of risk and stigma impacts.

placed at risk by these management failures. Eligible buyers either will not buy at all or will only buy when prices are discounted or when they are ignorant of the risk involved.

Hunsperger (1997) designed a study to examine the possible price impacts of the Rocky Flats, Colorado nuclear weapons plant on real estate in two nearby communities. The study determined that properties in the two communities experienced losses in value compared to other housing in the Denver metropolitan area. This finding was based upon (a) interviews with planners and local real estate experts, (b) review of analogous cases dealing with hazardous facilities and conditions, (c) examination of real estate records and transactions, and (d) a multiple regression analysis of property sales data. In collaboration with this study, we examined news media reports and conducted a survey to see if stigmatization based upon perceptions and behaviors in response to the Rocky Flats weapons facility may have accounted for some of the property value losses.

This research used the social amplification of risk model as the basis for an examination of risk, media, and stigma in connection with Rocky Flats. The basic hypothesis is that dramatic and highly visible stories about Rocky Flats mismanagement and the potential contamination of the adjacent communities stigmatized residential property in the subject communities. This process would provide a reasonable explanation for diminution of property values. We focused on events associated with the June 6, 1989 raid at Rocky Flats by the Federal Bureau of Investigation. This raid was an extraordinary event that created widespread news media coverage. News accounts of the raid and its consequences were reported over the following years with major stories up to the present time.

News coverage of an event, such as this FBI raid, may produce stigma impacts by (a) initiating awareness of a danger, (b) increasing perceptions of a known danger, (c) stimulating recall for people with latent negative reactions that have atrophied with time, and (d) increasing the number and geographical locations of people with knowledge about the danger.

The general public response to radioactive contamination and the ability of responsible management agencies to control exposure of residents near nuclear facilities provide a general background to the specific events surrounding the FBI raid. Together with our understanding about the role of the news media and the potential for technological stigma, we can structure three propositions to guide our research:

1. Media coverage after the FBI raid (compared to before) will communicate conditions consistent with stigma such as high risk perceptions including dread, distrust in facility managers, and aversion to the purchase of residential property associated with Rocky Flats.
2. These characterizations of Rocky Flats will continue to exist in 1995 news media presentations and in the evaluations of Denver area residents.
3. Stigma responses and behaviors will be more prominent for those residents who have heard of Rocky Flats events compared with those who have not heard of these events.

Background on Rocky Flats

The DOE's Rocky Flats facility, located in Jefferson County approximately 16 miles northwest of Denver, Colorado, began operations in 1953 to produce components for nuclear weapons. The plant produced "plutonium pits," the triggers for nuclear weapons, during the period 1953 to 1989 (DOE, 1995, pp. CO-11 to CO-39; CAO, 1992; U.S. Office of Technological Assessment, 1991; Colorado Council on Rocky Flats, 1993). Dow Chemical was the management contractor beginning in 1952 and Rockwell International operated the facility from 1975 through 1989.

On June 6, 1989, agents from the FBI and the EPA raided Rocky Flats with a warrant charging violations of environmental laws. These federal agents impounded a large volume of records and in August 1989 the U.S. Department of Justice impaneled a Federal Grand Jury to investigate the evidence and possibly authorize charges against Rockwell and its employees. The Grand Jury considered the case well into 1992 and prepared a report. Before any indictments were returned, Rockwell agreed in 1992 to plead guilty to ten violations of environmental law and paid a fine of $18.5 million. The Justice Department agreed not to prosecute any Rockwell employees (Abas, 1992).

The FBI raid was unprecedented—no other plant in the nuclear weapons complex (with 126 facilities) before or since has been raided—and the story was covered by national news media, including the major television networks. Subsequently, numerous stories about the raid, the Grand Jury, and conditions at Rocky Flats were communicated to residents of the Denver metropolitan area.

Rocky Flats was closed as a weapons plant in 1992 and designated as the Rocky Flats Environmental Technology Site. The new mission at the facility is to clean up and restore contaminated buildings, facilities, soils, and ground water while also managing the radioactive and chemical hazards that remain from the production period. Rocky Flats contains the largest inventory of plutonium in the DOE complex, about 60% of the total amount (NRC, 1996; Consortium for Environmental Risk Evaluation, 1995).

The FBI raid was not in and of itself a hazardous event. It did not create nor mismanage any hazard that existed at Rocky Flats. What it did was create public interest in the facility and a continuing scrutiny of its management. Subsequent newspaper stories encouraged distrust of the government and its contractors, prompting concerns about past, continuing, and future hazards at the facility. For example, one story reported how, in 1990, the Federal government resorted to a deliberate deception of state and local officials to maintain the option of future nuclear weapons production (Obmascik, 1994).[3]

This complex series of events generated questions about the risks to people off-site, especially in neighboring communities. Safety assurances from DOE and its contractors had lost credibility. One result of the FBI raid and the subsequent information about Rocky Flats was a major class action suit claiming damages to public health and property values.

A motion for a class certification of plaintiffs was filed in June, 1993. Geographical boundaries for health and property claims were based on exposure during the operating period to radioactive isotopes, such as plutonium, in an area east and south and approximately 90 degrees downwind from the Rocky Flats plant. The population claimed for the medical monitoring class area was about 40,000 persons (in 1990) and the property value class area contained about 15,000 properties ("Cook et al.," 1993).

Overview of the Studies

The studies reported here were conducted on behalf of the plaintiffs. One was an analysis of Rocky Flats newspaper coverage during the period 1989 to 1996 with a special focus on the 1989 FBI raid. The second was a survey of Denver metropolitan area residents' attitudes toward Rocky Flats and associated real estate purchase decisions. The purpose of these studies was to examine the issues of perceived risk from Rocky Flats, media coverage of events and stories related to Rocky Flats, and to increase understanding of possible stigma effects.

In Study 1, we examined the information communicated to the public about Rocky Flats in the time periods just prior to and after the raid with respect to stigma conditions such as attitudes and feelings of dread and distrust. In Study 2, a 1995 survey of experienced and/or potential Denver home buyers, we directly examined home buyers' attitudes toward Rocky Flats and home purchases in adjacent communities. We attempted to assess the impact of knowledge about the Rocky Flats events separately for those who had heard of the events (and therefore may show stigma impacts resulting from the FBI raid) and those who had not (and therefore may not show these stigma impacts).

Study 1

The newspaper analysis focused on three periods: (a) the year 1989, before and after the FBI raid (the "event" in the social amplification of risk model); (b) the period from 1990 through 1995; and (c) the period immediately preceding and during the 1995 survey data collection.

BACKGROUND ON DENVER NEWSPAPERS

In 1989, Denver was one of the major metropolitan areas in the nation. The 1990 U.S. Census placed the Denver-Boulder Standard Metropolitan Statistical

---

[3]Obmascik (1994) reports on the exit interview from the DOE of John Tuck, former undersecretary of energy, who was instrumental in negotiating a 1990 compliance agreement between DOE, EPA, and the State of Colorado. Tuck said that DOE could not meet its part of the agreement and knew it at the time they signed but made the agreement because they thought this was the best way to maintain weapons production as an option at Rocky Flats.

Area (SMSA) as the 22nd largest in the nation with a population of 1,848,319. When limited to only the Denver metropolitan area, the 1990 population was recorded as 1,622,980. This population is served by two major local newspapers, *The Denver Post (The Post)* and *The Rocky Mountain News (The News)*. In the 6-month period ending in March, 1991 *The Post* had an average net paid circulation of 248,493 (413,343 on Sundays) while *The News* averaged a net paid circulation of 374,009 (432,502 on Sundays; *The 1992 Information Please Almanac*, 1992).[4]

During the time periods in which we were interested, *The Post* and *The News* kept index systems of printed articles. In 1989 (the year of the raid), articles from *The Post* were indexed consistently into general categories. Articles on Rocky Flats could be located by finding the appropriate general categories (e.g., nuclear weapons, radiation) and reading article abstracts to determine their relevance to Rocky Flats events. The index system for *The News* was changed on June 1, 1989. Prior to this date, articles were filed by the Denver Library under general categories. Beginning with this date, *The News* articles were contained on a CD-ROM file with an electronic index system. The text of all published articles could be searched to select all articles that mentioned Rocky Flats (or Flats) at least once. A comparable index to the original *The News* system or to *The Post* system was not available.

FULL-TEXT ARTICLE ANALYSIS FOR TIME PERIODS A: 1989 AND C: JULY-OCTOBER, 1995

A full-text analysis was completed for the available Rocky Flats articles published in 1989 in both newspapers ($N = 609$). For *The Post* this database included 46 articles published before the FBI raid and 203 published after the raid. For *The News* it included 17 pre-raid articles and 343 post-raid articles. The same full-text analysis was completed for articles for the period July 1995 through October 1995 published in *The News* ($N = 38$). Articles that mentioned Rocky Flats, but not as a main point of the article (e.g., an obituary that mentioned earlier employment at Rocky Flats), were not included in this analysis.

The authors developed the coding protocol, tied to the theory of the social amplification of risk (Kasperson & Kasperson, 1996; Kasperson et al., 1992; White, Edwards, & Emani, 1990), and a coding sheet, which was refined during training sessions for research assistants who did the coding.[5] Coders were graduate students at the University of Oregon who were recruited for this task.

After reading each article, coders recorded general information about the article (e.g., date, article length), then indicated on a 7-point scale (–3 = *strongly negative* to +3 = *strongly positive*) their judgment of the general attitude that was conveyed about events at Rocky Flats, and the most important reasons for that attitude. Coders also indicated whether the article discussed exposures to radiation and whether the article indicated any issues of

[4]A third paper (*Boulder Camera*) did not have a comprehensive index, and its home delivery was limited to the Boulder County area. It was not included in the *1992 Information Please Almanac* which included all cities with at least one paper having an average net paid circulation of more than 75,000.

[5]Coding instruments and directions are available from the authors.

trust or blame in the Rocky Flats management. Harms and benefits to people and groups were indicated and, finally, coders marked any demands made or actions taken with respect to Rocky Flats.

ANALYSIS OF CITATIONS WITH HEADLINES FOR THE PERIOD B: 1990 - 1995

A brief analysis was made of all headlines of articles in which Rocky Flats appeared from 1990–1995 in the CD-ROM files of *The News*. For the period January 1990 through December 1995, the files provided 1,727 citations of Rocky Flats (i.e., Rocky Flats was mentioned at least once in each full-text story). The large volume of articles over this time period required a brief and efficient overview that we addressed by doing an analysis of the headlines only. In addition, a headlines analysis was completed for all articles from 1989 subjected to a full text analysis ($N = 609$) in order to check on the reliability of coding headlines only as compared to coding the full text of the same articles.

A separate and simplified coding sheet was used to analyze the headline items. On this simplified coding sheet, coders indicated their judgment of the attitude or affective tone conveyed by the headline on a 3-point scale (–1 = *negative* to +1 = *positive*). From the headline alone, it was often difficult to determine how the article might pertain to events at Rocky Flats. These articles were coded 99 = *can't tell*. In addition, coders indicated the general content of the headline (e.g., whether it mentioned Rocky Flats, environmental cleanup).

THE RELIABILITY OF NEWSPAPER ANALYSES

Twenty percent of the 609 articles from 1989 were double-coded to monitor coder reliability. Each pair of double-coded articles was compared, and the authors resolved any differences and calculated the reliability of each article. The standard for reliability of full-text articles was set at a conservative value of 75% agreement between independent analyses. Coders averaged 82% reliability, considerably above the standard.[6] Reliability was calculated as a simple percent by counting the number of correct responses and dividing by the total number of correct and incorrect responses. A response was correct only if the coder circled a correct response (e.g., the coder circled –2 for the attitude and the correct answer was –2). While it was also correct not to circle a response that should not have been circled, this was not included as a correct response (e.g., the coder did not circle "distrust of management" as a main reason for the attitude, and it was correct not to circle that item). A response was incorrect if a coder did not circle a correct response and if the coder circled an incorrect response. Note that this is a conservative estimate because the coding sheet frequently allows for multiple responses on a single line or table, and the coder could circle all or none of the responses. The coding reliability for the 1989 database is estimated at 86% with 120 double-coded articles with 100% reliability (after the authors corrected any mistakes) and 489 articles with 82% reliability based upon the double-coded data. A similar process and outcome was recorded for the full-text 1995 articles.

[6]Every fifth article was double coded in order to check reliability. Each coder was paired against every other coder in order to check reliability between each pair. Reliability articles were corrected, tabulated, and given back to each individual coder for review in order to continue training and avoid coder drift on the content analysis of articles.

For the full-text articles from 1989, the reliability of the coder's attitude judgment was calculated as well. The pairs of coders agreed on an exact attitude rating 70.8% of the time (172 articles). Pairs of coders differed by more than 1-point on the 7-point scale only 6 times (2.4% of the articles). Coders rated the attitudes in the same direction (i.e., positive, negative) or agreed that the attitude was neutral 92.6% of the time (225 articles).

The reliability procedure was the same for the headlines analysis, with 20% of the citations double-coded. The coding reliability for the headlines database was about 90% (i.e., 501 citations with 100% reliability after the authors corrected any mistakes and 2,004 citations with 87% reliability).

NEWSPAPER ANALYSES RESULTS

This section presents the results of the newspaper analysis chronologically in order to examine the FBI raid as an event in terms of the social amplification of risk model.

A full-text analysis of newspaper articles from 1989 was conducted for the periods before and after the FBI raid. This was followed by an analysis of the headlines only in the six years after the raid (1990–1995) and a full-text analysis for newspaper stories about Rocky Flats for the time immediately before and during the 1995 survey (July-October, 1995).

Articles from 1989

The Post was indexed consistently during the entire year 1989 and was therefore the best source of information to track any changes from before to after the FBI raid on June 6, 1989. Table 20.1 shows the publication of articles on Rocky Flats, by month, for the calendar year 1989. While the period from June 7, 1989, when the first article on the FBI raid was published, to December 31, 1989 makes up 57% of the year, the proportion of articles published in *The Post* during this same post-raid period was significantly higher (81.5%, $p < .01$). See Table 20.2 for a census of articles. In fact, *The Post* rated the FBI raid as the number one Colorado story of the year and as the number two story for the decade of the 1980s (Engdahl, 1989).

The articles were judged for overall attitude toward perceptions of Rocky Flats events on a scale ranging from +3 (*Strongly positive*) to –3 (*Strongly negative*). By averaging across articles appearing in a particular time period, we can calculate the average attitude conveyed by the newspaper in that time period. In *The Post* articles, the average attitude after the FBI raid was not significantly more negative than before the raid, suggesting that the potential for stigma was there before the raid. The raid, however, markedly increased the volume of negativity. There were more than eight times the number of negative articles than positive or neutral articles after the raid and the number of negative articles increased substantially (from 42 articles before the raid to 181 after the raid).

Not only did many more negative articles appear, but the representation of Rocky Flats changed following the raid when the hazards of the facility were more likely to be presented as something to be dreaded (a dreaded risk included characteristics such as the potential for catastrophic, uncontrolled, or extremely serious and involuntary risks; Slovic, 1987). These dreaded characteristics were associated with potential exposures of humans and the environment to radiation and/or chemicals. A greater proportion and number of articles

Table 20.1 Rocky Flats Articles Published in Two Denver Newspapers During 1989

		The Denver Post (January-December)		*Rocky Mountain News* (June-December)	
	Total articles	Positive/ neutral	Negative	Positive/ neutral	Negative
January	13	2	11	–	–
February	12	1	11	–	–
March	9	0	9	–	–
April	5	1	4	–	–
May	4	0	4	–	–
June	3	0	3	–	–
Pre-raid total	46	4	42		
6/6/89 FBI raid	**189**	**5**	**72**	**11**	**101**
July	74	3	26	13	32
August	65	4	23	4	34
September	72	5	21	6	40
October	65	2	16	3	44
November	49	2	12	5	30
December	32	1	11	1	19
Post-raid total	546	22	181	43	300
Total articles	592	26	223	43	300

after the FBI raid, compared to before the raid, communicated a condition of dreaded risk associated with Rocky Flats ($p < .01$, see Table 20.3).

Table 20.3 also clearly shows the increase in reports indicating distrust and blame of Rocky Flats management after the raid ($p < .05$). Interestingly, the management contractor seemed to bear the brunt of these issues of blame and trust since after the raid the articles were somewhat (although not significantly) more likely to indicate that the Department of Energy should be trusted more ($p > .35$). This difference may reflect attempts by James Watkins, then Secretary of Energy, to address hazardous conditions at Rocky Flats.

Post-Raid Comparison of The Post and The News

Articles from *The Post* were selected in order to compare pre- and post-raid periods in 1989 because the *Post* was indexed consistently throughout that year.

Table 20.2 A Census of *Denver Post* Articles in the Pre- and Post-raid Periods

	Pre-raid		Post-raid	
	%	No.	%	No.
Days of the year	43.0	157	57.0	208
Articles published	18.5	46	81.5	203**
Paragraphs published	16.2	613	83.8	3,168**

**$p < .01$.

Table 20.3 A Comparison of *Denver Post* Article Attributes, Pre- Versus Post-raid

Average attitude	Pre-raid −1.33		Post-raid −1.45	
	%	No.	%	No.
Number of positive and neutral articles	8.7	4	10.8	22
Number of negative articles	91.3	42	89.2	181
Potential exposure to radiation	50.0	23	68.5	139*
Dreaded potential exposure to radiation	43.5	10	70.5	98**
Increased trust (contractor)	4.4	2	4.4	9*
Decreased trust (contractor)	39.1	18	59.1	120
No blame in contractor	15.2	7	16.3	33*
Blame contractor	28.3	13	47.8	97
Increased trust (DOE)	10.9	5	16.3	33
Decreased trust (DOE)	60.9	28	49.8	101

*$p < .05$. **$p < .01$.

The index system for the *Post* was not ideal, however, because articles were indexed by hand under general categories, and it was therefore not possible to retrieve and examine any article that mentioned Rocky Flats. It is unknown how many relevant articles were excluded due to the less efficient nature of this index system. *The News*, while not indexed consistently throughout 1989, is a better database for article selection starting June 1, 1989 when it was entered onto CD-ROM. This change in format allows searches for all articles that mentioned Rocky Flats at least once as opposed to being limited (as with *The Post*) to articles that were indexed under general categories such as "nuclear weapons." Starting at about the time of the June 6 FBI raid and continuing up to the time of the telephone survey, it was possible to examine *The News* articles or headlines for articles that mentioned Rocky Flats at least once. Before drawing any conclusions that rely on data from these two different sources, however, we must examine the comparability of articles from *The Post* with articles from *The News*.

Table 20.1 shows the frequency of articles and their positive/neutral or negative values appearing in *The News* after the raid. Not surprisingly, more articles were coded from *The News* based upon an electronic word search for Rocky Flats as opposed to the manual search required for *The Post*. The general pattern of articles, however, remained similar. There were no statistical differences between *The Post* and *The News* in the average attitude conveyed by the post-raid articles in 1989 (−1.5 and −1.3, respectively; $p > .15$). Other sections of the content analysis also were compared across the two newspapers during this time period. No statistical differences were found on issues of trust or blame in the management contractor or DOE (all $p > .1$). *The News*, however, did seem to be somewhat less negative in its treatment of three issues: trust as a reason for the overall attitude; dread toward a suspected or actual exposure to radiation or chemicals; and manageability of the same exposure ($p < .10$; see Table 20.4).

The newspaper analyses that follow come from *The News*. Based on the post-raid comparison of *The Post* and *The News* above, *The News* is expected to

Table 20.4 Comparison of *The Post* and *The News*, Post-raid, 1989

Average attitude	*The Post* −1.5		*The News* −1.3	
	%	No.	%	No.
Distrust of management as reason for attitude	72	147	58	200**
Dread toward potential exposure to radiation	70.5	98	61.2	156
Potential radiation exposures are manageable	44.9	62	57.6	147*

*$p < .05$. **$p < .01$.

be somewhat less negative toward the management of Rocky Flats and in their assessments of potential exposure to radiation or chemicals.

1990-1995: Citations and Headlines Analysis

Headlines from stories reporting about Rocky Flats in *The News* files for the period 1990 to 1995 were coded for attitude and general content. Attitudes toward Rocky Flats, based upon newspaper headlines, were more difficult to assess than was the case for the full-text articles for two reasons. First, the authors did not examine the full text of the articles to assess their relevance to Rocky Flats events so that irrelevant articles, such as obituaries of former employees, were included in the analysis. Second, even when their relevance could be determined, the headlines contained much less information. In order to compare the ratings of the two data sources, *The News* headlines and full-text articles were analyzed separately for 1989. In general, the headlines conveyed a less negative attitude compared to the full-text articles (1989 headline ratings averaged −.64; 1989 full-text ratings averaged −.80; $p < .0001$). Still, the headlines often gave strong clues. The majority of the headlines matched the text (62.4%). An additional 28.6% of the headlines from stories that were negative in the full-text analysis were classified as neutral (21.6%), positive (2.0%), or "can't tell" (4.9%) in the headlines analysis. These comparisons suggest that the headlines database will understate the negative message contained in the articles.

A total of 1,727 *The News* headlines were assessed for the period 1990–1995. Of these, 55.5% ($n = 959$) were coded as "can't tell" for the attitude because the headlines did not state how the articles related to events at Rocky Flats;[7] 22.2% ($n = 384$) were coded with a negative attitude, 18.3% ($n = 315$) were coded as neutral, and 4.0% ($n = 69$) were coded as positive. Of those articles where an attitude could be assessed, the average overall attitude communicated to readers was negative (−.41 on a scale of −1 = negative to +1 = positive).

[7]Of the 959 headlines coded "can't tell," 923 (96%) did not mention Rocky Flats in the headline. Occasionally ($n = 36$), Rocky Flats was mentioned in the headline but coders still could not determine the article's relevance to conditions at Rocky Flats (e.g., "Flats grand juror McKinley ponders running for Congress").

From the time of the June 6, 1989 FBI raid through 1995, a total of 2,070 articles that mentioned Rocky Flats at least once appeared in *The News*. This number was at its highest in the seven months of 1989 following the raid ($n = 343$) and steadily decreased from 1989 to 1995 although the number remained high ($n = 225$) even in 1995.

The overall attitude conveyed by the headlines (for those headlines where an attitude could be determined) remained consistently negative throughout the time period although it seems to be somewhat less negative in 1995 (average attitude = –.23) compared to the earlier time periods (average attitudes range from –.40 to –.46, all $p < .06$). As shown in Table 20.5, a consistent negative attitude pattern persisted over the years following the FBI raid.

Study 2 and Discussion

The newspaper analysis clearly showed that the FBI raid produced more articles, especially negative articles, concerning events at Rocky Flats. In particular, the event marked an increase in articles suggesting that Rocky Flats should be dreaded and its facility managers should not be trusted. The citations analysis showed that the average attitude conveyed by the headlines remained consistently negative even six years after the raid. The results of this examination of newspaper coverage over a six-year period (Study 1) indicate that the negative messages were available to support public stigmatization or shunning of Rocky Flats and its communities.

An analysis of newspaper coverage cannot specify, however, the public's views toward home purchases in communities near Rocky Flats. As a result, in Study 2 we conducted a survey of Denver area residents in order to assess public attitudes toward Rocky Flats and to examine whether knowledge of the potentially stigmatizing events of the FBI raid and its aftermath impacted attitudes toward home purchases in nearby communities. A more extensive newspaper analysis was also done for the period before and during the survey.

Table 20.5 Newspaper Article Data Base—*The News*

	1989	1990	1991	1992	1993	1994	1995	Total
Number of articles in data base	343[a]	434	304	319	210	235	225	2,070
Number of headlines with ratings	315	175	129	173	100	100	91	1,083
Average rating of headline attitude	–.64[b,d]	–.46[c,d]	–.40[c]	–.45[c,d]	–.44[c,d]	–.41[c]	–.23[c]	

[a]This is based on the seven months of 1989 after the June 6 FBI raid.
[b]The average attitude in 1989 based on the full text analysis was –.78 on same range of –1 to +1.
[c]The average attitude is significantly less negative than 1989, $p < .01$.
[d]The average attitude is more negative than 1995, $p < .05$.

MEDIA COVERAGE AROUND THE TIME OF THE 1995 SURVEY

Most of the telephone survey interviews took place during September 1995. In order to examine how Rocky Flats was represented just before and during the time that survey interviews were conducted, the number of headlines and attitude conveyed by headlines were compared for three 4-month periods: July-October 1994; March-June 1995; and July-October 1995. These results are shown in Table 20.6.

Table 20.6 indicates that headline attitudes appearing in the four-month period prior to and during the time when survey interviews took place were similar to the same time period a year earlier, in 1994 ($p < .75$). Fewer articles appeared in the survey time period compared to the previous four-month time period (March-June 1995) although the headlines were somewhat more negative ($p < .06$). The headlines from these time periods, however, remained less negative than in previous years (e.g., –.41 in 1994 and –.44 in 1993, $p < .06$).

A full-text analysis of those articles ($n = 38$) with Rocky Flats as a primary or secondary point was made for the four months preceding survey completion (July-October 1995). Comparison of these 38 articles to primary/secondary articles from *The News* that appeared in 1989 after the FBI raid ($n = 343$) yielded some interesting results (see Table 20.7).

The percentage of articles reporting an exposure to chemicals or radiation from Rocky Flats as a main issue decreased, as did the proportion of articles mentioning a harm to health or the environment in the local community. However, the proportion of articles indicating dread and manageability of the risks from a human or environmental exposure remained the same ($p > .90$).[8] In other words, the percentage of articles with human or environmental exposure to radiation or chemicals as a main issue decreased, but of the articles with an exposure as a main issue, the majority of articles in both years portrayed the exposures as dreaded and manageable.

To believe that the risks are manageable and to have trust and confidence in the existing management can be two different things. Issues of trust and blame are interwoven throughout the history of Rocky Flats. A comparison of newspaper articles from 1989 and 1995 suggests that distrust and blame decreased (while trust increased) from the time in 1989 when Rockwell was the

Table 20.6 Comparison of Newspaper Headlines

	July-October 1994	March-June 1995	July-October 1995
Number of articles	64	82	58
Average attitude[a]	–.26	+.04[c]	–.3[b]

[a]The average attitude is based on the 26-27 headlines in each period for which coders could identify the article's relevance to Rocky Flats events.

[b]The average attitude based on the full text analysis was –.89 on a scale that ranged from –3 to +3. After rescaling the attitude based on the full-test analysis to the same –1 to +1 scale used in the headlines analysis, the average attitude was –.31.

[c]The March-June 1995 attitude was more positive than the July-October 1995 period, $p = .06$. No other comparisons approached significance.

[8]Dread and manageability of the risk was assessed only in those articles mentioning a contamination.

Table 20.7 Comparison of Full-text Analyses of Post-raid, 1989 with Pre-survey, 1995 (*The News* only)

	Post-raid, 1989		Pre-survey, 1995	
	%	No.	%	No.
Radiation/chemical exposure is a main issue	74.3	255	52.6	20**
Of these:				
Exposure is dreaded	61.2	156	60.0	12
Exposure is manageable	57.6	147	60.0	12
Harm to health or environment	71.7	246	52.6	20*
Distrust in management is a main reason for attitude	58.3	200	34.2	13*
Decreased trust in management contractor	50.7	174	31.6	12
Increased trust in management contractor	4.4	15	23.7	9***
Blame management contractor	40.2	138	21.1	8**
Increased or decreased trust in DOE	58.0	199	18.4	7***

*$p < .05$. **$p < .01$. ***$p < .001$.

management contractor and the time of the 1995 survey after five years of EG&G as the management contractor and just after Kaiser-Hill had taken over (see Table 20.7). Specifically, a larger proportion of the post-raid 1989 articles compared to the pre-survey 1995 articles indicated distrust of management in general and decreased trust in the management contractor in particular ($p < .05$).

Overall, *The News* articles on the subject of Rocky Flats during the period July-October, 1995—immediately before and during the survey interviews—focused on cleanup issues and were less negative than those published in the aftermath of the 1989 FBI raid.

THE 1995 SURVEY METHODOLOGY

A total of 604 persons were interviewed during the period August 31, 1995 to October 1, 1995. Targeted respondents were people who were knowledgeable and experienced in the Denver area residential housing market. These respondents were defined as people who had an interest, knowledge, and/or recent experience in considering or making a home purchase. The formal criteria were residents 18 years of age and older, currently residing in the Denver metropolitan area who have: (a) bought a home in the last five years; (b) plans to purchase a home in the next couple of years; or (c) been actively in the market for a home during the past five years. At the 95% level of confidence, the margin of error for the Denver area sample of 416 was 5.0% and for the Arvada/Westminster sample of 188 it was 7.2%. The response rate was conservatively estimated at 54%.

All interviews were conducted by telephone from the University of Maryland Survey Research Center at the College Park, Maryland campus. The Survey Research Center also developed weights to correct for sample bias and account for conditions such as the number of household telephones, adult household size, and eligibility for the sample. The questions asked for evaluations of the major downwind communities (Arvada and Westminster) and two comparable communities located in other areas of the Denver metropolitan area.[9] Respondents were asked about the advantages and disadvantages of Arvada and Westminster. Each respondent was asked about knowledge of the location of Rocky Flats, perceptions of the plant, and an evaluation of how the plant affected the desirability of homes in Arvada and Westminster. A question set was used to inquire about how distance from Rocky Flats and housing costs might affect the desirability and acceptability of a house.[10]

SURVEY RESULTS

This section provides a brief summary of key findings from the survey. The focus is on: (a) Impressions of Rocky Flats; (b) Effects of Rocky Flats on the desirability of homes in Arvada and Westminster; (c) Effects of distance and price incentives on willingness to buy in the Arvada and Westminster areas; (d) Recollection or knowledge of the 1989 FBI raid; and (e) Evaluations of the legacy and safety of Rocky Flats.

Impressions of Rocky Flats

Survey respondents were asked to identify Rocky Flats (as opposed to the Rocky Mountain Arsenal) and to locate it in the Denver metropolitan area. Those who had heard of Rocky Flats were asked to provide up to three images or word associations of Rocky Flats and they were then asked to provide an overall impression of Rocky Flats using a five-point scale from "strongly positive" to "strongly negative." For the Denver metropolitan area respondents, 7.0% provided positive responses, 28.3% were neutral (neither positive nor negative), and 63.8% were negative. Only 0.9% didn't know.

Effects on Home Desirability in Arvada and Westminster

Denver metropolitan area respondents were asked if Rocky Flats would make a house located in Arvada or Westminster more or less desirable (using a 5-point scale). Only 1.4% said that Rocky Flats would make a house "somewhat" or "considerably" more desirable, while 64.2% said the plant would make a house "somewhat" or "considerably" less desirable. A third of the sample, 32.8%, said Rocky Flats would "make no difference."

Effects on Willingness to Buy in the Arvada and Westminster Areas

Denver metropolitan area respondents ($n = 416$) were presented with a set of four progressive scenarios to measure the effects of distance and price on their willingness to buy a house in the target area. The order of the presentation was to move from the least costly to the most costly, from the seller's point of view. The order and the outcome of the questions is shown in Figure 20.2. If

---

[9]The two comparable communities, Ken Caryl and Highlands Ranch, were selected by an experienced Denver appraiser and are located southwest and south in the Denver metropolitan area. Arvada and Westminster are located adjacent to each other northwest of the Denver metropolitan area.

[10]A copy of the survey instrument and/or the coding protocols for full text and headlines analysis is available from the authors.

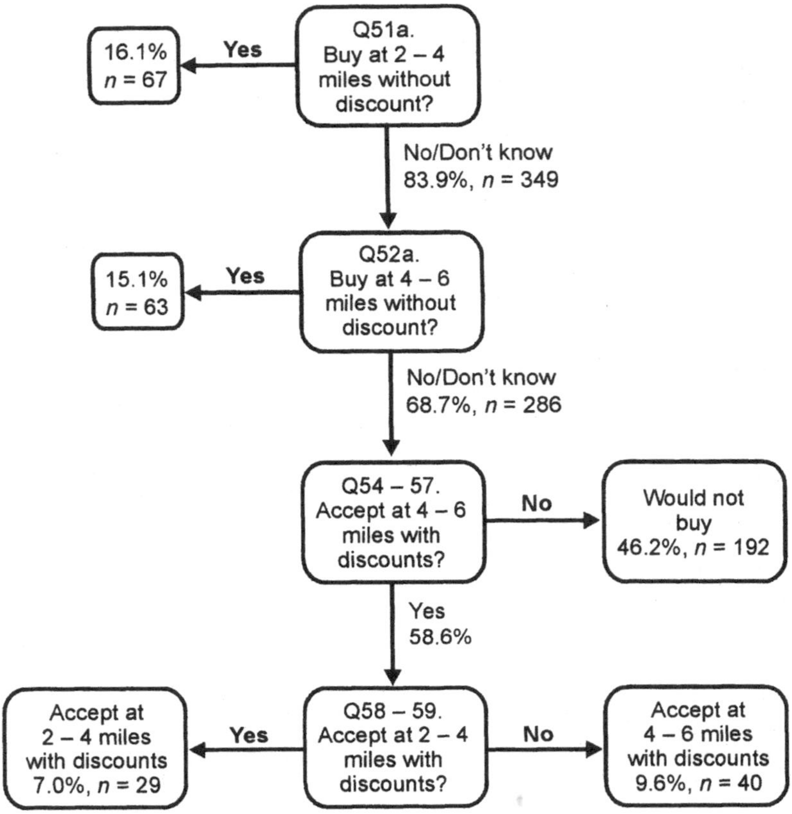

Notes:
- *N/n* = number of respondents.
- Twenty-five people (6.0%) said they didn't know or could not answer.
- Data for the Denver metropolitan area do not include Arvada/Westminster.
- Due to rounding, figures may not always sum to the total.

Figure 20.2 Design and distribution of results from a question set with distance and discount options applied to a hypothetical house purchase in the Arvada/ Westminster communities. Denver metropolitan area respondents ($n = 416$).

the respondents accepted the conditions at any point in the question set, this answer was recorded as their final outcome decision.

The first two mitigation questions asked people if they would accept an otherwise desirable house located 2-4 miles from Rocky Flats, to which 16.1% said "yes." Those who said "no" were then asked about a house at a distance of 4-6 miles, which an additional 15.1% accepted.

The remaining two-thirds of the sample were then offered price discounts (of $5,000, then $10,000, and finally at a price of their own choosing) for a house at the 4-6 mile distance. A total of 16.6% of the sample accepted a combination of the 4-6 mile distance and a price discount. These respondents were then asked if they would accept the same conditions or an even better discount (of their own choosing) for a house in the 2-4 mile range, which 7.0% percent did. Therefore, the final outcome for distance and price acceptance was that 7.0%

accepted at the 2-4 mile distance and 9.6% accepted at the 4 to 6 mile distance. The final outcomes for the distance and cost incentives are shown in Table 20.8.

A substantial proportion (46.2%) rejected offers of distance and price at the 4-6 mile distance. In other words, they stated they would not purchase a home in that area regardless of the price discount.

Recollection of the 1989 FBI Raid

Respondents were asked a single question about whether they had heard of the FBI raid and the subsequent $18.5 million fine paid by Rockwell. A substantial majority (61.3%) of the Denver metropolitan area respondents said they had heard of these events and 80% thought values were decreased by association with Rocky Flats. A comparison of those who knew about the FBI raid with those who did not is presented in Table 20.9.

Respondents who recalled the raid were more familiar with Arvada and Westminster, were more likely to say that property in these two communities would be more desirable without Rocky Flats, and to think that Rocky Flats had adversely affected property values. They were more likely to say Rocky Flats is a health risk, less likely to say they would buy a house within 6 miles of Rocky Flats even with a combination of distance and price incentives, and less likely to buy without any discount at either of the two distances specified in the questionnaire.

In some cases, the respondents who recalled the FBI raid and those who had not heard of it, recorded similar responses. Both groups were equally likely to have negative impressions of Rocky Flats, and to believe that Rocky Flats is unsafe or probably unsafe. This suggests that respondents who did not recall the FBI raid were influenced by other accounts of Rocky Flats. On the question about very small levels of contaminants in residential areas, both groups of survey respondents recorded similar proportions of moderate or high risk responses. This question indicates that those who had heard of the raid and those who had not were similar in their basic risk values. What stands out in these data is the greater proportion of negative evaluations of property near Rocky Flats on the part of those with knowledge of the FBI raid.

Evaluations of the Legacy and Safety of Rocky Flats

The FBI raid and its subsequent events, such as the investigations of a federal grand jury and the large fine paid by Rockwell, produced substantial news coverage. Other events, not directly connected with the raid, were reported before, during, and after this high-profile story was in the news. Almost half the Denver metropolitan area respondents (48.9%) said they had heard of pos-

Table 20.8 Price and Distance Discounts

	Accept without discount	Accept with discount, mean value	Mean value with and without discount
2–4 miles	$n = 67$	$n = 29$ $30,740	$n = 96$ $9,270
4–6 miles	$n = 63$	$n = 40$ $15,747	$n = 103$ $6,084

Table 20.9 Knowledge or Recollection of the 1989 FBI Raid

	YES, respondent had heard of raid		NO, respondent had not heard of raid	
	No.	%	No.	%
Total respondents (n = 408)	255	61.3	153	36.9
Familiar with Arvada	195	76.4	84	54.5***
Familiar with Westminster	196	77.0	100	65.0**
Negative impression of Rocky Flats	144	63.5	57	62.4
Believe Arvada/Westminster would be more desirable without Rocky Flats	133	73.4	52	54.6**
Believe Rocky Flats is a health risk	170	66.8	88	57.3*
Thought Rocky Flats had affected values of property located within 6 miles	229	89.7	102	66.7***
Thought Rocky Flats was unsafe or probably unsafe	201	78.8	116	75.9
Believe that contaminates in residential areas, at very small levels, would be a moderate or high risk	146	70.7	101	67.4
Summary outcomes for buying a house near Rocky Flats:[a]				
Would not buy at all within 6 miles of Rocky Flats	128	53.2	62	42.5
Would buy either with a discount or without any discount	112	46.8	83	57.5

[a] $x^2 = 4.1, p < .05$.
*$p < .05$. **$p < .01$. ***$p < .001$.

sibly hazardous events before the FBI raid. When confronted with two statements, one arguing for the present safety of Rocky Flats and the other claiming that hazardous conditions exist at the plant, over three-quarters (77.4%) of the Denver metropolitan area respondents said they thought the facility was unsafe (31.3%) or probably unsafe (46.1%). This finding is consistent with other findings of a negativity bias (Pratto & John, 1991; Skowronski & Carlston, 1989).

Conclusions

The research reported here examined newspaper reports about Rocky Flats beginning with January 1989 and continuing through to 1995. This research focused on the intense news media interest in the FBI raid of June 6, 1989 and the subsequent treatment of Rocky Flats as an ongoing story. Consistent with the conceptual model entitled "the social amplification of risk," the messages from the news media conveyed the contamination from Rocky Flats, distrust of the plant operators and managers, and serious concern about potentially harmful health effects in nearby communities. These characterizations of the risks associated with Rocky Flats also were found in a survey of the Denver metropolitan area target population that made up a second part of the research. The survey found considerable knowledge of the FBI raid and responses of aversion and avoidance with regard to buying real property in two nearby communities. These perceptions, attitudes, and responses are consistent with stigmatization of property in these communities and provide an explanation for the negative differential in property values, which was documented in the Hunsperger (1997) studies. The responses of people who recalled or knew about the FBI raid were more negative than the responses of people who did not have specific recall of this event.

Over the period 1989-1995, the DOE made a number of changes in the operations at Rocky Flats, attempting to improve conditions there and modify public perceptions of the facility. Within a few months following the FBI raid, they replaced the long-term operator, Rockwell International. In 1992, they terminated the facility as a weapons-production plant and designated it as the Rocky Flats Environmental Technology Site. Numerous programs to clean up the environmental problems at the facility were initiated during the 1990s. Despite these efforts, the Rocky Flats plant and nearby residential communities are viewed with concern as places with significant risks to public health. The survey supports a finding that the FBI raid played a key role in the stigmatization of the target communities within the Denver metropolitan area. While negative impressions toward Rocky Flats were similar for those who recalled or knew about the FBI raid and those who did not, respondents who recalled the raid exhibited more negative responses toward the communities and housing values associated with Rocky Flats.

Acknowledgments

An earlier version of this article was presented as a paper at the Risk, Media, and Stigma Conference held at the Annenberg School of Communications, The University of Pennsylvania, Philadelphia on March 23-24, 1997. We would like to thank the conference hosts Kathleen Hall Jamieson, Dean of the Annenberg School, and Howard Kunreuther, Co-Director of the Center for Risk Management and Decision Processes at the Wharton School. Data collection and analysis was funded by the law firm of Berger and Montaque, Philadelphia. We would like to thank Daniel Berger, Merrill G. Davidoff, Peter Norberg, and Jonathan Auerbach. We appreciate the help of James Williams, Planning Information Corporation, Denver, Colorado who obtained the newspaper files upon which that portion of the analysis was based and Wayne Hunsperger, Hunsperger and Weston, Denver, who provided access to his professional appraisal studies of property values in the Rocky Flats area. Janet Douglas and Leisha Mullican at Decision Research prepared the manuscript.

Coping with Stigma

This final section considers what can be done to cope with stigma. Kunreuther and Slovic identify four strategies for coping with stigma. Walker, in a response to Kunreuther and Slovic, questions the assumption that society ought to adopt the normative values of reducing stigma through a variety of strategies or techniques. Fischhoff proposes a definition of stigma and draws a number of inferences including that "stigma reflects a moral statement of what constitutes unacceptable behavior." When this is the case, Fischhoff says that attempts to change or adjust perceptions, evaluations, or behaviors that stigma produces, presume a moral superiority on the part of those applying strategies to cope with stigma. In a final comment, he notes that this "may not be a comfortable, or appropriate, role for social scientists."

21 Coping with Stigma: Challenges and Opportunities

Howard Kunreuther and Paul Slovic

Introduction

How should we deal with stigma and its impacts? This question would probably seem absurd to an ancient Greek, about to brand someone with a visible mark to signify that this person was immoral or dangerous and thus undesirable, someone to be denigrated and avoided. Stigmatization in ancient Greece was a form of risk management. Even today, stigmatization can be a positive force for risk reduction. Food manufacturers or restaurants with lax safeguards against bacterial contamination, for example, deserve to be stigmatized, and the economic costs associated with that stigma may serve as a deterrent or punishment. In these situations stigmatization and its consequences may thus be a good thing for society.

But if stigma were purely beneficial, we would not be writing this paper. Stigma is a powerful force in our modern industrial society because science, technology, and communications media often interact with the idiosyncrasies of human cognition, perception, and emotion to produce extreme disruption in the lives of industries, products, communities, and people. In many instances we sense that the social and economic response is exaggerated, even unwarranted, leading to impacts far more serious than the initial threat. In such cases, we face the challenge of how to manage stigma and reduce the vulnerability of important products, industries, and institutions to its effects.

The challenge of managing stigma is nowhere better illustrated than in the case of British beef. The possibility that eating beef might lead to a fatal brain disease had a catastrophic impact upon the beef industry in Britain and threatened the very economic and political stability of the country. Following a series of accounts in the press in the fall of 1995, beef consumption in the United Kingdom fell 20% by Christmas 1995. Approximately 1.4 million British households stopped buying beef and thousands of schools took beef off the menu (Powell, this volume). The widespread publicity given to the deaths of a dozen young people from a mysterious disease that looks similar to BSE in cattle and CJD in humans led to the destruction of hundreds of thousands of cattle. If this action and the avoidance of British beef by consumers in the United Kingdom and throughout Europe prevented an epidemic of this gruesome and fatal human disease, then the response could be considered quite appropriate—a public health miracle perhaps. If there was really little or no danger from eating British beef, the response could be considered exaggerated and destructive.

Episodes of stigmatization such as the BSE scare are noteworthy because they are textbook examples of what has been called the "social amplification of risk" (Kasperson et al., 1988) and illustrate a new form of societal vulnerability. Whereas human health was the primary vulnerable commodity in the past, increasing technical and medical sophistication, combined with hypervigilant monitoring systems to detect incipient problems, make such scourges less likely now. But the price of this vigilance, based in no small part upon the incredible ability of modern media to "spread the word," is the impact that this information itself has upon social, political, industrial, and economic systems. Thus we live in a world in which information, acting in concert with the vagaries of human perception and cognition, has reduced our vulnerability to pandemics of disease at the cost of increasing our vulnerability to massive social and economic catastrophes. Is this latter vulnerability inevitable? What might be done to reduce it without losing the benefits of hypervigilant warning systems? This paper examines the causes of stigma and explores ways to manage it better.

Stigma: A Conceptual Framework

THE ROLE OF IMAGERY AND AFFECT

To create and evaluate strategies for dealing with the destructive effects of stigma we must understand something of its nature. Building upon the work of a large number of behavioral scientists, we propose a model in which stigma is based upon negative imagery that has become associated with places, products, technologies and, of course, people.

The eminent learning theorist, Hobart Mowrer (1960a, b), for example, concluded that human behavior is guided and controlled in a generally sensible and adaptive manner by conditioned emotional responses to images that could be viewed as "prospective gains and losses." More recently Damasio (1994) argues that human thought is made largely from images, broadly construed to include perceptual and symbolic representations. Through experience, these images become "marked" by positive and negative feelings (Mowrer and other learning theorists would call this conditioning) and these feelings motivate action. When a negative marker is linked to an image it sounds an alarm and motivates avoidance. When we think of the prime targets for stigmatization in our society, members of minority groups, the aged, homosexuals, drug addicts, and persons afflicted with physical deformities and mental disabilities, we can appreciate the affect-laden images that, rightly or wrongly, are associated with such individuals (Goffman, 1986).

Empirical support for the proposed relationship between images, affect, decision making and stigma has come from a program of research at Decision Research (Slovic, Flynn, et al., 1991). This work was motivated by a practical question: "What is the potential for a high-level nuclear waste repository at Yucca Mountain to stigmatize the city of Las Vegas and the State of Nevada, thus producing significant economic impacts in those places?" Building upon previous research linking images and behavior, the studies were designed to develop a measure of environmental imagery, assess the relationship between imagery and choice behavior, and describe economic impacts that might occur as a result of altered images and choices.

Specifically, research was designed to test the following three propositions: (a) Images associated with environments have diverse positive and negative affective meanings that influence preferences (e.g., in this case, preferences for sites in which to vacation, retire, find a job, or start a new business); (b) A nuclear-waste repository evokes a wide variety of strongly negative images, consistent with extreme perceptions of risk and stigmatization; and (c) The repository at Yucca Mountain and the negative images it evokes will, over time, become increasingly salient in the images of Nevada and of Las Vegas. If these three propositions are true, it seems quite plausible that, as the imagery of Las Vegas and of Nevada becomes increasingly associated with the repository, the attractiveness of these places to tourists, job seekers, retirees, and business developers will decrease and their choices of Las Vegas and Nevada within sets of competing sites will decrease.

Table 21.1 illustrates the results obtained by asking one respondent to associate to each of four cities and, later, to rate each image on an affective scale ranging from –2 (very negative) to +2 (very positive). An overall affective score is obtained for each stimulus city by summing the individual ratings. The cities in this example produce a clear affective ordering with San Diego being perceived most favorably and Los Angeles most negatively. The research showed that image scores such as these were highly predictive of expressed preferences for living, working, and vacationing in different places. In one study we found that the image score predicted the location of actual vacations taken over the next 18 months.[1]

In sum, the research pertaining to Yucca Mountain supported the three propositions that the study aimed to test: Images of cities and states, derived from a word-association technique, exhibited positive and negative affective meanings that were highly predictive of preferences for vacation sites, job and retirement locations, and business sites (Proposition 1). The concept of a nuclear-waste storage facility evoked extreme negative imagery (Proposition 2) indicative of stigmatization. Dominant associations to a repository were "dangerous," "death," "pollution," "bad," "undesirable," and "somewhere else." The nuclear-weapons test site, which has been around far longer than the Yucca Mountain nuclear-waste project, was found to have led to a modest amount of nuclear imagery becoming associated with the state of Nevada. This provided indirect evidence for Proposition 3, which asserts that nuclear-waste related images will also become associated with Nevada and Las Vegas if the Yucca Mountain Project proceeds. Moreover, nuclear imagery, when present in a person's associations to Nevada, was found to be linked with much lower preference for Nevada as a vacation site, indicative of a stigmatization response.

Nuclear facilities are not the only technologies with highly stigmatizing imagery. The image of chemical technologies is so stigmatized that when you ask college students or members of the general public to tell you what first comes to mind when they hear the word "chemicals," by far the most frequent

[1]The role of imagery in relocation decisions is illustrated by a Seattle friend's comment to one of the authors in a letter: "Ed's saying Montana may be the answer to middle age contentment. I think Montana has too many movie stars and militia, and too many creeps in cabins. I refuse to cross state lines in an Easterly direction."

Table 21.1 Images, Ratings, and Summation Scores for Respondent 132

Image no.	San Diego		Denver		Las Vegas		Los Angeles	
	Image	Rating	Image	Rating	Image	Rating	Image	Rating
1	Very nice	2	High	2	Rowdy town	–2	Smoggy	–2
2	Good beaches	2	Crowded	0	Busy town	–1	Crowded	–2
3	Zoo	2	Cool	2	Casinos	–1	Dirty	–2
4	Busy freeway	1	Pretty	1	Bright lights	–1	Foggy	–1
5	Easy to find way	1	Busy airport	–2	Too much gambling	–2	Sunny	0
6	Pretty town	2	Busy streets	–2	Out of the way	0	Drug place	–2
Totals		10		–1		–7		–9

Note: Based on these summation scores, this person's predicted preference order for a vacation site would be: San Diego, Denver, Las Vegas, and Los Angeles. Source: Slovic, Layman, Kraus, et al. (1991, p. 690).

response is "dangerous" or some similar response (e.g., toxic, hazardous, poison, deadly, cancer); beneficial uses of chemicals are rarely mentioned. National surveys have found that up to 75% of the public agree with the statement, "I try hard to avoid contact with chemicals and chemical products in my everyday life" (Krewski, Slovic, Bartlett, Flynn, & Mertz, 1995).

The salience of certain images may help explain why a product, place, or industry is likely to be stigmatized even when only one expert views a situation as being dangerous and points out what can happen from an adverse incident. There is considerable empirical evidence suggesting that individuals often dwell on the consequences associated with specific events, particularly if they are characterized in a salient way. In such situations little attention is given to the likelihood of those consequences (Camerer & Kunreuther, 1989). Consider the following example of siting liquified natural gas (LNG) facilities in California.

In the 1970s California was considering building a terminal for receiving LNG and converting it into a gas in order to distribute it through pipelines. Three locations were proposed for the LNG terminal. One of the proposed areas was Oxnard, where there was considerable opposition to the facility by citizen groups. One of the opposition groups focused attention on a series of worst-case scenarios showing the number of fatalities from accidents and vapor cloud explosions.

A series of maps of the community were drawn which depicted the impacts of these vapor cloud explosions should the wind be blowing from different directions. Any resident of Oxnard could find their house covered by a vapor cloud on at least one of the maps. This study was used by citizen groups to claim that a LNG facility would stigmatize Oxnard and decrease property values, even though experts asserted that the probabilities of these events occurring were extremely low.[2] The impact of the opposition was sufficiently strong that the California State legislature passed a siting act that redefined eligible siting areas and excluded Oxnard because it was viewed as too dangerous (Kunreuther & Linnerooth, 1982).

STIGMA AND THE SOCIAL AMPLIFICATION OF RISK

Although stigmatizing images are created through direct experiences such as bad odors, ugly landscapes, accidents, illnesses, etc., the greatest contributor to stigma, by far, is the news media through the process known as social amplification of risk. Social amplification, as schematized in Figure 21.1, is triggered by the occurrence of an adverse event such as a major or minor accident, a discovery of pollution, an incident of sabotage, and so on.

Risk amplification reflects the fact that the adverse impacts of such an event sometimes extend far beyond the direct damages to victims and property and may result in massive indirect impacts such as litigation against a company or loss of sales, increased regulation of an industry, and, of course, stigmatization. In some cases, all companies within an industry are affected, regardless of which company was responsible for the mishap. Thus, the event can be thought of as a stone dropped in a pond. The ripples spread outward, encompassing first the directly affected victims, then the responsible company or agency, and, in the extreme, reaching other companies, agencies, or industries. In addition to the British beef incident mentioned earlier, other well-known examples of events resulting in extreme higher-order impacts include the chemical manufacturing accident at Bhopal, India, the disastrous launch of the space shuttle Challenger, the nuclear-reactor accidents at Three Mile Island and Chernobyl, the adverse effects of the drug Thalidomide, the Exxon Valdez oil spill, and the adulteration of Tylenol capsules with cyanide. An important feature of social amplification is that the direct impacts need not be large to trigger major indirect impacts. The seven deaths due to Tylenol tampering resulted in more than 125,000 stories in the print media alone, and inflicted losses of more than $1 billion upon the Johnson & Johnson (J&J) Company, due to the damaged image of the product (Mitchell, 1989; see chapter 12). J&J was able to turn this negative event into a more positive one for them by introducing new packaging devices to avoid future tampering. Other firms in the industry followed suit and the company was eventually able to restore its image and become the industry leader again.

It appears likely that multiple mechanisms contribute to the social amplification of risk. First, extensive media coverage of an event can contribute to heightened perceptions of risk, propagation of stigmatizing images, and amplified impacts (Burns et al., 1990). Second, a particular risk or risk event may enter into the agenda of social groups, or what Mazur (1981) terms the partisans, within the community or nation. The public relations attack on "Alar" by the Natural Resources Defense Council demonstrates the high media profile and important impacts that special interest groups can trigger (Moore, 1989).

A third mechanism of amplification arises out of the interpretation of unfortunate events as clues or signals regarding the magnitude of the risk and the adequacy of the risk-management process (Burns et al., 1993; Slovic, 1987). The informativeness or signal potential of a mishap, and thus its potential social impact, appears to be systematically related to the perceived characteris-

---

[2]The Federal Energy Regulatory Commission, the principal body at the federal level determining whether a proposed LNG project is in the public interest, indicated that the risk associated with one of the worst case scenarios projecting 130,000 deaths in Oxnard was 7.1×10^{-55}.

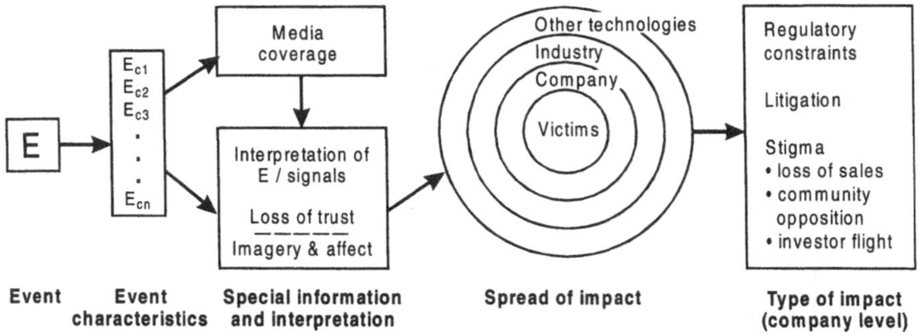

Figure 21.1 A preliminary model of social amplification of risk and stigma impacts. Development of the model will require knowledge of how the characteristics (E_c) associated with a hazard event interact to determine media coverage and the interpretation or message drawn from that event. The nature of the media coverage and the interpretation is presumed to determine the type and magnitude of ripple effects (adapted from Slovic, 1987).

tics of the hazard. An accident that takes many lives may produce relatively little social disturbance (beyond that caused to the victims' families and friends) if it occurs as part of a familiar and well-understood system (e.g., a train wreck). However, a small accident in an unfamiliar system (or one perceived as poorly understood), such as a nuclear-waste repository or a recombinant-DNA laboratory, may have immense social consequences if it is perceived as a harbinger of future and possibly catastrophic mishaps.

The concept of accidents as signals helps explain our society's strong response to mishaps involving nuclear power and nuclear wastes. Because the risks associated with nuclear energy are seen as poorly understood and catastrophic, accidents anywhere in the world may be seen as omens of disaster everywhere there are nuclear reactors and wastes, producing extensive media coverage (e.g., a minor valve failure in a reactor in England was described in a story in the San Francisco *Examiner*) and "ripple responses" (e.g., increased regulation, stigma) that carry large social and economic costs.

INNUENDO

Besides reporting certain hazard or risk stories in considerable detail, the news media are often accused of covering them in a biased or sensationalist way. This may sometimes be true, but the fact that the media see themselves as having a responsibility to warn the public about dangers (high signal events) can explain and rationalize much of the coverage. A number of studies have also shown that the media often do a good job of reporting a story as it is given to them by scientists or regulators, although they sometimes try to balance a story by giving equal time to opposing views, even when one view is held by relatively few persons. An intriguing social-psychological study titled "Incrimination through innuendo: Can media questions become public answers?" (Wegner et al., 1981) suggests that even the mildest hint of trouble can stigmatize a person (and probably a product or technology as well). The study found that indirect statements or assertions about people, in the form of innuendoes (e.g., "Ginger is not using drugs") were as destructive to a person's image as direct, incriminating statements. It may be that even subtle and indirect associations in the media between products or technologies and undesirable characteristics or events can induce stigma. If subtle hints do this, so might airing the views of a single outlier scientist or a report of a study concluding that the risk of some product or technology is high, as in the case of the consulting report for the citizens group in Oxnard regarding the dangers of siting an LNG facility in their community.

CARCINOGENS AND STIGMA: A SPECIAL PROBLEM

Many substances, activities, and technologies have become stigmatized in our society because of their proven or suspected associations with cancer. Chemicals, radiation and radiation technologies (e.g., nuclear power, nuclear-waste storage), asbestos, and electromagnetic fields come quickly to mind in this regard. We shall briefly discuss why this is so and, in subsequent sections, point out its implications for managing stigma.

The pain and drawn-out suffering of cancer make it one of the most dreaded diseases in many societies. The very name "cancer" or the label "carcinogen" evokes stigmatizing images and avoidance responses. Avoidance responses to carcinogens are amplified when exposure is perceived as involuntary, when the risks are unfamiliar or viewed as poorly understood, when the victims are young and innocent, and when the benefits of the activity are small or distributed inequitably across society.

In addition there are several major problems with risk-assessment studies of carcinogens. First, it is now widely believed that animal studies overpredict cancer because the high doses used in such studies kill cells and overpower natural defense mechanisms in ways that would not happen in humans exposed to far lower doses of the chemical in question. Yet the public believes that chemicals that cause cancer in animals (at any dose) also cause cancer in humans, even if scientists do not believe this (Kraus, Malmfors, & Slovic, 1992; Slovic et al., 1996; Slovic, Malmfors, Mertz, Neil, & Purchase, 1997). Similar problems occur with epidemiological studies of risk which often seem better able to frighten the public than to reduce scientific uncertainty (Doll, 1997; Taubes, 1995).

Second, in order to be "conservative" and provide maximum protection for the public, cancer risk assessments are based on assumptions designed to minimize the chance that the risk will be underestimated. Data from animals exposed to chemicals at high doses is extrapolated to low doses using a linear, no-threshold model. No exposure is said to be without risk. A very small exposure to a carcinogen may be estimated as producing, say, 1 case of cancer per 100,000 people, but if one million people were exposed, this leads to an estimated 10 cases of cancer. In fact, such low risks could well be zero.

High signal potential and extensive media coverage typically accompany events or scenarios involving exposure to carcinogens. One of the most dramatic examples of media-amplified stigmatization of a product occurred in the spring of 1989, when millions of consumers stopped buying apples and apple products after CBS ran a news story on "60 Minutes" stating that the chemical Alar could cause cancer. The assertion that Alar was carcinogenic was based upon animal studies that were considered suspect because the doses used had been so large as to have been acutely toxic. Moreover, there was no evidence from epidemiological studies showing Alar to be a human carcinogen. Nevertheless, the public reaction was extreme. The apple growers claimed losses in excess of 100 million dollars and these losses would undoubtedly have been even greater had they not stopped using Alar soon after the CBS program was aired.

Within a year of the Alar scare, a minute amount of Benzene (a known human carcinogen) was found in water bottled by Perrier. The contamination did not originate in the mineral water itself, but from dirty filters used in the process of adding carbon dioxide to the water to give it its fizz. Although the health risk was judged by the FDA to be nil, every bottle of Perrier worldwide was withdrawn from the market in response to consumer pressure. Perrier estimated its short-term losses at $79 million (Reuters Library Report, 1990). During the year following the incident, Perrier lost half of its share of the U.S. imported mineral water market as a result of the stigmatization of its product (Brown, 1991).

These strong reactions to Alar and Benzene are understandable in light of risk-perception studies that have found a high percentage of the public (70% or more) believing that any contact with a "carcinogen" (no matter how small or how brief) is likely to lead to cancer (Kraus et al., 1992). As a result of this belief, people want absolutely no contact with substances that are believed to be carcinogenic. Weinstein (1988), for example, found that 85% of the respondents in his survey of New Jersey residents agreed that "If even a tiny amount of a cancer-producing substance were found in my water, I wouldn't drink it." This "mental model of carcinogenesis" stands in stark contrast to the way that scientists think about the risks from chemicals. The fundamental principle of toxicology is the fact that "the dose makes the poison."[3] Scientists believe that even very toxic substances, including carcinogens, may pose little risk if the dose is small enough.

Psychological and anthropological research provides insight into why people are so concerned about even minute exposures to carcinogenic substances. Frazer (1890/1959) and Mauss (1902/1972) described a belief, widespread in many cultures, that things that have been in contact with each other may influence each other through transfer of some of their properties via an "essence." Thus, "once in contact, always in contact," even if that contact (exposure) is brief. Rozin and his colleagues (see chapter 3) have shown that this belief system, which they refer to as a "law of contagion," is common in our present culture. The implication of this research is that even a minute amount of a toxic substance in one's food will be seen as imparting toxicity to the food; any amount of a carcinogenic substance will impart carcinogenicity, and so on.

The "essence of harm" that is contagious is typically referred to as contamination. Being contaminated has an all-or-none quality to it—like being alive, or being pregnant. When a young child drops a sucker on the floor, the brief contact with "dirt" may be seen as contaminating the candy, causing the parent to throw it away rather than washing it off and returning it to the child's mouth. This all-or-none quality irrespective of the degree of exposure is evi-

---

[3]This maxim of toxicology dates back to the observation by Paracelsus in the 16th Century that, "All things are poison and nothing is without poison. It is the dose only that makes a thing not a poison" (as quoted in Ottoboni, 1984).

dent in the observation by Erikson (1990) that, "To be exposed to radiation or other toxins…is to be contaminated in some deep and lasting way, to feel dirtied, tainted, corrupted" (p. 122). The effect of contagion or contamination in the public's view is obviously very different from the scientist's model of how contact with a chemical induces carcinogenesis or other kinds of harm. In the scientific model, exposure (and the resulting risk) is viewed as a continuum, rather than as an "all or none" phenomenon.

The stigmatization and avoidance of such products as red apples treated with Alar and Perrier water containing Benzene, thus appears both understandable and to some extent predictable in light of research on perceived risk and social amplification of risk.

Characterizing Stigma Effects

When a place or a product is stigmatized, its economic value may decrease to levels below what one would expect when one takes into account the risk associated with it. In this section we will characterize the economic impacts of stigma as the difference between the market price and the risk-adjusted price. Consider a product or piece of property which has no risk associated with it and sells at a price B on the market. Suppose scientific experts agree that there is a probability p that users of the product or owners of the property will suffer an injury or illness that results in a discounted loss of L over the life of the product/property. Then a risk-neutral person, who believes the experts and computes the expected discounted loss (pL), will want to purchase the product at a lower price ($P = B-pL$) to reflect this risk.

The market price of the product or property may be below P for several reasons. Some buyers may be risk-averse rather than risk-neutral and some may perceive the risk to be more serious than the scientific experts. If the buyer overestimate the probability that the item is contaminated and/or conclude that the consequences are likely to be worse than the scientists think it will be, then the resulting price is likely to be lower than P. When experts disagree, then a person may give greater weight to the views of those who have higher than average estimates of the risk. Those individuals will also want to pay a price below P.

There may be some individuals who want absolutely nothing to do with the product or property because they believe that it is too risky for them. By withdrawing from the market, this group may reduce the demand for the product so that the market price falls sharply, as it did with British beef or Alar. In this case the economic effect of the stigma is the lost revenue due to the decrease in demand. If demand is sufficiently reduced, then some firms in the industry will be forced to shut down their operations because they cannot cover their costs.

The reduction in demand for a product could be due to changes in the social norm with respect to the particular product (Sunstein, 1996). The media coupled with notable individuals can play a key role in changing certain norms and creating norm bandwagons, a form of social amplification. These occur when small shifts lead to large ones as people join the "bandwagon." Boycotting of certain products which eventually leads to a general ban would be an example of such a bandwagon effect. As Sunstein suggests, the impact of a social norm on behavior can be viewed as a tax should it discourage the purchase of a particular product.

Property values reflect pricing dynamics similar to products with one major difference. When one purchases property, there is a concern not only about its consumption while the property is being used but its investment value should the owner decide to sell it (Adams & Cantor, this volume). If a prospective buyer believes that the local area is perceived by others to be more risky than in the past because of an adverse event (e.g. a chemical accident in a nearby plant), then this will lower the price that he or she is willing to pay. This is due not only because of the personal anxiety created by this event while one occupies the property, but also because of the perceived increased difficulty in selling it in the future. If enough people refuse to live in the area for these reasons, the market price of property in the region will likely plummet.

To illustrate how stigma can affect property values, consider the case of the Rocky Flats Nuclear Weapons Facility. It was raided by the FBI in June, 1989 with its operator, Rockwell International, pleading guilty to environmental crimes from accidents and releases involving toxic substances. In the same year Rocky Flats was added to the National Priority List for cleanup under the Superfund program. A survey of 416 residents from the Denver metropolitan area revealed that 46.2% of the respondents would not buy a house at any price within 6 miles of Rocky Flats. The estimated mean discount in selling price required to attract those still willing to live in a 4 to 6 mile range was $15,747 (Hunsperger, this volume).

A related measure of the economic effects of stigma is offered by Chalmers and Jackson (1996) who examined the impact of contamination on the types of loans provided by banks to purchase property. They concluded that the required risk-adjusted equity yields were as much as 15% higher when the property was contaminated than when it was pristine. At a more general level Chalmers and Jackson suggest that if the risk characteristics of the property increase the cost of capital, this difference from a normal interest rate can be used to measure the economic impacts of stigma.

Strategies for Dealing with Stigma

From the analyses described above, we can see why stigma is so powerful a phenomenon and so hard to manage. It arises from fundamental psychological processes of imagery, affect, and learned associations between the two, amplified by the media and by the use of risk-assessment studies whose results are often uncertain or distrusted. It can lower demand for the affected goods or property, so that residents trying to sell their homes and businesses trying to sell their products are harmed economically, sometimes severely.

In this section we discuss a number of potential strategies for reducing vulnerability to stigma. As outlined in Table 21.2, these include efforts to:

- Prevent the occurrence of stigmatizing events;
- Reduce perceived risk;
- Reduce the number of stigmatizing messages and their social amplification;
- Reduce the impacts of stigma.

Table 21.2 Strategies for Coping with Stigma

1. Prevent stigmatizing events	• Model stigma impacts as explicit costs in decision analysis
2. Reduce perceived risk	• Create and maintain trust • Inform, educate, and desensitize the public • Educate scientists about how risk studies and quantitative risk assessments breed fear
3. Reduce social amplification	• Educate the media about stigma • Educate the government about stigma
4. Reduce stigma impacts	• Provide insurance • Guarantee property values • Provide compensation

PREVENT STIGMATIZING EVENTS

One implication of signals, ripples, and stigma is that effort and expense beyond that indicated by the expected losses from direct impacts (the inner circle in Figure 21.1) might be warranted to reduce the frequency of occurrence of high-signal, stigmatizing events. For example, in the event of another "contained" core-damaging accident in a nuclear reactor such as the one that took place at Three Mile Island, the major costs of such an accident would not be those from immediate loss of life, latent cancers, and direct economic expenses (e.g., property damage, repairs, cleanup), important as these may be. Instead, the dominant costs might arise from secondary impacts such as public reaction, perhaps leading to long-term interruption or even shutdown of the industry. The resulting higher-order consequences of this suspension or shutdown (e.g. dependence on more costly and less reliable energy sources) could total tens or hundreds of billions of dollars.

These sociopolitical and long-run economic impacts must be considered when determining how much should be spent on the facility to reduce the probability of a core-damaging accident. In other words, the design of nuclear safety criteria might be phrased in terms of the question: "Given the cost of making a facility safer and the economic impacts of an accident, what probability of a core-damaging accident is tolerable?"

This notion calls for a more comprehensive modeling of the overall social costs (including stigma impacts) of nuclear accidents and the benefits of reducing the risk. If even small and contained (but frightening) accidents are likely to have immense costs, this would imply the need for strict criteria, even at great expense, to make the probability of such accidents smaller than it currently is. Similar logic might argue in favor of remote siting of hazardous facilities, dedicated trains for transporting hazardous materials, tamper-resistant packaging on products, expensive safety precautions in blood banks, and other measures to prevent stigma producing events from taking place.

Before taking such steps one needs to analyze the trade-offs between the additional expenses required to lower the risk of an accident and the expected benefits from taking this action. This requires us to better understand how to model and predict stigma impacts as well as to determine how certain protective measures will affect the likelihood and consequences of future events.

With respect to the stigma impacts, we already know a lot about qualities of hazards and their victims that trigger media coverage, ominous signals, stigmatized images, and strong avoidance behavior and this information should be incorporated into impact projections (for a related discussion of modeling and predicting potential future sources of toxic tort litigation see Foran, Goldstein, Moore, & Slovic, 1996).

REDUCE PERCEIVED RISK

Create and Maintain Trust

Reducing perceived risk should decrease stigma, but altering risk perception is not easy. One key link to perception is through trust. If trust in experts, managers, and policy makers increases, perceived risk will decrease and so will stigma. Unfortunately, trust in risk management is difficult to achieve and maintain (Slovic, 1993). Trust is fragile. It is typically created rather slowly, but it can be destroyed in an instant—by a single mishap or mistake. Once trust is lost, it may take a long time to rebuild it to its former state. In some instances, lost trust may never be regained.

The fact that trust is easier to destroy than to create reflects certain fundamental mechanisms of human psychology that Slovic (1993) called "the asymmetry principle." When it comes to winning trust, the playing field is not level. It is tilted toward distrust for each of several reasons. First, negative (trust-destroying) events are more visible or noticeable than positive (trust-building) events. Negative events often take the form of specific, well-defined incidents such as accidents, lies, discoveries of errors, or other mismanagement. Positive events, while sometimes visible, more often are fuzzy or indistinct. For example, how many positive events are represented by the safe operation of a nuclear power plant for one day? Is this one event? Dozens of events? Hundreds? There is no precise answer. When events are invisible or poorly defined, they carry little or no weight in shaping our attitudes and opinions.

Second, when events do come to our attention, negative (trust-destroying) events have much greater weight on our opinions than do positive events. Adding to the asymmetry is yet another idiosyncrasy of human psychology—sources of bad (trust-destroying) news tend to be seen as more credible than sources of good news. These various trust-destroying tendencies are, of course, amplified by the news media. Just as individuals give greater weight and attention to negative or trust-destroying events, so do the media (Lichtenberg & MacLean, 1992). The implication of all this for managing stigma is that, again, great efforts and costs may be warranted to prevent the occurrence of events that could fuel distrust. In other words, loss of trust is one of the ripple effects referred to earlier that needs to be modeled and valued when making decisions about risk management.

One way to generate trust is to encourage public participation as an integral part of the decision-making process. For example, in siting new facilities it is important to hear the concerns of the affected public and respond to them. As part of the monitoring and control procedures for making certain the facility remains safe, a committee could be established to inspect the facility at regular intervals and report its findings back to the local community. If such a process helps to establish trust and confidence between the developer and the affected parties, it could reduce perceived risk and stigmatization.

A set of guidelines for a fairer, wiser, and more workable siting process—the Facility Siting Credo—was developed during a National Facility Siting Workshop in 1990. A questionnaire based on the Credo was completed by stakeholders in 29 waste facility siting cases, both successful and unsuccessful, across the United States and Canada. Using an independent determination of outcome (success), a preliminary rank of the importance of various Credo principles was obtained. The data revealed that establishing trust between the developer and host community was an important factor in facilitating the siting process. The siting process was most likely to be successful when the community perceived the facility design to be appropriate and to satisfy its needs. Public participation also was seen to be an important process variable, particularly when it led to the view that the facility does a good job of meeting community needs. (Kunreuther, Fitzgerald, & Aarts, 1993).

Inform, Educate, and Desensitize the Public

It is natural to turn to informing and educating people about risk in order to reduce stigma impacts by calming what technical experts may view as "exaggerated fears." However, the affective nature of stigma limits the influence of quantitative risk information. Consider phobias, for example, which are strong affective and aversive reactions to stimuli such as spiders, snakes, airplane travel, the outdoors, and so on. It is well known that risk communication (i.e., statistics showing that the risks are small or nonexistent) is virtually useless in treating phobias. We can expect the same to be true with stigma. What works with phobias is systematic desensitization or counterconditioning to reduce the negative affect. For example, a subject who developed an inability to take pills following the Tylenol poisonings in 1981 was "cured" by starting with relaxation training and then being exposed very gradually to stimuli increasing in similarity to medicine pills (first, simply imagining pill taking, then drinking wine, then taking a vitamin pill, etc., see Pbert & Goetsch, 1988). Similar counterconditioning may be necessary to deal with the conditioned negative affect that drives stigmatization and avoidance behavior.

In the difficult area of dealing with minute exposures to carcinogens, it may help people to learn of Bruce Ames' (1983) studies showing that many common fruits and vegetables contain natural carcinogens in far greater amounts than the small exposures they are concerned about (hopefully this will not stigmatize fruits and vegetables). It may also help if people understand that the body has defenses against the cell damage that might lead to cancer and if they understand the essence of current theories of carcinogenesis, so they can appreciate that minute exposures may pose little or no risk.

Educating about the benefits of such exposures (e.g., the great benefits of water chlorination, a procedure that may also expose people to very small amounts of carcinogens) may also reduce negative affect as there is evidence to believe that the positive affect associated with benefits can partially offset the negative affect associated with risks (Alhakami & Slovic, 1994; Margolis, 1996).

Educate Scientists: Risk Studies Breed Stigma

Risk assessment, as currently practiced and communicated, is part of the problem of stigma. The practice of quantitative risk assessment has steadily increased in prominence during the past several decades as government and industry officials have sought to develop more effective ways to meet public demands for a safer and healthier environment. Ironically, as society has expended great effort to make life safer and healthier, many in the public have become more, rather than less concerned about risk. This is particularly true for involuntary exposure to chemicals, which the public associates to a remarkable extent with danger, cancer, and death.

The linear, no-threshold model of cancer risk assessment has long been the subject of debate and criticism (see, e.g., Nichols & Zeckhauser, 1980). Recently, Purchase and Auton (1995) described an alternative model in which the lowest dose at which the critical effect has been observed is identified and used to define the No Observed Adverse Effect Level (NOAEL) for that effect. Purchase and Auton show that one cannot distinguish empirically between the linear, no-threshold model and a model in which the NOAEL is divided by a safety factor. Thus, for example, the linear model used by the EPA regulates any lifetime risk in excess of 1 chance in one million which can be shown to be equivalent to the NOAEL divided by a safety factor of about 250,000.

Use of a nonthreshold linear model to express risk in probabilistic terms leads to higher perceived risk than does the safety-factor format based upon the same test results (Purchase & Slovic, 1999). Thus with data implying a 1 in one million excess cancer risk, risk assessors and risk communicators are faced with a choice between making one of the following two scientifically accurate statements which are framed differently:

> STATEMENT 1: The probability you will develop cancer from exposure to Chemical Y is about 1 chance in 1,000,000.
>
> STATEMENT 2: Chemical Y has been observed to cause cancer in animals but only at doses more than 250,000 times greater than what you will ingest. At doses less than 250,000 times greater than what you will ingest, no cancer in animals has been found.

According to Purchase and Slovic (1999) the latter statement may be less likely to contribute to the stigma-causing view that any exposure to this carcinogen is likely to cause cancer in the exposed person.

More generally, the point we are trying to make here is that the rise of quantitative risk assessment, with its proliferation of high-dose animal studies and reliance on conservative extrapolation methods, may be a strong contributor to destructive stigmatization involving chemical products. This poses a dilemma for risk managers who cannot and should not abandon animal studies and risk assessment. However, they must recognize that stigma is a side-effect of such efforts and consider ways to offset its damaging impacts without jeopardizing public health and safety.

REDUCE SOCIAL AMPLIFICATION

Another strategy would focus on possibly altering the number and content of stigma producing messages reaching the public by educating the media and the regulatory community about the effects their messages may have.

Educate the Media

The media will not easily be persuaded to change their way of "reporting on risk." They believe they are providing an important duty to society in warning of potential threats, and this is true. Moreover they are well protected by the First Amendment of the Constitution. Although apple growers in Washington State sued CBS for making numerous false statements about the risks of apples treated with Alar, the suite was dismissed by a federal judge who argued that "Even if CBS' statements were false, they were about an issue that mattered, cannot be proven as false and therefore must be protected" (Auvil vs. CBS *60 Minutes*, 1993). The judge further affirmed the right of a news-reporting service to rely upon a scientific government report (which CBS did) and communicate the report's results.

However, we suspect that CBS, in their reporting on Alar, did not intend to harm the apple growers. They wanted to motivate EPA to take action against Alar and did not anticipate the massive stigma response they created. Thus one strategy would be to educate the media about the nature and potency of stigma and their responsibility to anticipate and weigh potential stigma losses when deciding what information to present about risk and how they should frame the data.

In response to the Alar incident, Florida, Washington, and other states have proposed (and in some cases passed) legislation allowing producers of agricultural commodities to recover damages for the disparagement of any agricultural commodity (where disparagement is defined as "dissemination to the public . . . of any false information regarding the application of any agricultural chemical or process to agricultural commodities that is not based on reliable scientific data, that the disseminator knows or should have known to be false, and that causes the consuming public to doubt the safety of any agricultural commodity."

Although "anti-disparagement legislation" may discourage and punish some wanton attacks on products, it seems unlikely to penetrate the First Amendment defense of the news media. A recent case in point is the rejection of the suit filed by the Texas Beef Group against Oprah Winfrey for uttering critical remarks about the feeding of cattle and about eating hamburgers.

Educate the Government

In an earlier section, we argued that government-endorsed methods of risk assessment may contribute to exaggerated fears of chemical carcinogens and the destructive stigmatization of certain chemicals (and perhaps the entire chemical industry).

In its attempts to carefully regulate public exposures to harmful substances, government does other things besides promulgating risk assessments that place products in jeopardy. Like the media, the government communicates risk information to the public in ways that may be destructive as well as helpful. A case in point is the *Biennial Report on Carcinogens*, a list of known or suspected cancer-causing substances prepared by the National Toxicology Program (NTP). At the lowest (least dangerous) level on the list is a category labeled "reasonably anticipated to be a carcinogen." The degree of evidence necessary to "earn" this listing is far less than what is needed to be listed as a "probable" or "known" human carcinogen. This "reasonably anticipated" listing is meant as the very first step in the hazard-identification process, a red flag so to speak. It carries no legal penalty but, as should be obvious from the preceding discussions, it has the potential to inflict massive losses to manufacturers who have products labeled as having potential carcinogens because of the stigma associated with the asserted link to "cancer."

This issue came to attention of the glass-wool (fiberglass) industry around 1990 when the government proposed placing glass-wool insulation in the "reasonably anticipated to be a carcinogen" category. The industry objected strenuously, claiming that the animal data had little relevance to human health because they were obtained from studies that surgically implanted massive quantities of fibers into the body cavities of test animals, bypassing normal bodily defense mechanisms. Moreover, the relevant human exposure pathway is inhalation. Fearing that the industry would be "needlessly stigmatized" and irreparably harmed by the listing, the glass-wool manufacturers challenged the nature of the research methods used by the NTP as well as the criteria used to evaluate data and label degrees of cancer risk. Subsequently, the NTP revised its criteria to provide a greater role for scientific expert judgment in classifying substances (*Federal Register*, 1996). This change represents an improvement over the former, rigid criteria that required a listing whenever there were two positive carcinogenicity studies in animals, regardless of the relevance of the animal studies to human exposures and regardless of how many negative studies had been reported.

This case shows that it is not only the media who need to be sensitive to the nature and impacts of stigma but researchers and government regulators as well as they seek to compile, evaluate, and communicate research findings. The glass-wool episode forces us to consider how much evidence is necessary to warrant the issuing of a very early (and thus possibly wrong) warning, given that the consequences of the warning itself could be very damaging (not unlike the question of deciding when to yell fire in a crowded theater). We have no easy solution or answer to this question except to note that the reality of stigma should force a reexamination of research methods and criteria for identifying and labeling carcinogens.

REDUCE STIGMA IMPACTS

The economic impacts of some stigmatizing events may possibly be mitigated through insurance and compensation mechanisms which may convince the public that property is likely to be safer than they may have anticipated.

Insure Brownfields

Property may be stigmatized when there is a concern that contamination of the land will require a very expensive cleanup under Superfund (administered by the EPA). Several uncertainties confront the property owner: the probability that contamination will be found on the land, the cost of the resulting cleanup and who will be responsible for covering all or part of this expense. In many cities large parcels of land may have no value at all, even though the technical analyses conclude that their potential benefits may exceed the expected cost of cleanup (i.e., the probability that there is contamination multiplied by the average cleanup cost for land of this type).

The term "brownfields" has been used to characterize property that was previously developed but is currently idle because it is either known or thought to be contaminated. The GAO has estimated that there are 130,000 to 450,000 contaminated sites around the country and that it will cost an estimated $650 billion to remediate these parcels of land (GAO, 1993). Many cities are concerned with their future development due to the stigma associated with the large amount of industrial land that is vacant or inactive within their boundaries. The cleanup and re-utilization of brownfields is envisioned by the EPA as being the key to revitalization of America's inner cities, and as the beginning of an urban renewal process that can result in a cleaner environment, new jobs, an enhanced tax base, and a sense of optimism about the future (EPA, 1996).

One of the reasons that the brownfields problem exists is because of great concern among buyers of property that they may be forced to incur large cleanup expenses if the land that they purchase is found to be contaminated. In addition, the financial institution that issues a mortgage is concerned that it may have to cover the costs of cleanup if the buyer goes bankrupt. Many areas in the inner cities are stigmatized because there is great uncertainty regarding the risk (probability and consequences) that the land is contaminated with toxic wastes.

One way to deal with the stigma threatening the development of certain parcels of land is to provide property transfer insurance so that buyers know that they are protected against future losses. In order for insurance to be marketable, there needs to be a more precise estimate of the risk associated with any particular piece of property than is normally found through title searches. Audits and inspections, such as a pre-acquisition site assessment by an engineering consulting firm to determine whether the property is contaminated, are needed to clarify the nature of the risks. If, in addition, federal and state environmental protection agencies are willing to issue well specified standards for cleanup, then property owners and potential insurers will even have a better idea on what the costs are likely to be should an audit reveal contamination (Bray, Freeman, & Kunreuther, 1998).

Currently several insurers are marketing property transfer coverage using the following procedure:

1. Require that the prospective buyer undertake an audit or inspection of the property. Phase 1 audits usually cost approximately $8,000 and provide a good indication of whether there is contamination on the land. If there is some concern with contamination then a Phase 2 audit will be necessary (a much more thorough audit) to determine the extent of the contamination. This costs of this audit normally ranges from $25,000 to $30,000.[4]

2. Even after undertaking a Phase 1 audit and declaring "no contamination" there is a 10% chance that the audit is incorrect. It is here that insurance is very useful. If the insurer can determine the range of cleanup costs associated with contamination when a Phase 1 audit fails, then a premium reflecting risk can be charged. An examination of the costs of federal and state mandated cleanups at more than 3,000 sites revealed that the average costs of remedial action ranged from $102,000 for underground storage tanks to just over $33 million for federal national

[4]In theory, a buyer who undertakes audits is not held liable using the "innocent landowner purchase" defense as part of recent amendments to Superfund (CERCLA) legislation. However, to our knowledge this defense has not been tested in the courts (M. Gerrard, personal communication, February 1997.)

priority list sites (in 1991 dollars; Freeman & Kunreuther, 1997). These data and related information enable insurers to set a premium based on risk where the amount charged will vary with the type of land, the nature of the audit and the terms of the contract (e.g. deductibles, upper limits of coverage).

3. Banks will often require insurance as a condition for a mortgage to protect their investments should the buyer go bankrupt, much as they do with homeowners coverage today. Potential buyers may find the property attractive if they know they are protected against losses.

Suppose that an audit has been undertaken on the property and reveals that there is no contamination on the land. Recognizing that there is still some chance that the audit has incorrectly diagnosed the situation, the buyer decides to purchase insurance or is forced to do so by the bank issuing a loan and is now protected from losses due to contamination. The purchase price will now reflect the cost of insurance and the audit. Should the buyer want to sell the now-insured property in the future, there should be no decrease in value due to stigma because the existing coverage would protect the new owner from any cleanup costs.

Insure New Facilities

Residents in a community may be concerned that locating a noxious facility (e.g. landfill, hazardous waste disposal plant, or radioactive waste repository) in their backyard may stigmatize the area because of the perceived risks associated with the wastes that are stored there. This concern may be associated both with potential negative health effects from exposure to radiation or toxic substances and with projected negative economic impacts such as decreases in property values or reduced business activity.

Even if analysts offer scientific evidence that the risks are very low, these concerns are unlikely to be allayed. For example, residents living near the OII landfill in the Los Angeles, California metropolitan area were told by experts that the health risks associated with odors from the landfill were harmless. A survey conducted by the California Department of Health found no statistical differences in mortality or incidences of cancer and liver disease for residents in the OII area and control communities. Yet many of those residing where the OII landfill was located, believed that the facility posed serious health risks to them. This concern adversely affected property values in the community (McClelland et al., 1990).

If one is siting a hazardous facility in an area, then the type of insurance that the developer or company purchases may reduce the discrepancy between the experts' and public's views of the risk. The idea is a simple one: if an insurer is willing to offer coverage against an accident then it must be confident that the risk associated with the facility can be quantified. The lower the premium charged by the insurer, the safer the facility is likely to be. In this sense insurance can serve as a signal of relative safety, to the extent that the information on premiums is publicly available. Furthermore the public can view the insurance premium as a surrogate for the risk and hence better appreciate how risky it is.

Insurance can be viewed as a signal for safety in much the same way that past accidents or near-misses can be viewed as a signal for danger (Slovic,

1987; March, Sproull, & Tamuz, 1991). One reason that a low premium should be viewed as a legitimate signal is that the insurance industry is a competitive one. Hence no company can charge an unusually high rate and not be undercut by someone else who can make a profit. Otherwise insurers would have no incentive to treat the risk as smaller than what the public perceives it to be. Of course, if the new facility is hazardous, higher premiums or inability to obtain insurance would signal that too.

However, insurance may sometimes send mixed or confusing signals about safety. For example, consider the Price Anderson Act, which was passed in 1957 to encourage the utilities and industry to develop nuclear power as a source of energy. This piece of legislation provided insurance to the utilities with a $560 million limit on liability. This relatively low figure could be interpreted to mean that the insurance industry was so concerned about the risk that they did not want to provide more than this amount of protection. Alternatively one could interpret this low dollar value as saying that the catastrophic potential was not high. When the limit was raised a few years ago to $6.2 billion, the signal may also have been mixed for the same reasons.

Guarantee Property Values

One way to address the concern that residents in a community have with respect to the economic impact that siting a noxious facility will have is for the developer to provide property value guarantees. For example, in 1990 Champion International Corp. established a program to protect the property values of residents within two miles of an industrial landfill they sited. The company monitors changes in the sales prices of property in the county over a 10-year period and pays residents who sell their homes for any decrease in property value that is attributed to the presence of the landfill (Ewing, 1990).

As with insuring a new facility, this type of guarantee serves as a signal for safety since the developer is willing to cover costs associated with the facility. If it felt there would be large expenses then it would *not* offer this type of compensation. It is clearly designed to allay any concerns that citizens in the area may have with respect to the risk.

The challenge in developing property value guarantees is to develop an index that will effectively measure normal changes in comparable structures in areas which are not subject to the risks associated with the new facility. Champion's approach seems to make sense, but it would be interesting to know what their experience has been since it instituted the program.

Provide Compensation

Empirical evidence indicates that compensation can prove effective in gaining public acceptance for siting facilities on the benign end of the spectrum (e.g., landfills, prisons), but it is subject to serious limitations when it comes to facilities that the public regards as particularly risky or of questionable legitimacy such nuclear wastes. These require creative mitigation measures such as independent inspections of the facility and local authority to monitor and shutdown the facility. Even then they may be viewed as too risky to be acceptable with or without compensation (Kunreuther & Easterling, 1996).

With respect to relatively benign facilities, as shown in Table 21.3, a local landfill was acceptable to 30% of the Bacot, Bowen, and Fitzgerald (1994) sample and to 25% of the Jenkins-Smith and Kunreuther (2000) sample when compensation was not made part of the package. In both cases the rate of acceptance approximately doubled with the introduction of compensation.

The survey by Jenkins-Smith and Kunreuther (2000) also investigated the impact of compensation on acceptance in the case where the facility being sited was a hazardous-waste incinerator and a medium-security prison. Although these two facilities differed markedly in the absolute level of acceptability (15% versus 29% in the no-compensation case), the introduction of benefits produced similar levels of increased acceptance (17 percentage points for the incinerator, 22 percentage points for the prison). In situations where residents view the facility as sufficiently hazardous compensation will be unable to overcome opposition as in the case of high- level nuclear waste (HLNW).

Judged acceptability of compensation across six surveys is presented in Table 21.4. Studies 5 and 6 found that offering payments led respondents to be less accepting of a radioactive waste repository, perhaps because they viewed it as a bribe. In the Dunlap and Baxter (1988) survey, 60% of the sample voted in favor of a HLNW repository at Hanford without benefits, while only 51% voted in favor when tax rebates were offered. In the survey of attitudes toward a low- and mid-level waste repository by Frey, Oberholzer-Gee, and Eichenberger (1996), acceptance was cut in half—from 51% to 25%. Moreover, the size of the compensation did not significantly influence the acceptance rate. Overwhelmingly, those who refused compensation reported that they could not be bribed.

A striking illustration of actual rejection of compensation comes from North Dakota. In 1990, three county commissioners in sparsely populated Grants County applied for a nonbinding grant to study the possibility of hosting an MRS facility for temporarily storing high level radioactive waste. The three commissioners who initiated the process were all voted out of office in a recall election because they accepted the grant even though the grant was not binding in any way (Kunreuther, Linneroth, & Fitzgerald, 1996).

Even if residents in the host community are willing to accept compensation, it may provoke very strong negative reactions and stigmatization because

Table 21.3 Effect of Compensation Measures in Increasing Acceptance of Facilities

	Landfill for municipal waste		Hazardous waste	
	Study 1[a]	Study 2[b]	incinerator[b]	Prison[b]
Acceptance without incentives	30%	25%	15%	29%
Acceptance with economic benefits		50%	32%	51%
Rebates on property tax	63%			
State money for schools	62%			
State money for roads	56%			

[a]Bacot et al. (1994). Sample of 844 Tennessee residents. The 30% figure for acceptance without incentives was derived from the reported result that 70% opposed the landfill; 30% is an upper bound on the actual figure. The authors do report the proportion in favor under the incentives conditions.

[b]Jenkins-Smith and Kunreuther (2000). Total sample of 1200 U.S. residents. Each condition has n = 150.

Table 21.4 Limited Effectiveness of Compensation the Case of Nuclear Waste Repositories

	Study 1[a]	Study 2[b]	Study 3[c]	Study 4[d]	Study 5[e]	Study 6[f]
Acceptance without incentives:	22%	10%	27%	24%	60%	51%
Acceptance with economic benefits:						25%?
"substantial payments"	26%					
"economic benefits"		14%				
$1,000/yr for 20 years			26%	23%		
$3,000/yr for 20 years			30%			
$5,000/yr for 20 years			30%			
$100–$900/yr for 20 years					51%	25%

[a]Carnes et al. (1983). 1980 survey of 420 Wisconsin residents.

[b]Jenkins-Smith and Kunreuther (2000). Total sample of 1,200 U.S. residents. Each condition has $n = 150$.

[c]Kunreuther, Easterling, Desvouges, and Slovic (1990). 1987 survey of 1,001 Nevada residents ($n = 498$ answered compensation questions).

[d]Herzik (1993). 1993 survey of 1,212 Nevada residents.

[e]Dunlap and Baxter (1988). 1987 survey of 658 persons living near Hanford, Washington.

[f]Frey et al. (1996) 1993 survey of 305 persons living in Wolfenschiessen, Switzerland.

some view it as morally wrong. Elster (1992) suggests that people may view health and safety as inherent rights that should never be traded off for material goods. An illustration of this point is provided by the German city of Bergkamen where concerned citizens were persuaded by the prospective operator to accept a power plant in exchange for money. The German press objected, claiming that this exchange creates incentives for groups to protest a facility under the expectation that they will eventually be bought off by the developer (Kunreuther & Linnerooth, 1982). Since that time direct monetary compensation has not been utilized in Germany in connection with the siting of facilities.

Taiwan offers another example of a negative response to monetary compensation. In this case, villagers forced 23 petro-chemical firms in an industrial park to close in 1988 after an overflow of waste water from the treatment plant polluted nearby streams and adversely affected fishing in the area. The Minister of Economic Affairs responded by offering substantial amounts of monetary compensation to residents of the area who accepted the funds in return for reopening the facilities. This action produced an outcry throughout the country and led the legislature to pass the Pollution Conflicts Resolution Act which explicitly prohibits this kind of individual compensation in the future (Shaw, 1996).

Conclusion

The varieties of risk-induced stigma described in this chapter and in this book are noteworthy because they exemplify what has been called the "social amplification of risk" and illustrate a new form of societal vulnerability. Although stigma has been studied by social psychologists, sociologists, and other social scientists as it applies to people, there is very little empirical research or stigmatization of places, products, industries, and technologies. It seems obvious that the better we understand the dynamics of these forms of stigmatization, the better we can forecast and manage in any given situation. The challenge before us is to learn how to manage stigma and reduce the vulnerability of important products, industries, and institutions to its effects, without suppressing the proper communication of risk information to the public.

22 Defining and Identifying "Stigma"

Vern R. Walker

With "Coping with Stigma" (chapter 21 of this volume), we are beginning to address the normative dimension of our subject. We are discussing what we *ought* to *do* about stigma. This normative focus is somewhat different than the one we have had until now. The focus before this point has been on the normal scientific problems of predicting and explaining stigma-related phenomena: that is, on the problems associated with developing models for the dynamic processes surrounding stigma, with identifying causal factors that increase or decrease stigma formation, and with devising effective methods for modifying stigma phenomena. We have been discussing, for example, different metrics for measuring negative affective response, such as a decreased willingness-to-pay. These are inquiries in which scientists routinely engage.

With regard to this prior empirical inquiry, I would like to suggest three scientific disciplines to include in future discussions of stigma: semiotics, the science of signs and symbols; anthropology, which looks at human phenomena in a cultural context; and the advertising sciences, which know a great deal about stigma formation—witness the political consultants who devise "negative attack ads." I think these additional scientific disciplines could have some valuable insights on how this process of stigma works.

But the discussion has now moved out of the purely descriptive and explanatory phase, and into the normative phase. The topic before us now is whether and how we ought to reduce the incidence or severity of stigma by "de-sensitizing" the public and by educating the scientists, the media, and government. The topic is our *justification* for modifying the risk perceptions of others. The soft assumption we have been making is that we probably *should* do so if we *can*, because we are, after all, dealing with *stigma*. What I am going to urge is that our justification for intentionally modifying another person's perception of risk depends upon our being able to define and correctly identify "stigma," as contrasted with a person's accurate perception of "real risk." How do we know "stigma" when we see it? I will confine my comments to this one normative precondition for attempting risk-perception modification, although justification surely has additional requirements as well.

My comments are premised on a fundamental background principle. The principle is that we ought to have an open marketplace of ideas, for libertarian reasons as well as for efficiency. I assume that we are all in basic agreement with the libertarian policies underlying a constitutional right of free speech. Our personal right to decide what to think is inherent in our concept of being a free human individual. Complementing this libertarian ideal is an efficiency argument that is somewhat Darwinian in nature. If our society has a productive engine of variability for producing ideas, then the selective mechanisms of society can draw from a wealth of ideas and beliefs. Our society should be more adaptable to changing circumstances if we have a lively source of free expression, which is promoted by minimizing social and political control over any individual's concepts, beliefs, values, and personal decision rules. This principle of an open marketplace of ideas should be the anchor as we begin to talk about what we *ought* to *do* about stigma. I suspect that everyone would be more nervous about the prospect of an effective technology for modifying risk perception or behavior if we thought there was any chance of it succeeding. But quite apart from our ability to succeed in modifying behavior, if this principle of freedom of expression is our common starting point, then we should be in agreement that any intrusive engineering of any individual's beliefs needs to be justified.

This leads to the next question, which I will pose this way: When are we *warranted* in calling something "stigma," with the result that we might be justified in eliminating it? What are the criteria for "stigma," and how do we know "stigma" when we see it? I think that we must address this question before launching any normative crusades. My pre-conference intuition about the normal use of the word "stigma" was that stigma entails an *un*warranted level of avoidance behavior. Stigma represents a *mis*conception, a *mis*perception of risk, an *over*reaction to some thing. And in the last day and a half we have definitely heard such words used to describe stigma. After hearing the earlier papers at the conference, I have refined this account somewhat, and now I suggest a conceptual structure something like the following.

In listing criteria for a "risk-induced stigma-phenomenon" (this observable phenomenon that we are trying to identify), I would suggest starting with the semiotic, community-based notions introduced here yesterday. I start with four factors on which I think we would probably agree. First, there must be a communication. Second, the communication occurs through a marker. Third, this marker is associated with or attached to a person, a place, a product, or an industry. And fourth, perception of the marker triggers a negative affective response or some avoidance behavior in the perceiver. I think that perhaps we all now agree on that part of the construct.

These four characteristics, however, are true of many phenomena we should hesitate to call stigma. For example, a "skull-and-crossbones" or other such symbol on a bottle of poisonous chemical is presumably not a stigma. So what subset of that broad type of phenomenon constitutes "risk-induced stigma"? I suggest two further features. One is that the effectiveness of the communication derives from a negative affective response that originally occurred in some other context, but which has been *transferred* to the new context. There is a transference of this negative affective response by means of the marker. The avoidance behavior was originally associated with one thing, but now the stigmatizing process transfers that response to something else, and is able to do so because the affective response remains associated with the marker. In that sense a stigmatizing process is a parasitic phenomenon, utilizing an emotional charge that is already there. Moreover, the marker must have suitable characteristics or structure, such that the affective response is transferable with it.

The second additional feature of stigma is that the degree of negative affective response that the transferred marker generates is not fully explained by what I will call "real risk," but which others here have called "assessed risk." The assessed risk is the risk as determined by the currently dominant scientific paradigm or methodology for predicting future adverse effects. This dependency upon current scientific theory is emphasized by the phrase "assessed risk." The concept of "assessed risk" reminds us that there are subjects assessing that risk. On the other hand, the phrase "real risk"has some advantages as well. Talking about the "real risk" emphasizes that the current paradigm for predicting adverse effects might not be completely correct, and that it might overestimate or underestimate the real dangers. There is still a real world full of real dangers, and it often eludes our attempts to assess it. When I use the phrase "real risk," I am trying to capture all of these connotations.

I think that we define stigma *against* the real risk posed by the thing to which the marker is attached. For all practical purposes, we define it against whatever we adopt as the criterion method for predicting adverse effects. Stigma has to do with our inability to explain a person's degree of negative affective response on the basis of the real risk as assessed by that paradigm. That is, we posit stigma to account for the *difference* between the avoidance behavior we actually observe and the avoidance behavior we would predict based on the current paradigm's assessment of the true danger. We use the concept of stigma to help explain the degree of response that we actually observe. In that sense, the stigma communication "adds something" by explaining why we observe an otherwise unpredicted degree of avoidance behavior.

This also explains why scientists regard stigma phenomena as inherently irrational. To say that a product is stigmatized *means* that some communicative marker has become associated with the product and has transferred to it a negative affective response that is *unwarranted* given the current scientific theories about the real hazard posed by the product. A stigma reaction is by definition out of line with what is warranted, which we determine on the basis of the best available science. Stigma, then, shares the scientific opprobrium normally heaped upon superstition. Just as superstition is the natural enemy of science, so is stigma. That is why, within the gathering of scientists at this conference, we have softly assumed that *if* certain behavior *is* a response to stigma, then of course it would be a good thing to modify it. But now I want to say: "That may be true *if* it is *in fact* stigma-induced. But before we are justified in modifying it, we ought to be somewhat certain that it *is* in fact stigma-induced."

The importance of this point did not become concrete for me until talking with Terre Satterfield yesterday about how an anthropologist might define a "taboo." It seemed to me that a taboo situation is one in which we might find a paradigmatic instance of stigma. We can imagine what might be occurring in a group or community when a taboo is transgressed. For sake of discussion, imagine that it is taboo in our community to mix the cooked meat and the mashed potatoes on our plates at dinner. In this culture, you are not supposed to do it. What happens, however, if someone does do it? This might cause three kinds of immediate problem for the rest of the community—for example, for everyone else at the dinner table. First, there is the social problem of what to do about the transgressor. There is a social norm that has been violated, and now a social problem that must be solved. It might be solved by some kind of purification or reconciliation ritual, because we have a transgressor out of harmony with the community. This is a social problem in the sense that the community probably has it in its power to fix it, however difficult it might be to do so because of the dynamics of human psychology.

Let me call the second problem, without any irreverence, "how to appease the gods." What drives this second problem is the notion that when someone violates the taboo, they have set in motion some external chain of events that is going to hurt us, the community. We can have interesting sub-cases: Does it hurt all of us? Does it hurt only the transgressor's relatives? But unlike the first problem, this problem is believed to be not purely social. The transgression sets in motion some objective chain of events *outside* the community that has the potential to bring about real harm, and we are collectively afraid of that harm. So part of the reaction to the transgression is a state of fear for what has been done to us. In contrast with the first problem, we now need "engineers" who can actually modify the objective causal chain, for our mere forgiveness of the transgressor is probably not a solution to this second problem.

The third immediate problem is that the transgression brings to the surface, and challenges us to reaffirm, our most deeply held beliefs about the world and our relationship to the universe. For we are now called upon to punish a transgressor whom we might know or love, and to take steps to ward off an evil event that might or might not be real, and both of these difficult tasks have the potential to make us question the need for doing them. It requires a new vote of confidence in our received worldview.

Each of these problems is created as soon as the taboo is violated. At least in taboo-like circumstances, these problems are so immediate and so serious that the creation of a stigma might make a great deal of sense. We know that we will have all of these problems *if* the taboo is violated. One strategy to keep the taboo from being transgressed is to establish a stigma, a warning marker effective in modifying behavior, in order to help people not to violate the taboo. This use of stigma is a risk-reduction, avoidance strategy to decrease the likelihood of taboo transgression. We engage in stigma formation to get our children and others in the community to keep from doing those things that would cause us to have these three kinds of problems. In view of this function of stigma within a risk-reduction strategy, we ought to be very sure, before we eliminate any stigma produced by the community over time, that the fear of external evil that might befall the community is indeed an irrational fear. One person's gods are another person's myths, and one person's worldview is another person's superstition. Whether a "skull-and-crossbones" symbol is a stigma or an appropriate warning depends upon the facts about the true risks.

I have given a taboo situation merely as an example where the development of a stigma makes sense and why a community would engage in stigmatizing. But we can undoubtedly have attenuated forms of the phenomenon. A few of us have been talking about the apple growers, those owners of small orchards who were producing apples for the community when the Alar debate broke in the press. There are analogies to a taboo situation. The community may react by thinking that it has been let down, that these apple growers have transgressed. First problem: what should we do about those transgressors, to remove their transgressor status? Second problem: what should we do about the objective causal chain set in motion by selling Alar-treated apples, and about the exposure our children have had to this chemical? Third problem: what should we think about our "scientific" risk assessment methodology, with all of its uncertainties, and about the pesticide industry and our regulators? All of these problems are created by selling the treated fruit.

Let me summarize the direction of my thoughts to this point. We need to justify any serious efforts at education or behavior modification, including the elimination or diminution of negative affective behavior. Even if we are *able* to reduce negative affective behavior, we need to be able to explain why it *ought* to be reduced. We need to justify our actions not only because our general normative principles, such as freedom of thought, require it, but also because the use of stigma may itself be a beneficial risk-reduction social strategy. It may be justifiable to eliminate stigma provided we can verify it to be *stigma*. So we need to operationalize the concept of stigma in order to justify eliminating instances of it.

There is also a scientific motive for operationalizing a concept of stigma. On our current paradigm of science, it is generally accepted that we are not doing "science" until we operationalize a concept for identifying or measuring what we are studying. We need operational criteria for stigma in order to study it scientifically. But how could we conduct a two-day scientific conference without defining the central concept? I think that, to the extent that we have not possessed an agreed operational definition of "stigma," we have been largely using the word in its ordinary meaning. Frankly, I think that we have been using the word "stigma" as itself a stigmatizing marker. A scientist can easily use the word "stigma" in its pejorative sense precisely because the word ascribes unwarranted negative affective behavior to others who lack scientific data to support their perceptions of risk. At the same time, the use of the word by a scientist *generates* negative affective behavior among other scientists because the word connotes irrationality. The meaning of the word "stigma" therefore helps to explain how we here could engage in two days of discussions about a concept we have not operationalized. We have tacitly agreed that stigma is associated with unscientific behavior patterns that really should be discouraged through "education." And I agree—provided it is indeed an instance of stigma, and not just another occasion when nonexpert perceptions of risk differ from those of the experts, and when experts are labeling all dissenters as "irrational" and "stigmatizing." Who is justified in educating whom, in other words, is a function of who is in fact correct about the real risks.

So in addition to the scientific methodological reasons for operationalizing the concept of stigma, we should define it because the concept *does* play a normative role in our society. When negative affective behavior is in fact warranted because the risk is accurately perceived, society (especially the state) is unjustified and inefficient in trying to modify those perceptions. We might be acting wrongfully, and probably unsuccessfully and counterproductively, if we engage in efforts to engineer preference formation in situations where the real risk is uncertain. We have discussed examples of the tragedies of airbags and BSE and blood supply and dioxin, and these case studies suggest that we need to get our facts straight about risk. This means assessing the risk using our currently dominant scientific paradigms for predicting adverse effects, and determining how much public reaction is due to stigma. But it also means continually reassessing the validity of that currently dominant risk-assessment paradigm. Our legitimacy, as well as our effectiveness and efficiency, is at stake.

That brings me to my final point. It might be useful for our methodologists and social scientists to examine the capabilities of the fact-finding institutions in our society for distinguishing real risk from stigma, facts from fiction. I have been talking about why it is important to identify stigma. We should also look at how such fact-finding can be accomplished effectively and efficiently. I am using a single word, "fact-finding," although social scientists might prefer the phrase "institutional fact-finding process." We should study fact-finding as an institutional process that has epistemic goals (such as descriptive accuracy, uncertainty characterization and reduction, and a transparent evidentiary basis), but has those goals in combination with institutional constraints and objectives (such as administrative efficiency, fairness to parties, and separation of governmental powers).

Where we find a fact-finding process in a societal institution, it is always a blend of these different types of goals. The optimal blend of such diverse goals for each institution constitutes its particular model of the fact-finding process. The question I pose is: What is the best process or blend of goals for a fact-finding institution when it comes to identifying "stigma"? This is an empirical research question. To the extent that such an ideal fact-finding process would use the currently dominant scientific paradigm for predicting adverse effects and would use it correctly and successfully to characterize risks, then such a process would bring us closer to producing a legitimate basis for societal intervention to eliminate stigma.

If we were to list the private and public management structures in our society that engage in fact-finding about risk, we would see that there are quite a few of them. Scientists engage in fact-finding in many settings, not only as individuals but also in professional associations. There are institutions such as the National Academy of Sciences, Bell Labs, and Greenpeace. Fact-finding about risk is performed in universities and in private companies, and by insurers, the press, legislatures, courts, and regulatory agencies. One agency well worth examining, I think, is the National Transportation Safety Board. Perhaps that institution comes close to being a model as a stigma identifier—even after such difficult cases as Flight 800 or the rudder control problems on the Boeing 737s, where I think the fact-finding capabilities of the NTSB have been severely tested. But something wonderful and quite effective goes on when we witness a daily public narrative produced by the NTSB about dredging the ocean floor for wreckage from Flight 800, about the percentage of debris brought up, and about the modes of analysis being performed. It is a narrative with the potential to build public confidence in the fact-finding process. It is a sustained effort to determine the real risk in future air travel and to legitimate a sustained campaign to combat stigma. Without a fact-finding process that builds public confidence that the real risk is being determined, we have no means of identifying stigma, and no justification for changing people's minds about the real risks in the world.

In conclusion, we might do well to evaluate and compare the fact-finding processes of the different institutional structures in our society, and to examine better and worse processes for distinguishing real risk from stigma. Such an empirical inquiry should supplement our continued conceptual investigations into how we know stigma when we see it. Both are needed if we are to develop better justifications and more effective techniques for dealing with stigma. That is, assuming we can ever agree on what we *mean* by the word "stigma." Perhaps the definition I have proposed can help us on our way: a "stigma" is a communicative marker that has become associated with a thing, and which has transferred to that thing a negative affective response originally generated in another context, which response is unwarranted in its new context given the current scientific theories about the real hazards posed by that thing. This definition suggests many scientific investigations into the incidence and role of stigma in our society, and suggests the importance of a fact-finding process for accurately identifying stigma before attempting to eliminate it.

23 Defining Stigma

Baruch Fischhoff

Stigma, like any other social science term, is most useful if it can be restricted to specific, well-defined phenomena that can, in turn, be related to other well-defined phenomena. Settling on a restrictive definition is inherently controversial. It will conflict with the definitions held by some investigators. It might even exclude some of those investigators from the community of stigma researchers. Incurring that risk must be justified by avoiding the competing risk, namely, that a broadly defined notion of "stigma" will have so little explanatory power that the research community will collapse, as a result of being unable to produce stable, converging results.

The proposal that follows comes from an admiring bystander, intrigued by the work stimulated by interest in stigma, but puzzled about what that concept's unique value is. It begins by proposing a definition of stigma. It then illustrates this definition by applying it to characterizing the role of stigma in several complex social processes. It concludes with thoughts on the implications of this definition for two practical tasks faced by stigma researchers: measuring stigma's effects on property values and reducing stigma. No attempt is made to relate the proposed definition to competing ones advanced by various stigma researchers. That would require greater familiarity with the stigma literature, and with the web of associations that its contributors assume, than is possible for a visitor to the area.

Stigma Is . . .

Stigma is demonstrated by *principled refusal to engage in an act that would otherwise be acceptable*. It happens when an individual feels that the act is *just not done*. In this view, stigma is a dichotomous variable. A stigmatized act is unacceptable whatever the associated benefits. If deciding what to do involves any cost-benefit calculus, then the resulting choice is not stigma driven.

Observing others' behavior can shape whether to invoke a stigma rule: If other people won't perform the act, perhaps they know something about it. However, those observations induce stigmatization only if they convince the observer that the act is a wrong thing to do. That can occur when the others' behavior either (a) demonstrates general rules of stigma that the observer might adopt or (b) teaches how to apply already adopted rules in a specific setting. It is not stigma when people mimic others' behavior for instrumental reasons, such as avoiding the penalty for doing something that another person stigmatizes (but without rejecting that act in principle).

Stigma may be attached to an act under all circumstances, or just in specific contexts. For example, an act might be legitimate when one is requested to perform it, but not when one provides the initiative, or vice versa. (Some sexual practices fall into these categories, for some people.) There are also acts whose acceptability depends on who is making the request (e.g., colleague, relative, student).

Individuals can be stigmatized as a result of their own voluntary acts—marking them as people who have done things that one does not do. They may also be marked by involuntary acts—as sometimes happens to the victims of rape, incest, or smear campaigns. Shunning such victims reflects that stigma, however, only if it is felt to be the right thing to do. It is not stigma if one enjoys victims' company less because they are less lively companions, if one feels guilty about failing to protect them, or if one wants to avoid being seen with people whom others have stigmatized.

Stigmatization can occur in risky or risk-free situations. For example, one may believe that it is wrong to incur an epsilon chance of losing one's life under frivolous circumstances, hence stigmatize someone who does so. Adopting that rule would be a matter of principle. Applying it would depend on the attendant risk perceptions. Overestimating risks would mean applying the rule in more circumstances than one intended. However, deciding whether a stigma rule could apply (i.e., the circumstances are frivolous) is logically independent of determining whether the risk passes the threshold of opprobrium.

Stigma can also be a source of risk misperceptions. People who avoid an act because they stigmatize it might be misinterpreted as avoiding it because they see it as too risky. Those individuals might encourage that misattribution by explaining their behavior in risk terms, if they feel uncomfortable espousing their stigma principles. Once invoked, those risk-related explanations might begin to convince both those who hear them and those who speak them. For example, people who keep kosher view pork and shellfish as unclean, as a matter of principle. They might also claim that those foods tend to be unhygienic, providing a practical side benefit for believers. Those claims might lead some nonbelievers to avoid those foods as well, but for risk reasons, not stigma ones.[1]

The present definition of stigma is close to that for *taboo*. However, taboos are associated with specific acts, whereas stigma arises from violating general principles of proscribed behavior. Thus, stigma requires interpretation. As a result, it should be less predictable (and more malleable) than taboo. Stigma becomes increasingly important, relative to taboo, as the variety of potential acts increases, so that relatively fewer acts are governed by specific rules. Thus, stigma (as defined here) is particularly important in modern societies, whereas taboos are more central for traditional ones.

---

[1]Some people like to find instrumental collaboration of their principled beliefs. If so, then they might be good sources of selective samples of risk information (e.g., vegetarians who know about the latest salmonella outbreaks in red meat).

The potential malleability of stigma creates a challenge for developing predictive theories of stigma, and an opportunity for devising interventions that increase or decrease it. As a result, we need theoretical work, creating general accounts of stigma, and descriptive work, providing rigorously characterized examples, capable of guiding and testing the theories.

Stigma in General

The definition offered above is ostensibly quite simple: Stigma means rejecting an otherwise acceptable action, as a matter of principle. However, its application can be quite complicated when analyzing actual situations. Figure 23.1 summarizes the relationships between stigma and other processes with which it might be confused.

Stigmatic avoidance involves interpreting an action as unacceptable in principle. It arises from observing an initiating event that leads one to invoke a stigma rule proscribing a possible action (link a). Such avoidance may be prompted—or confirmed—by observing others' stigmatic avoidance of that action (link b), which shows them applying a general stigma rule to that specific situation (link c). The same initiating event can also change one's perceptions of the risks and benefits associated with the act (link d). Those *instrumental* concerns could lead to avoidance behavior, but for non-stigma reasons (link e). Stigma can be determined by assessments of risk, although they would have to be performed in a way that evokes no cost-benefit calculus (e.g., when one never does "that kind of thing" if it has any risk) (link f). If the initiating event changes other people's risk and benefit perceptions (link g), it can lead them to either non-stigmatic (link h) or stigmatic avoidance (link i). Other people's non-stigmatic avoidance behavior may erroneously be interpreted as stigmatic avoidance, leading to more of the same (link j). It can also affect one's own perceptions of risk and benefit (link k), thereby affecting the chances of stigmatic (link f) or nonstigmatic avoidance (link e). Finally, one's own perceptions of risk and benefit can be influenced by those of other people (link l).[2]

The next section applies the scheme of Figure 23.1 to several archetypal cases, often cited as representing stigma.[3]

Stigma in Context

CRIME SCENES

People are often reluctant to live in a house that was the scene of a violent crime. Such behavior would be a direct effect of stigmatization (link a), if the house were otherwise attractive, but is not even being considered for reasons like "people don't live in houses visited by tragedy," "it would be violating the victim's memory to live there," or "we can't just go on with life as usual."

---

[2]Of course, one's actions and beliefs could also influence other people. That could be handled in the figure either by adding arrowheads or by reversing the roles of "own" and "others."

[3]The complement of stigmatization might be called "veneration": doing or liking or adopting something just because it is right, regardless of the associated instrumental benefits. An analysis comparable to that in Figure 23.1 might help to disentangle veneration-motivated attraction from related processes (e.g., dietary practices adopted for religious or nutritional reasons, adulation that comes from reverence or sycophancy).

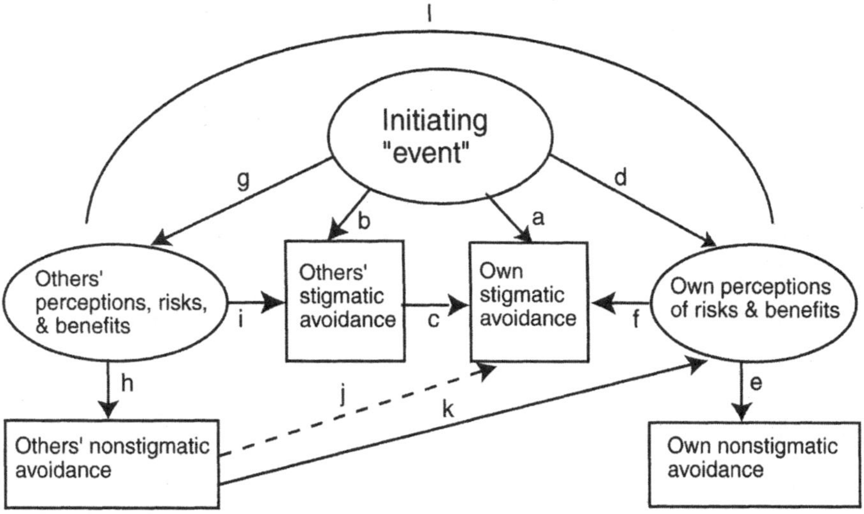

Figure 23.1 A general model of stigma and stigma-related avoidance.

It would not, however, be stigma if people avoided the house for instrumental reasons (link e), such as believing that it would be hard to sell the house because of others' stigmatization or thinking that "there must be some hidden defect if it's been on the market this long." Although caused by stigma, such interpretations fit directly into a cost-benefit calculus, hence affect the chances of non-stigmatic avoidance behavior. However, they do not override that calculus, as stigma would.

Similarly, it would not be stigma if the house lost value because people anticipated that living there would evoke unpleasant reminders of the crime—from the house itself or from others familiar with its history ("Oh, you live in the Johnson house"). Nor would it be stigma if the availability of crime-related cues, or the belief that criminals return to the scenes of their crimes increased the perceived probability of additional crimes. Stigma could play an indirect, but still instrumental role, if potential buyers believed that others would stigmatize the house's inhabitants ("How could you choose to live there, just because you knew that others' squeamishness made it really cheap?").

COCKROACHES

Paul Rozin's classic studies have shown experimental subjects' unwillingness to drink juice in which sterilized cockroaches have been dunked. That reluctance reflects direct stigmatization for individuals who have an overriding belief that it is wrong to engage in such strange acts. Stigma can have indirect instrumental effects for subjects who are not bothered by the act itself, but believe that others might stigmatize them for doing it. (Those "others" could include the experimenters who staged the study, and would know who drank what.) Preserving one's reputation can be a powerful motivator even when one sees no merit to others' stigma principles (i.e., link c does not exist). Individuals with abiding prejudices can still invoke powerful sanctions, justifying instrumental avoidance. On the other hand, inadvertent experimenter communication could signal what stigma means in this (unusual) situation. That could happen even with experimenters whose demeanor is one of practiced neutrality. Such impassivity might itself seem sufficiently unusual to prompt thoughts of "What do they mean by that?" Strange situations evoke a search for cues, whether intended or not.

Avoiding the act would not be stigma driven, if subjects declined the drink because they expected to experience recurrent reminders of having performed a questionable act with little benefit. Those reminders would reduce their future quality of life, making the act less attractive in instrumental terms. Nor would stigma be involved if subjects doubted the investigators' claim of cockroach purity, hence saw some risk. Even if one accepted the investigators' sincerity, one might wonder how thoroughly the investigators had done their work and how well they understood the integrity of cockroach corpses in (acidic?) apple juice. After all, the researchers are psychologists, not microbiologists or entomologists.

Another unwelcome role for the investigators arises if subjects start to reflect on the kind of person who would create such unusual tasks. If that act is viewed as a wrong thing to do, then the investigators themselves might be stigmatized. Should that happen, then subjects' evaluations of the drinking act would be confounded by their evaluations of the person presenting it ("Who accepts drinks from someone like that?")

The web of experiments surrounding Rozin et al.'s demonstrations is intended to control for such alternative explanations and, indeed, strongly suggests that direct stigmatization (as defined here) causes some subjects to decline the beverage. Those subjects who stigmatize the act might also see it as having an unfavorable cost-benefit ratio. However, the stigma alone would dissuade them.

ALAR

U.S. apple growers suffered tremendous financial losses as the result of public claims about the health risks associated with a common spray, Alar. Particularly affected were purchases of apple juice for consumption by children.[4] Such increases in avoidance behavior could be explained without recourse to stigma if consumers perceived (a) increased risk from apple juice consumption; (b) reduced benefit from that consumption (e.g., from worrying about safety, from having to argue with those who saw consumption as questionable); or (c) acceptable costs for shifting to substitutes (e.g., apple-flavored Kool Aid—which, roughly speaking, costs less, tastes similarly, and has the same nutritional value as pasteurized apple juice).

It would be stigma if consumers avoided apple juice *despite* believing that the risks were no greater than for otherwise equivalent beverages. Their stigma rule could be risk-based, for example, "one does not feed children foods whose risks are the subject of controversy or were created in an immoral way—even if the magnitude of those risks would otherwise be acceptably small, relative to the benefits." People who endorse that rule should have no respect for apple growers who performed the stigmatized act of feeding their children apple juice during the crisis, in order to "show the world" how safe the juice was.

[4]There are no reports on its popularity as a marinate for sterile cockroaches.

They might even stigmatize the act of buying food products from "people like that."

STIGMATIZED RESEARCHERS

Any putative behavioral truth should apply to one's own behavior, when the appropriate circumstances arise. Thus, it should be possible to apply this scheme to cases in which one scientist deliberately avoids another scientist's work. Consider, for example, a scientist who had maligned an esteemed colleague, with (what are seen as) charges based on lies and half-truths. Unreliable scholarship undermines the trust upon which the academic community is based, hence might be stigmatized. One possible response is not to believe or cite anything that the attacker writes. If the reason for such avoidance is not being able to trust the attacker's citations without double-checking the original sources, that would, however, reflect prudence, not stigma. It would also reflect instrumental behavior, if one avoided citing the attacker wherever possible, so as not to provide that individual with satisfaction, recognition, citation counts, etc. Nor would stigma be involved, if one declined invitations to debate the attacker, simply because that task was too unpleasant.

It would be stigma, though, if one let the attacker's charges remain unchallenged because one "does not stoop to deal with that type of person," "will not be seen on the same stage," or "will not prostitute oneself to engage in such debates." The price that one will pay to maintain that distance (e.g., the risk that silence will be treated as assent or cowardice) reflects the power of that stigma.

Estimating and Preventing Stigma

ESTIMATING THE COSTS OF STIGMA

Housing prices may have dropped around Superfund sites, like Rocky Flats. Stigma would be directly involved for potential buyers who felt that one does not voluntarily live "where weapons of mass destruction were created," "where one has to deal with the Department of Energy (or Defense)," or "where the integrity of nature has been destroyed by unnatural acts." Instrumental concerns are involved, however, for potential buyers who worry about being censured by people who have stigmatized the area, or who (mis)interpret those individuals' stigma-induced avoidance as reflecting perceived risk (and not principled rejection). Cost-benefit concerns might include the expectation of dwelling on past events ("This was once an atomic weapons factory"), the perception of residual risk, or the fear of depressed future real-estate prices. The level of perceived risk needed to trigger stigma-induced avoidance might not tip the balance of an instrumental decision ("The risk is tiny, but I can't countenance the process that produced it"). As elsewhere, these perceptions might (or might not) be accurate, and might (or might not) incorporate beliefs about the sources of others' stigmatizing behavior.

Figure 23.2 applies the general scheme of Figure 23.1 to the specific case of real estate prices near Rocky Flats. That picture could be converted to a formal pricing model, following its logic: Where stigma arises, some fraction of prospective buyers reject otherwise attractive offers because something is "just not right" about the houses or the neighborhood in question. As a result, those individuals will be absent from the set of potential buyers, thereby reducing

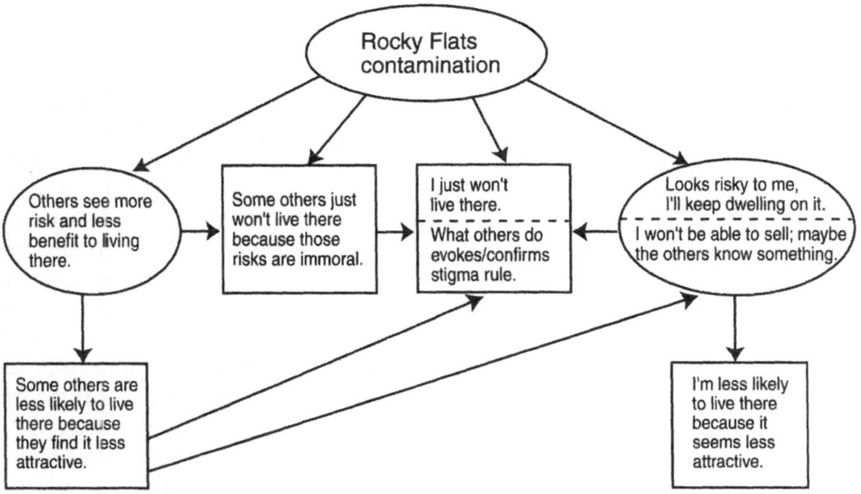

Figure 23.2 Stigma-related processes at Rocky Flats.

demand, hence prices. In addition to such direct stigma effects, there will be indirect ones. Well informed buyers who have no principled objection to purchasing these homes should still be willing to pay less if they expect future stigmatization to reduce future demand. (Reductions in current demand should be reflected in today's already reduced prices.) A more poorly informed potential buyer might inappropriately infer that others' instrumental avoidance is motivated by perceived risk, rather than by stigma. Whatever their source, perceptions of risk should reduce people's willingness to pay in proportion to the magnitude of the perceived risk.[5]

Fitting this model and estimating this aspect of the cost of stigma would involve the following steps: (a) Create a conventional model for property valuation in that kind of market, ignoring the potential for stigma; (b) reduce demand by removing the number of people for whom such a purchase would be stigmatized—either by direct assessment of their numbers or by prediction from an appropriate theory; (c) identify potential buyers' theories for interpreting others' avoidance behavior in such situations; (d) determine how those theories affect the perceived value of properties—by creating expectations of risk to future sales or to life and limb; and (e) devalue properties by the appropriate amount and run the model to predict how the market will behave.

STIGMA REDUCTION

According to the present account, stigma reflects a moral statement of what constitutes unacceptable behavior. It points to three routes to reducing stigma. One is exercising moral suasion in order to change people's stigma rules. The second is convincing people that their general stigma rules do not apply to a specific situation. The third is providing information that changes how the

[5]For instrumental choices, the value of a house for individual potential buyers should decline in direct proportion to perceived risk. For stigma-related choices, the magnitude of the perceived risk should determine the probability of that risk being above the threshold of principled unacceptability.

situation is perceived, so that even if the stigma rule applies in principle, it does not apply in practice.

Those who hope to "treat" stigma face both a scientific and an ethical challenge. The former is to understand the determinants of stigma and the opportunities for reducing it. The latter is to decide on the kinds of influence that they are willing to exert. For example, when will they try to help people who claim that they (or their products or organizations or activities) have been illegitimately stigmatized? Those claims can be true. However, they can also be self-serving, if they imply that the stigmatizers are irrational or uncompromising. One way to evaluate the quality of such claims is to measure the accuracy of the perceptions upon which the judgments of stigma are based. If stigmatizers view the situation accurately, then trying to get them to remove the stigma means trying to get them to abandon the stigma rule that they are applying. Doing so means choosing sides in a political, ethical struggle. That may not be a comfortable, or appropriate, role for social scientists. Providing better information is a more neutral role. Depending on the facts of the matter, it might lead others to greater understanding of the stigmatized or it might lead the stigmatized to a better understanding of themselves, and why they are so disliked.

References

The 1992 Information Please Almanac. (1992). New York: Houghton Mifflin.

Abas, B. (1992, December). The Rocky Flats cover-up continued. *Harper's Magazine,* 19–23.

Ablon, J. (1981). Stigmatized health conditions. *Social Science and Medicine, 15,* 5-9.

Agency for Toxic Substances and Disease Registry (ATSDR). (1990, June). *Case studies in environmental medicine: Arsenic toxicity.* Washington, DC: U.S. Department of Health and Human Services.

Agency for Toxic Substances and Disease Registry (ATSDR). (1996, March 6). *Exposure investigation CR #40W1.* Atlanta, GA: U.S. Department of Health and Human Services.

Alhakami, A.S., & Slovic, P. (1994). A psychological study of the inverse relationship between perceived risk and perceived benefit. *Risk Analysis, 14*(6), 1085–1096.

Allard wants EPA to reduce Superfund areas in Leadville. (1997, July 29). *Denver Post,* p. 4B.

Allport, G. (1954). *The nature of prejudice.* Reading, MA: Addison-Wesley.

Altman, L.K. (1997, January 3). FDA proposal would ban using animal tissue in feed. *The New York Times,* p. A7.

American Automobile Manufacturers Association. *Air Bag Safety: Hearings Before the U.S. Senate Committee on Commerce, Science, and Transportation,* 104th Cong., 2d Sess. (1996, March 7) (testimony of Richard Klimisch).

American Psychological Association (APA). (1989, March). [1989 annual meeting announcement]. *American Psychologist,* p. 583.

Ames, B. (1983). Dietary carcinogens and anticarcinogens. *Science, 221,* 1256–1264.

Ames, B.N., Magaw, R., & Gold, L.S. (1987). Ranking possible carcinogenic hazards. *Science, 236,* 271–280.

Anderson, A. (1997). *Media, culture and the environment.* London: UCL Press.

Anderson, C. (1983). Abstract and concrete data in the perseverance of social theories: When weak data lead to unshakeable beliefs. *Journal of Experimental Social Psychology, 19,* 93-108.

Appadurai, A. (1981). Gastro-politics in Hindu South Asia. *American Ethnologist, 8,* 494–511.

The Appraisal Institute. (1992). *Measuring the effects of hazardous materials contamination on real estate values: Techniques and applications.* Chicago: Author.

Arizona Department of Environmental Quality (ADEQ). (1995). *South Phoenix waste minimization survey project.* Phoenix, AZ: Author.

Associated Press (AP). (1980, May 30).

Associated Press (AP). (1996a, October 24). Federal report blames air bag in death of child wearing a belt. *The Boston Globe,* p. A6.

Associated Press (AP). (1996b, October 24). Properly belted child is killed by air bag. *The New York Times,* p. A17.

Atkinson, A. (1992). Environmental inequity—An emerging concern for government. *Maryland Journal of Contemporary Legal Issues, 5*(1), 81–103.

Auvil v. CBS *60 Minutes,* 800 F. Supp. 928 (E.D. Wash. 1993).

Bacot, H., Bowen, T., & Fitzgerald, M. (1994). Managing the solid waste crisis: Exploring the link between citizen attitudes, policy incentives, and siting landfills. *Policy Studies Journal*, 22(2), 229–244.

Baker, E.J., Moss, D.J., West, S.G., & Weyant, J.K. (1977). *Impact of offshore nuclear generating stations on recreational behavior at adjacent coastal sites* (NUREG-0394). Washington, DC: U.S. Nuclear Regulatory Commission.

Bandura, A. (1983). Self efficacy determinants of anticipatory fears and calamities. *Journal of Personality and Social Psychology*, *45*(2), 464-469.

Barr, M., Jr. (1981, February). Environmental contamination of human breast milk. *American Journal of Public Health*, *71*(2), 124-6.

Barthes, R. (1984). *Camera lucida: Reflections on photography*. London: Fontana Paperbacks.

Barzel, Y. Measurement cost and the organization of markets. (1982, April). *Journal of Law and Economics*, 27–48.

Bassett, G.W., & Hemphill, R.C. (1991). Comments on perceived risk, stigma, and potential economic impacts of a high-level nuclear waste repository in Nevada. *Risk Analysis, 11*, 697–700.

Basso, K.H. (1996). *Wisdom sits in places: Landscape and language among the western Apache*. Albuquerque: University of New Mexico.

Baum, A., Fleming, R., & Singer, J. (1983). Coping with victimization by technological disaster. *Journal of Social Issues*, *39*(2), 117-138.

Beck, K., & Frankel, A. (1981). A conceptualization of threat communications and protective health behavior. *Social Psychology Quarterly*, *44*(3), 204-217.

Beck, R.J., & Wood, D. (1976). Cognitive transformation of information from urban geographic fields to mental maps. *Environment and Behavior, 8*, 199–238.

Been, V. (1994a). Locally undesirable land uses in minority neighborhoods: Disproportionate siting or market dynamics? *The Yale Law Journal, 103*, 1383–1422.

Been, V. (1994b, Spring). Unpopular neighbors: Are dumps and landfills sited equitably? *Resources, 115*, pp. 16–19.

Bell, A. (1991). *The language of news media*. Oxford: Blackwell.

Benjamin, D.K. (1978, July). The use of collateral to enforce debt contracts. *Economic Inquiry*, 333–359.

Benjamin, D.K., & Mitchell, M.L. (1989). *Quality-assuring price premia: Classic evidence from the real thing*. Working paper, Securities and Exchange Commission.

Beron, G.L. (1992). Measuring the effects of proximity to a uranium processing facility. In *Measuring the effects of hazardous materials contamination on real estate values: Techniques and applications* (pp. 67–100). Chicago: The Appraisal Institute.

Beyer, B.K., & Hicks, E.P. (1968). *Images of Africa: A report of what secondary school students know and believe about Africa south of the Sahara*. Pittsburgh, PA: Carnegie Mellon University, Project Africa.

Biddle, W. (1982, November 10). 10 years later, Missourans find soil tainted by dioxin. *The New York Times*.

Bisconti, A. (1989, May). *Polling an inattentive public: Energy and U.S. public opinion*. Washington, DC: U.S. Member Committee of the World Energy Conference.

Bisconti, A. (1991, July/August 1991). Public attitudes about nuclear waste. *Nuclear Plant Journal*, pp. 49ff.

Black Elk, W., & Lyon, W.L. (1990). *Black Elk: The sacred ways of a Lakota*. San Francisco: Harper and Row.

Blair, J.M. (1994). *Issues raised by highway traffic noise and the performance of Phoenix's freeways*. Master's thesis, Arizona State University, Tempe.

Blood safety director stresses importance of restoring public confidence in blood supply. (1996, October 18). *ABC [America's Blood Centers] Newsletter, 40*, 1–2.

Boerner, C., & Lambert, T. (1994). *Ethical land use: Principles of policy and planning*. Baltimore: John Hopkins Press.

Boholm, Å. (1998). Visual images and risk messages: Commemorating Chernobyl. *Risk Decision and Policy, 3*(2), 125–143.

Boholm, Å., Löfstedt, R., & Strandberg, U. (1988). Tunnelbygget genom hallandsåsen:

Lokalsamhällets dilemman (*CEFOS Rapport 12)*. Göteborgs Universitet.
Boulding, K.E. (1956). *The image*. Ann Arbor: University of Michigan Press.
Brandner, S., Isenmann, S., Raeber, A., Fischer, M., Sailer, A., Kobayashi, Y., Marino, S., Weismann, C., & Aguzzi, A. (1996). Normal host prion protein necessary for scrapie-induced neurotoxicity. *Nature, 379*, 339–343.
Brandt, A., & Rozin, P. (Eds.) (1997). *Morality and health*. New York: Routledge.
Bray, S., Freeman, P., & Kunreuther, H., (1998, May). *Environmental insurance as facilitator in brownfields transactions*. Paper Presented at the American Enterprise Institute Conference on "Challenges in Private-Public Partnerships in Managing Environmental Risk," Washington, DC.
Brooke, J. (1995, May 3). Tourist site springs from nuclear horror story. *The New York Times*, p. A4.
Brown, J. (1996, April 4). Vegetarian monks change habits and tuck into beef. *PA News*.
Brown, N.J., Barton, J.A., & Bjork, K.E. (1996, October). *Rocky Flats community needs assessment: Final report*. Denver, CO: University of Colorado Health Sciences Center.
Brown, W. (1991, January 4). Perrier's market share fizzles in the aftermath of its recall. *Washington Post*, p. F3.
Brown, W. (1996, November 2). Risks spark search for better airbag. *Washington Post*, p. A12.
Brown, W., & Ottaway, D. (1997, June 1). From life saver to fatal threat, how the U.S. automakers and a safety device failed. *Washington Post*, pp. Al, A16–A18.
Bryan, R.H. (1987). The politics and promises of nuclear waste disposal: The view from Nevada. *Environment, 29*(8): 14–17, 32–38.
Bryant, B., & Mohai, P. (1992). *Race and the incidence of environmental hazards: A time for discourse*. San Francisco: Westview Press.
Bullard, R.D. (1993). The threat of environmental racism. *Natural Resources and Environment, 7*(3), 23–26.
Bullard, R.D. (1994a). *Dumping in Dixie: Race, class, and environmental quality*. Boulder, CO: Westview Press.
Bullard, R.D. (1994b). *Unequal protection: Environmental justice and communities of color*. San Francisco: Sierra Club Books.
Burgess, J.A. (1978). *Image and identity: A study of urban and regional perception with particular reference to Kingston upon Hull* (Occasional Papers in Geography, No. 23). UK: University of Hull.
Burnham, D. (1974, August 10). Scientist urges Congress to bar any use of the pesticide 2,4,5-T. *The New York Times*.
Burns, W., Slovic, P., Kasperson, R., Kasperson, J., Renn, O., & S. Emani, S. (1990). Social amplification of risk: An empirical study (Report No. NWPO-SE-027-90). Carson City: Nevada Agency for Nuclear Projects, NWPO.
Burns, W.J., Slovic, P., Kasperson, R.E., Kasperson, J.X., Renn, O., & Emani, S. (1993). Incorporating structural models into research on the social amplification of risk: Implications for theory construction and decision making. *Risk Analysis, 13*, 611–623.
Camerer, C., & Kunreuther, H. (1989). Decision processes for low probability events: Policy implications. *Journal of Policy Analysis and Management, 8*, 562–592.
Campbell, A. (1981). *The sense of well-being in America*. New York: McGraw Hill.
Canadian Council of Ministers of the Environment. (1993). *Dioxins and furans: The Canadian perspective* (PN 1050) [Brochure]. Winnipeg: Manitoba Statutory Publications.
Cantrell, S., Maloney, M.T., & Mitchell, M.L. (1989). *On estimating the variance of abnormal stock market performance*. Working paper, Securities and Exchange Commission.
Carnes, S.A., Copenhaver, E.D., Sorensen, J.H., Soderstrom, E.J., Reed, J.H., Bjornstad, D.J., & Peelle, E. (1983). Incentives and nuclear waste siting: Prospects and constraints. *Energy Systems and Policy, 7*(4), 324–351.
Carter, L.J. (1987). *Nuclear imperatives and public trust: Dealing with radioactive waste.*

Washington, DC: Resources for the Future.
Cerra, F. (1980, August 2). Garbage recycling plant stays closed over dioxin. *The New York Times*.
Chalk, A. (1986, January). Market forces and aircraft safety: The case of the DC-10. *Economic Inquiry*, 43–60.
Chalk, A. (1987, September). Market forces and commercial aircraft safety. *Journal of Industrial Economics*, 61–82.
Chalmers, J.A., & Jackson, T. (1996). Risk factors in the appraisal of contaminated property. *The Appraisal Journal*, *64*(1), 44–58.
Chalmers, J.A., & Roehr, S.A. (1993). Issues in the valuation of contaminated property. *The Appraisal Journal*, *61*(1), 28–41.
Cheron, E., & Ritchie, J.R.B. (1982). Leisure activities and perceived risk. *Journal of Leisure Research*, *14*(2), 139-154.
Christensen-Szalanski, J., Beck D., Christensen-Szalanski, C., & Koepsell, T. (1983, May). Effects of experience and expertise on risk judgments. *Journal of Applied Psychology*, *68*(2), 278-284.
City of Arvada (1989). *Arvada citizen attitude survey*. Arvada, CO: Author.
City of Santa Fe v. Komis, 845 P.2d 753 (NM 1992).
Claybrook, J. (1996, December 1). The auto industry, the air bag. *Washington Post*, p. C-7.
Cohen, B.L. (1983). *Before it's too late: A scientist's case for nuclear energy*. New York: Plenum.
Cohen, B.L. (1985). A simple probabilistic risk analysis for high-level waste repositories. *Nuclear Technology*, *68*, 73–76.
Collin, R.W. (1992). Environmental equity—A law and planning approach to environmental racism. *Virginia Environmental Law Journal*, *11*, 195–546.
Colorado Council on Rocky Flats. (1993). *The handbook on Rocky Flats*. Golden, CO: Author.
Colwell, P.F. (1990). Power lines and land value. *The Journal of Real Estate Research*, *5*(1), 117–126.
Consortium for Environmental Risk Evaluation. (1995). *Health and ecological risks at the U.S. Department of Energy's nuclear weapons complex: A qualitative evaluation* [CERE Interim Risk Report]. New Orleans, LA: Author (see chapter 5 on Rocky Flats).
Cook et al., plaintiffs v. Rockwell International Corp. and Dow Chemical Co., defendants (Plaintiff's Motion for Class Certification). (1993, March 29). *Civ. A. No. 90-K-181*. U.S. District Court, CO.
Council for Agricultural Science and Technology (CAST). (1994). *Foodborne pathogens: Risks and consequences*. Ames, IA: Author.
Covello, V.T. (1983). The perception of technological risks: A literature review. *Technological Forecasting Social Change*, *23*, 285–297.
Covello, V. T. (1991). Risk comparisons and risk communication: Issues and problems in comparing health and environmental risks. In R. E. Kasperson & P. J. M. Stallen (Eds.), *Communicating risks to the public: International perspectives*. The Netherlands: Kluwer Academic.
Coyer, B.W., & Schwerin, D. (1981). Bureaucratic regulation and farmer protest in the Michigan PBB contamination case. *Rural Sociology*, *46*(4), 703-723.
Craik, K.M. (1970). Environmental psychology. In T. Newcomb (Ed.), *New directions in psychology* (pp. 1–121). New York: Holt, Rinehart, and Winston.
Crisis public relations. (1982, August). *Dun's Business Month*, p. 50.
Cutter, S. (1995). Race, class and environmental justice. *Progress in Human Geography*, 19(1), 111–122.
Dake, K. (1991). Orienting dispositions in the perception of risk: An analysis of contemporary worldviews and cultural biases. *Journal of Cross-Cultural Psychology*, *22*, 61–82.
Dalmotas, D.J., Hurley, R.M., & German, A. (1995). *Air bag deployments involving*

restrained occupants (SAE Technical Paper No. 950868). Warrendale, PA: Society of Automotive Engineers.

Damasio, A.R. (1994). *Descartes' error: Emotion, reason, and the human brain.* New York: Avon.

Deane, J. (1997, June 27). Chancellor warned over rising cost of BSE crisis. *PA News.*

Dickson, D. (1981, January). Limiting democracy: Technocrats and the liberal state. *Democracy, 1*(1), 61-79.

Dimson, E. (1979). Risk measurement when shares are subject to infrequent trading. *Journal of Financial Economics*, 197–226.

DOE. See U.S. Department of Energy.

Doll, R. (1997). Weak associations in epidemiology. *Radiological Protection Bulletin, 192,* 10–15.

Douglas, M., & Wildavsky, A. (1982). *Risk and culture: An essay on the selection of technical and environmental dangers.* Berkeley: University of California Press.

Downs, R.M., & Stea, D. (1973). *Image and environment: Cognitive mapping and spatial behavior.* Chicago: Aldine.

Drotman, D.P. (1983, March). Contamination of the food chain by PCB from a broken transformer. *American Journal of Public Health, 73,* 302-313.

Dunlap, R.E., & Baxter, R.K. (1988). *Public reaction to siting a high-level nuclear waste repository at Hanford: A survey of local area residents* (For Impact Assessment, Inc.). Pullman: Washington State University, Social and Economic Sciences Research Center.

Dunlap, R.E., Rosa, E.A., Baxter, R.K., & Mitchell, R.C. (1993). Local attitudes toward siting a high-level nuclear waste repository at Hanford, Washington (pp. 136-172). In R.E. Dunlap, M.E. Kraft, & E.A. Rosa (Eds.), *Public reactions to nuclear waste: Citizens' views of repository siting.* Durham, NC: Duke University Press.

Dupont, R.L. (1981, September 2). The nuclear power phobia. *Business Week,* pp. 8–9.

Easterling, D.V. (1997). The vulnerability of the Nevada visitor economy to a repository at Yucca Mountain. *Risk Analysis, 17,* 635–647.

Easterling, D.V., & Kunreuther, H. (1990). *The convention attendees survey: Frequency distributions* (Report prepared by the Wharton Center for Risk and Decision Processes). Carson City: NWPO.

Easterling, D.V., & Kunreuther, H. (1993). The vulnerability of the convention industry to the siting of a high-level nuclear waste repository (pp. 209-238). In R.E. Dunlap, M.E. Kraft, & E.A. Rosa (Eds.), *Public reactions to nuclear waste: Citizens' views of repository siting.* Durham, NC: Duke University Press.

Easterling, D.V., & Kunreuther, H. (1995). *The dilemma of siting a high-level nuclear waste repository.* Boston: Kluwer Academic.

Easterling, D.V., Morwitz,V., & Kunreuther, H. (1991). Forecasting behavioral response to a repository from stated intent data. In *High-level radioactive waste management: Proceedings of the second annual international conference* (pp. 1540–1547). La Grange Park, IL: American Nuclear Society and American Society of Civil Engineers.

Edelstein, M.R. (1982). *The social and psychological impacts of groundwater contamination in the Legler section of Jackson, New Jersey* (Report to the law firm Kreindler and Kreindler).

Edelstein, M.R. (1987). Toward a theory of environmental stigma. In J. Harvey & D. Henning (Eds.), *Public environments* (pp. 21–25). Ottawa, Canada: Environmental Design Research Association.

Edelstein, M.R. (1988). *Contaminated communities: The social and psychological impacts of residential toxic exposure.* Boulder, CO: Westview Press.

Edelstein, M.R. (1991). Ecological threats and spoiled identities: Radon gas and environmental stigma. In S.R. Couch & J.S. Kroll-Smith (Eds.), *Communities at risk: Collective responses to technological hazards* (pp. 205–226). Worster: Peter Lang.

Edelstein, M.R. (1993). When the honeymoon is over: Environmental stigma and distrust in the siting of a hazardous waste disposal facility in Niagara Falls, New York. In W. Freudenburg & T. Youn (Eds.), *Research in social problems and public policy,*

Vol. 5 (pp. 75-96). Greenwich, CT: JAI Press.

Edelstein, M.R. (1997, March). *Crying over spoiled milk: Contamination, visibility, and expectation in environmental stigma.* Paper presented at the Annenberg Conference on Risk, Media, and Stigma, Philadelphia, PA. Reprinted as chapter 4 of this volume.

Edelstein, M.R., & Kleese, D.A. (1995, February). The cultural relativity of impact assessment: Native Hawaiian opposition to geothermal energy development. *Society and Natural Resources, 8*, 19-31.

Edelstein, M.R., & Kakofske, W.J. (1998). *Radon's deadly daughters: Science, environmental policy, and the politics of risk.* Lanham, MD: Rowman & Littlefield.

Edmonston, B. (1994, March). The trend you can't ignore. *American Demographics*, 16, 60–64.

Egan, T. (1988, April 6). On good days the odor is hardly noticeable. *The New York Times*, p. 16A.

Eldreidge, E. (1991, November 14). Officials rip ad campaign mounted for dump. *Reno Gazette-Journal.*

Elgin, D. (1981). *Voluntary simplicity*. New York: William Morrow.

Elgin, D., & Mitchell, A. (1977, summer). Voluntary simplicity (3). *The Co-Evolution Quarterly*, 4–19.

Elliot-Jones, M. (1992). Real estate value and toxic sites. *The Digest of Environmental Law*, 5(7), 89–93.

Elster, J. (1992). *Local justice: How institutions allocate scarce goods and necessary burdens.* New York: Russell Sage Foundation.

Engdahl, T. (1989, December 28). 1989: Top Colo. stories played on national stage. *The Post.*

Environment Canada (1996, December 10). News conference notes for the Honourable Sergio Marchi, P.C., M.P., on the occasion of the tabling of the new Canadian Environmental Protection Act. *National Press Theatre.*

EPA. See U.S. Environmental Protection Agency.

Ericson, R., Baranek, P., & Chan, J. (1987). *Visualizing deviance: A study of news organizations.* Milton Keynes: Open University.

Erikson, K. (1990). Toxic reckoning: Business faces a new kind of fear. *Harvard Business Review, 68*(1), 118–126.

Erikson, K. (1994). *A new species of trouble: The human experience of modern disasters.* New York: Norton.

Ewing, T.F. (1990, July 8). National notebook: Hamilton, Ohio; Guarantees near a landfill. *The New York Times Sunday National Edition,* Section 8, p. 1.

Eysenck, M.W. (Ed.). (1990). *The Blackwell dictionary of cognitive psychology*. Oxford, UK: Blackwell.

Fallon, A.E., Rozin, P., & Pliner, P. (1984). The child's conception of food: The development of food rejections with special reference to disgust and contamination sensitivity. *Child Development*, *55*, 566–575.

Farmer, B. (1995, December 7). Major seeks to ease beef fears. *PA News.*

FDA bans animal parts in feed. (1997).

Federal Register, 61 Fed. Reg. 50499-50500 (1996, September 26).

Ferguson, J., & Keating, M. (1980a, November 29). *The Globe and Mail.*

Ferguson, J., & Keating, M. (1980b, December 3). *The Globe and Mail.*

Ferguson, J., & Keating, M. (1980c, December 4). *The Globe and Mail.*

Ferguson, J., & Keating, M. (1980d, December 9). *The Globe and Mail.*

Ferguson, J., & Keating, M. (1980e, December 10). *The Globe and Mail.*

Ferguson, S.A., Braver, E.R., Greene, M.A., & Lund, A.K. (1996, September). *Preliminary report: Initial estimates of reductions in deaths in frontal crashes among right front passengers in vehicles equipped with passenger airbags.* Arlington, VA: Insurance Institute for Highway Safety.

Fick, Bob. (1996, November 28). Mall fender-bender kills one-year old girl. *The Cincinnati Enquirer*, p.Al.

Finsterbusch, K. (1980). *Understanding social impacts: Assessing the effects of public projects.* Beverly Hills, CA: Sage.
Fischhoff, B. (1977). Cognitive liabilities and product liability. *Journal of Products Liability, 1,* 213.
Fishbein, M. & Ajzen, I. (1975). *Belief, attitude, intention, and behavior: An introduction to theory and research.* Reading, MA: Addison-Wesley.
Fitchen, J.M. (1989). When toxic chemicals pollute residential environments: The cultural meanings of home and home ownership. *Human Organization, 48*(4), 313–324.
Flynn, J. (1992, April 15). How not to sell a nuclear waste dump. *The Wall Street Journal,* p. A20.
Flynn, J., Chalmers, J., Easterling, D., Kasperson, R., Kunreuther, H., Mertz, C.K., Mushkatel, A., Pijawka, K.D., & Slovic, P., with Dotto, L. (1995). *One hundred centuries of solitude: Redirecting America's high-level nuclear waste policy.* Boulder, CO: Westview Press.
Flynn, J., Kasperson. R., Kunreuther, H., & Slovic, P. (1992). Time to rethink nuclear waste storage. *Issues in Science and Technology, 8*(4), 42–48.
Flynn, J., Kasperson, R., Kunreuther, H., & Slovic, P. (1997). Overcoming tunnel vision: Redirecting the U.S. high-level nuclear waste program. *Environment, 39*(3), 6–11, 25–30.
Flynn, J., Mertz, C.K., & Slovic, P. (1991). *The Autumn 1991 Nevada state telephone survey.* Carson City: State of Nevada, NWPO.
Flynn, J., Peters, E., Mertz, C.K., & Slovic, P. (1998). Risk, media, and stigma at Rocky Flats. *Risk Analysis, 18*(6), 715–727. Reprinted as chapter 20 of this volume.
Flynn, J.H., Slovic, P., & Mertz, C.K. (1993). The Nevada initiative: A risk communications fiasco. *Risk Analysis, 13*(5), 497–502.
Flynn, J., Slovic, P., & Mertz, C.K. (1994). Gender, race, and perceptions of environmental health risks. *Risk Analysis, 14*(6), 1101–1108.
Flynn, J.H., Slovic, P., Mertz, C.K., & Toma, J. (1990). *Evaluations of Yucca Mountain: Survey findings about attitudes, opinions and evaluations of nuclear waste disposal and Yucca Mountain, Nevada* (Report No. NWPO-SE-029-90). Carson City, NV: NWPO.
Food Marketing Institute (FMI). (1997). *Trends in the United States: Consumer attitudes and the supermarket.* Washington, DC: Author.
Foran, J.A., Goldstein, B.D., Moore, J.A., & Slovic, P. (1996). Predicting future sources of mass toxic tort litigation. *Risk: Health, Safety & Environment, 7,* 15–22.
Fowler, C.S., Hamby, M., Rusco, E., & Rusco, M. (1990). *Native Americans and Yucca Mountain: A summary report.* Carson City, NV: NWPO.
Fox, W.F. et al. (1985). *An economic analysis of a monitored retrievable storage facility for Tennessee.* Knoxville: University of Tennessee Center for Business and Economic Research.
Frazer, J.G. (1959). *The new golden bough: A study in magic and religion.* New York: MacMillan (Original work published in 1890).
Freeman, P., & Kunreuther, H. (1997). *Managing environmental risk through insurance.* Washington, DC and Norwood, MA: American Enterprise Institute and Kluwer.
Freud, S. (1924). *Collected papers.* London: Hogarth.
Freudenburg, W.R. (1986). Social impact assessment. *Annual Review of Sociology, 12,* 451–478.
Freudenburg, W.R. (1993). Risk and recreancy: Weber, the division of labor, and the rationality of risk perceptions. *Social Forces, 71,* 909–932.
Freudenburg, W.R., & Pastor, S.K. (1992). Public responses to technological risks: Toward a sociological perspective. *Sociological Quarterly, 33,* 389–412.
Frey, B., Oberholzer-Gee, F., & Eichenberger, R. (1996). The old lady visits your backyard: A tale of morals and markets. Journal of Political Economy, 104(6), 1297-1314.
Gallup Poll. (1977, July 24). *Survey #977-K.* Princeton, NJ: The Gallup Organization.

Gallup Poll. (1984, July 8). *Survey #234-G; 236-G*. Princeton, NJ: The Gallup Organization.
Galton, F. (1880). Psychometric experiments. *Brain, 2*, 149–162.
GAO. See U.S. General Accounting Office.
Gardner, R.M. (1992). Stereotypes and media. *The Midwest Quarterly, 34*, 121–135.
Garling, R., Book, A., & Ergezen, N. (1982). Memory for the spatial layout of the everyday physical environment: Differential rates of acquisition of different types of information. *Scandinavian Journal of Psychology, 23*, 23–55.
Garthright, W.E., Archer, D.L., & Kvenberg, J.E. (1988). Estimates of incidence and costs of intestinal infectious diseases in the United States. *Public Health Report, 103*, 107–115.
Gelobter, M. (1994). The meaning of urban environmental justice. *Fordham Urban Law Journal 21*, 841–856.
Gerber ignores Tylenol textbook. (1986, March 10). *Advertising Age*, p. 3.
Gibbs, L.M. (1982). *Love Canal: My story*. Albany, NY: SUNY Press.
The Globe and Mail. (1980, December 9).
The Globe and Mail. (1981a, May 14).
The Globe and Mail. (1981b, July 17).
The Globe and Mail. (1981c, August 13).
The Globe and Mail. (1981d, October 27).
The Globe and Mail. (1982a, January 5).
The Globe and Mail. (1982b, January 6).
The Globe and Mail. (1982c, April 2).
The Globe and Mail. (1982d, June 25).
Goffman, E. (1963). *Stigma: Notes on the management of spoiled identity*. Englewood Cliffs, NJ: Prentice-Hall.
Golledge, R.G., & Stimson, R.J. (1987). *Analytical behavioural geography*. London: Croom Helm.
Goodey, B. (1973). *Perception of the environment* (Occasional Paper No. 17). UK: University of Birmingham, Centre for Urban and Regional Studies.
Goodey, B. (1974). *Images of place: Essays on environmental perception, communications, and education* (Occasional Paper No. 30). Birmingham, UK: University of Birmingham, Centre for Urban and Regional Studies.
Gottlieb, R. (1993). *Forcing the spring: The transformation of the American environmental movement*. Washington: Island Press.
Gould, P.R. (1966). *On mental maps* (Michigan Inter-University Community of Mathematical Geographers Discussion Paper No. 9). Ann Arbor: University of Michigan, Department of Geography.
Gowda, M.V.R., & Easterling, D. (1998). Nuclear waste and Native America: The MRS siting experience. *Risk: Health, Safety & Environment, 9*, 229–258.
Graham, J.D., Clemente, K.M., Glass, R.J., & Pasternak, N. (1999). Measuring public confidence in hazard claims: Results of a national survey. *Technology, 6*, 63–75.
Graham, J.D., Goldie, S.J., Segui-Gomez, M., Thompson, K.M., Nelson, T., Glass, R., Simpson, A., & Woemer, L.G. (1998). Reducing risks to children in vehicles with passenger airbags. *Pediatrics (e), 102*(1), 1–7.
Gregg, R. (1977, Summer). Voluntary simplicity (1). *The Co-Evolution Quarterly*, pp. 20-27.
Gregory, R., Flynn, J., & Slovic, P. (1995). Technological stigma. *American Scientist, 83*, 220–223. Reprinted as chapter 1 of this volume.
Gregory, R., & Mendelsohn, R. (1993). Perceived risk, dread, and benefits. *Risk Analysis, 13*, 259–264.
Gregory, R., Slovic, P., & Flynn, J. (1995). Risk perceptions, stigma, and health policy. *Health & Place, 2*, 213–220.
Gricar, B.G., & Baratta, A. (1983). Bridging the information gap at Three Mile Island: Radiation monitoring by citizens. *Journal of Applied Behavioral Science, 19*(1), 35-49.

Grinspoon, P. (1992, November 23). Atom and Eve—A love story: Nuclear power for women. *The Nation*, pp. 624-626.
Haidt, J., McCauley, C.R., & Rozin, P. (1994). A scale to measure disgust sensitivity. *Personality and Individual Differences*, *16*, 701–713.
Hannigan, J.A. (1995). *Environmental sociology: A social constructionist perspective*. London: Routledge.
Hansen, A. (1991). The media and the social construction of the environment. *Media, Culture and Society*, *13*(4), 443–458.
Hartley, J. (1992). *The politics of pictures. The creation of the public in the age of popular media*. London and New York: Routledge.
Hatcher, S.L. (1992). The psychological experience of nursing mothers upon learning of a toxic substance in their breast milk. *Psychiatry*, *45*, 172-181.
Healey, J.R., & O'Donnell, J.O. (1996). Poll shows adults know the dangers of airbags. *USA Today*, p. 2B.
Heberlein, T., & Black, J.S. (1981, November). Cognitive consistency and environmental action. *Environment and Behavior*, *13*(6), 717-734.
Heider, F. (1963). *The psychology of interpersonal relationships*. New York: John Wiley and Sons.
Heimer, C.A. (1988). Social structure, psychology and the estimation of risk. *Annual Review of Sociology*, *14*, 491–519.
Heitzmann, M., & Wilson, R. (1997, March). Low-dose linearity: The rule or the exception? *Belle Newsletter*, *6*(1), 2–8.
Herman, J. (1997). *Trauma and recovery: The aftermath of violence from domestic abuse to political terror* (Rev. ed.). New York: Basic Books.
Herzik, E. (1993, July). *Nevada statewide telephone poll survey data*. Report presented to Nevada State and Local Government Planning Group, University of Nevada, Reno.
Hewitt, K., & Burton, I. (1971). *The hazardousness of a place: A regional ecology of damaging events* (Department of Geography Research Paper No. 6). Canada: University of Toronto Press for the Department of Geography.
Hickox, K. (1992, February 6). DOE got media training. *Las Vegas Sun*.
Higgins, R.R. (1993). Race and environmental equity—An overview of the environmental justice issue in the policy process. *Polity*, *26*(2), 281–300.
Hilgartner, S. (1992). The social construction of risk objects. In J.F. Short & L. Clark (Eds.), *Organization, uncertainties and risk* (pp. 39-53). Boulder, San Francisco, & Oxford: Westview Press.
Hofmann, J., Liu, Z., Genese, C., Wolf, G., Manley, W., Pilot, K., Dalley, E., & Finelli, L. (1996). Update: Outbreaks of cyclospora cayetanensis infection—United States and Canada, 1996. *MMWR*, *45*(28), 611.
Hunsperger, W. (1991). Methane seepage from a nearbye landfill: Its effect on commercial real estate. *Environmental Watch*, *4*(3), 1–5.
Hunsperger, W. (1997, March). *The effects of the Rocky Flats nuclear weapons plant on neighboring property values*. Paper presented at the Annenberg Conference on Risk, Media, and Stigma, Philadelphia, PA. Reprinted as chapter 9 of this volume.
Ichheiser, G. (1970). *Appearances and realities*. New York: Jossey-Bass.
IFST. *See* U.K. Institute of Food Science & Technology.
Inglehart, R. (1971). The silent revolution in Europe: Intergenerational change in post-industrial societies. *American Political Science Review*, *65*, 991–1017.
Inglehart, R. (1981). Post-materialism in an environment of insecurity. *American Political Science Review*, *75*, 880–900.
Inglehart, R. (1990). *Culture shift in advanced industrial society*. Princeton, NJ: Princeton University Press.
Inmates in 60's test of a poison sought. (1981, January 18). *The New York Times*.
International Atomic Energy Agency (IAEA). (1988). *The radiological accident at Goiânia*. Vienna: Author.
International Commission on Radiological Protection (ICRP). (1991). *Recommendations*

of the ICRP (ICRP Publication 60, Ann ICRP 21, No. 1-3). London: Pergamon.
Jacob, G. (1990). *Site unseen: The politics of siting a nuclear waste repository*. Pittsburgh: University of Pittsburgh Press.
Janis, I. (1983). *Groupthink.* Boston: Houghton-Mifflin.
Janis, I.L., & Mann, L. (1977). *Decision-making*. New York: The Free Press.
Janoff-Bulman, R., & Frieze, I.H. (1983). A theoretical perspective for understanding reactions to victimization. *Journal of Social Issues, 39*(2), 1-17.
Jarrell, G., & Peltzman. S. (1985, April). The impact of product recalls on the wealth of sellers. *Journal of Political Economy*, pp. 512–536.
Jefferson County Economic Council. (1996). *Petition to the state of Colorado to create an enterprise zone*. Golden, CO: Author.
Jenkins-Smith, H. (1994). *Stigma models: Testing hypotheses of how images of Nevada are acquired and values are attached to them* (ANL/DIS/TM-17). Argonne, IL: Argonne National Laboratory.
Jenkins-Smith, H., Espey, J., Rouse, A., & Mulund, D. (1991). *Perceptions of risk in the management of nuclear waste: Mapping elite and mass beliefs and attitudes* (Report No. SAND90-7002). Albuquerque, NM: Sandia National Laboratories.
Jenkins-Smith, H., & Kunreuther, H. (2000). *Mitigation and benefits measures as policy tools for siting potentially hazardous facilities: Determinants of effectiveness and appropriateness.* Manuscript submitted for publication.
Johnson & Johnson suffers second blow in 5 months with Zomax's links to deaths. (1983b, March 7). *Wall Street Journal*, p. 7.
Johnson & Johnson unit sues 9 insurers for $117 million on Tylenol-recall costs. (1983a, January 12). *Wall Street Journal*, p. 6.
Johnson and Johnson. (1982, December 31). *Value Line Investment Survey*, p. 233.
Johnson and Johnson. (1983, April 1). *Value Line Investment Survey*, p. 233.
Johnson, E.J., & Tversky, A. (1983). Affect, generalization, and the perception of risk. *Journal of Personality and Social Psychology, 45*(1), 20-31.
Johnson, J.A., & Vogt, D.V. (1996). *"Mad cow disease" or bovine spongiform encephalopathy: Scientific and regulatory issues* (CRS Report for Congress, 96-644 SPR). Washington DC: Library of Congress, Congressional Research Service.
Johnston, B.R. (Ed.). (1994). *Who pays the price? The sociocultural context of environmental crisis.* Washington, DC: Island Press.
Johnston, B.R. (Ed.). (1997). *Life and death matters: Human rights and the environment at the end of the millennium.* Walnut Creek, CA: AltaMira Press.
Jones, E., Farina, A., Hastorf, A.H., Markus, H., Miller, D.T., Scott, R.A., & French, R.deS. (1984). *Social stigma: The psychology of marked relationships.* New York: W.H. Freeman.
Kahneman, D., & Tversky, A. (1979). Prospect theory: An analysis of decision under risk. *Econometrica, 47*, 263–291.
Kahneman, D., & Tversky, A. (1984). Choices, values, and frames. *American Psychologist, 39*, 341-350.
Kahneman, D., Slovic, P., & Tversky, A. (Eds.). (1982). *Judgment under uncertainty: Heuristics and biases.* New York: Cambridge University Press.
Kameron, J. (1975). *Man-nature value orientations.* Doctoral dissertation, City University of New York.
Kaplan, P. (1996, December 8). Air-bag backer steers clear of safety blame. *Washington Times National Weekly Edition*, p.3.
Kasperson, J.X., Kasperson, R.E., Perkins, B.J., Renn, O., & White, A.L. (1992). *Information content, signals, and sources concerning the proposed repository at Yucca Mountain: An analysis of newspaper coverage and social-group activity in Lincoln County, Nevada* (Report No. RP0136/MRDB). Worcester, MA: Clark University, Center for Technology, Environment, and Development (CENTED).
Kasperson, R.E. (1992). The social amplification of risk: Progress in developing an integrative framework of risk. In S. Krimsky & D. Golding (Eds.), *Social theories of*

risk (pp. 153–178). New York: Praeger.

Kasperson, R. E., Jhaveri, N., & Kasperson, J.X. (2000). Stigma and the social amplification of risk: Toward a framework of risk analysis. Chapter 2 of this volume.

Kasperson, R.E., & Kasperson, J.X. (1996, May). The social amplification and attenuation of risk. *The Annals of the American Academy of Political and Social Science, 545*, 95–105.

Kasperson, R.E., & Minghi, J.V. (Eds.). (1969). *The structure of political geography*. Chicago: Aldine.

Kasperson, R.E., Renn, O., Slovic, P., Brown, H.S., Emel, J., Goble, R., Kasperson, J.X., & Ratick, S. (1988). The social amplification of risk: A conceptual framework. *Risk Analysis, 8*, 177–187.

Kassarijan, H. (1982). Consumer psychology. *Annual Review of Psychology, 33*, 619-649.

Kearsley, G.W. (1985). Methodological change and the elicitation of images in human geography. *Journal of Mental Imagery, 9*, 71–82.

Kerr, J. (1990, October 26). Repository's effects on Vegas's economy, image raise concerns. *Las Vegas Review-Journal.*

Kinnard, W.N. (1992). Tools and techniques for measuring the effects of proximity to radioactive contamination on single-family residential sales prices. In *Measuring the effects of hazardous materials contamination on real estate values: Techniques and applications* (pp. 101–124). Chicago: The Appraisal Institute.

Kitchener-Waterloo Record. (1994, September 14).

Klein, B., Crawford, R.G., & Alchian, A.A. (1978, October). Vertical integration, appropriable rents, and the competitive contracting process. *Journal of Law and Economics*, pp. 297-326.

Klein, B., & Leffler, K. (1981, August). The role of market forces in assuring contractual performance. *Journal of Political Economy*, pp. 615–641.

Kohlhase, J.E. (1991). The impact of toxic waste sites on housing values. *Journal of Urban Economics, 30*(1), 1–26.

Kraus, N.N., Malmfors, T., & Slovic, P. (1992). Intuitive toxicology: Expert and lay judgments of chemical risks. *Risk Analysis, 12*, 215–232.

Krewski, D., Slovic, P., Bartlett, S., Flynn, J. & Mertz, C.K. (1995). Health risk perception in Canada II: Worldviews, attitudes and opinions. *Human and Ecological Risk Assessment, 1*(3), 231–248.

Kroll-Smith, J.S., & Floyd, H.H. (1997). *Bodies in protest: Environmental illness and the struggle over medical knowledge.* New York University.

Kuklinski, J., Metlay, D., & Kay, W. (1982). Citizen knowledge and choices on the complex issue of nuclear energy," *American Journal of Political Science, 26*, 615–642.

Kunreuther, H., Desvousges, W.H., & Slovic, P. (1988). Nevada's predicament: Public perceptions of risk from the proposed nuclear waste repository. *Environment, 30*(8), 16–20, 30–33.

Kunreuther, H., & Easterling, D. (1992). Gaining acceptance for noxious facilities with economic incentives. In D.W. Bromley & K. Segerson (Eds.), *The social response to environmental risk: Policy formulation in an age of uncertainty* (pp. 151-186). Boston: Kluwer Academic Press.

Kunreuther, H., & Easterling, D. (1996). The role of compensation in siting hazardous facilities. *Journal of Policy Analysis and Management, 15*, 601–622.

Kunreuther, H., Easterling, D., Desvousges, W., & Slovic, P. (1990). Public attitudes toward siting a high-level nuclear waste repository in Nevada. *Risk Analysis, 10*, 469–484.

Kunreuther, H., Easterling, D., & Kleindorfer, P. (1988). *The convention planning process: Potential impact of a high-level nuclear waste repository in Nevada* (Report No. NWPO-SE-021-90). Carson City, NV: NWPO.

Kunreuther, H., Fitzgerald, K., & Aarts, T.D. (1993). Siting noxious facilities: A test of the facility siting credo. *Risk Analysis, 13*, 301–318.

Kunreuther, H.C., Ginsberg, R., Miller, L., Sagi, P., Slovic, P., Borkin, B., & Katz, N. (1978). *Disaster insurance protection: Pubic policy lessons*. New York: Wiley.

Kunreuther, H., & Linnerooth, J. (1982). *Risk analysis and decision processes: The siting of LEG facilities in four countries*. Berlin, Germany: Springer Verlag.

Kunreuther, H., Linnerooth, J., & Fitzgerald, K. (1996). Siting hazardous facilities: Lessons from Europe and America. In P. Kleindorfer, H. Kunreuther, & D. Hong (Eds.), *Energy, environment and the economy: Asian perspectives*. Cheltenham, UK: Edward Elgar.

Kunreuther H., & Slovic, P. (1997, March). *Coping with stigma: Challenges and opportunities*. Paper presented at the Annenberg Conference on Risk, Media, and Stigma, Philadelphia, PA. Revised and printed as chapter 21 of this volume.

Kunreuther, H, & Slovic, P. (2000). Coping with stigma: Challenges and opportunities. Chapter 21 of this volume.

Las Vegas Convention and Visitors Authority. (1996). *Las Vegas marketing bulletin–1995 annual summary*. Las Vegas, NV: Author.

Leiss, W. (1996). Governance and the environment. In T. Courchene (Ed.), *Policy frameworks for a knowledge economy* (pp. 121-163). Kingston, Ontario: Queen's University, John Deutsch Institute.

Leiss, W., & Chociolko, C. (1994). *Risk and responsibility*. Montréal, Quebec: McGill-Queen's University Press.

Lenssen, N. (1992). Confronting nuclear waste. In L.R. Brown et al. (Eds.), *State of the world*. New York: Norton.

Levine, A. (1982). *Love canal: Science, politics and people*. Boston: Lexington Books.

Lichtenberg, J., & MacLean, D. (1992). Is good news no news? *The Geneva Papers on Risk and Insurance, 17*, 362–365.

Lifton, R. J. (1967). *Death in life: Survivors of Hiroshima*. New York: Random House.

Linthicum, L. (1995, March 10). Vote tears rift in tribe. *Albuquerque Journal*, pp. A1, A2.

Little, R.L., & Krannich, R.S. (1990). *Major sociocultural impacts of the Yucca Mountain high-level nuclear repository on nearby rural communities* (Report No. NWPO-SE-033-90). Carson City, NV: NWPO.

Löfstedt, R.E., & Boholm, Å. (1999). Off track in Sweden. *Environment, 41*(4), 16–20, 40–44.

Lund, A.K., Ferguson, S.A., & Powell, M.R. (1996). Fatalities in air bag-equipped cars: A review of 1989-93 NASS cases (SAE Technical Paper Series 960661). Warrendale, PA: Society of Automotive Engineers.

Luoma, J.R. (1990, May 15). Scientists are unlocking secrets of dioxin's devastating power. *The New York Times*.

Lyons, R.D. (1973, April 6). Contamination of fish in Vietnamese waters laid to U.S. defoliant. *The New York Times*.

MacGregor, D.G., Slovic, P., Mason, R.G., Detweiler, J., Binney, S.E., & Dodd, B. (1994). Perceived risks of radioactive waste transport through Oregon: Results of a statewide survey. *Risk Analysis, 14*(1), 5–14.

MacGregor, D.G., Slovic, P., & Morgan, M.G. (1994). Perception of risks from electromagnetic fields: A psychometric evaluation of a risk-communication approach. *Risk Analysis, 14*(5), 815–828.

MacInnis, D.J., & Price, L.L. (1987). The role of imagery in information processing: Review and extensions. *Journal of Consumer Research, 13*, 473–491.

Manning, M. (1992, March 30). Casks' safety in doubt. *Las Vegas Sun*.

March, J., Sproull, L., & Tamuz, M. (1991). Learning from samples of one or fewer. *Organization Science, 2*, 1–13.

Margolis, H. (1996). *Dealing with risk: Why the public and the experts disagree on environmental issues*. University of Chicago Press.

Marriott, M. (1968). Caste ranking and food transactions: A matrix analysis. In M. Singer & B. S. Cohn (Eds.), *Structure and change in Indian society*. (pp. 133–171.) Chicago: Aldine.

Masters, B.A. (1997, July 28). Tainted basil shows the challenges of tracking a microbe. *Washington Post*, p. B1.
Mauss, M. (1972). *A general theory of magic* (R. Brain, Trans.). New York: W. W. Norton. (Original work published 1902)
Mayr, E. (1960). The emergence of evolutionary novelties. In S. Tax (Ed.), *Evolution after Darwin: The evolution of life* (Vol. 1, pp. 349–380). University of Chicago Press.
Mazur, A. (1981). *The dynamics of technical controversy*. Washington, DC: Communications Press.
McClelland, G.H., Schulze, W.D., & Hurd, B. (1990). The effect of risk beliefs on property values: A case study of a hazardous waste site. *Risk Analysis, 10*(4), 485–497.
McEvoy, S. A. (1994). Double edged sword of Damocles: Utility companies' liability for diminution of property values due to electromagnetic fields. *Real Estate Law Journal, 23*(4), 109–122.
McNeil, D.G., Jr. (1978, December 22). 3 chemical sites near Love Canal possible hazard. *The New York Times*.
McQuail, D., & Windahl, S. (1993). *Communication models for the study of mass communication*. London and New York: Longman.
Meigs, A.S. (1984). *Food, sex, and pollution: A New Guinea religion*. New Brunswick, NJ: Rutgers University Press.
Melvin, J.W., Horsch, J.D., McCeary, J.D., Wideman, L.C., Jensen, J.L., & Wolanin, M.J. (1993). *Assessment of air bag deployment loads with the small females hybrid III dummy* (SAE Technical Paper No. 933119). Warrendale, PA: Society of Automotive Engineers.
Meredith, R. (1996, November 24). Hundreds call government phone line to ask about air bags. *The New York Times*, p.27.
Merleau-Ponty, M. (1962). *The phenomenology of perception*. London: Routledge and Kegan Paul.
Mertz, C.K., Flynn, J., & Slovic (1994). *The 1994 Nevada state telephone survey: Key findings* (Draft Report). Carson City, NV: NWPO.
Metz, W.C. (1992). Perceived risk and nuclear waste in Nevada: A mixture leading to economic doom? *Impact Assessment Bulletin, 10*, 23–42.
Metz, W.C. (1994). Potential negative impacts of nuclear activities on local economics: Rethinking the issue. *Risk Analysis, 14*, 763–770.
Metz, W.C. (1996). Historical application of a social amplification of risk model: Economic impacts of risk events at nuclaer weapons facilities. *Risk Analysis, 16*, 185–193.
Mitchell, M.L. (1989). The impact of external parties on brand-name capital: The 1982 Tylenol poisonings and subsequent cases. *Economic Inquiry, 27*, 601–618. Reprinted as chapter 12 of this volume.
Mitchell, M.L., & Maloney, M.T. (1989, October). Crisis in the cockpit? The role of market forces in promoting air travel safety. *Journal of Law and Economics, XXXII*, 329–355.
Mitchell, R.C., Payne, B., & Dunlap, R.E. (1988). Stigma and radioactive waste: Theory, assessment, and some empirical findings from Hanford, WA. In R.G. Post (Ed.), *Waste Management '88: Proceedings of the Symposium on Waste Management: Vol. 2. High-level waste and general interest* (pp. 95–102). Tucson: University of Arizona.
Monmonier, M.S. (1997). *Cartographies of danger: Mapping hazards in America*. University of Chicago.
Moore, J.A. (1989, May-June). Speaking of data: The Alar controversy. *EPA Journal, 15*, 5–9.
Morgan, M.G., & and Lave, L. (1990). Ethical considerations in risk communication practice and research. *Risk Analysis 10*(3), 355–358.
Morrison, D.G. (1979, Spring). Purchase intentions and purchase behavior. *Journal of Marketing 43*, 65–74.

Morrison, J.A. (1991, September 25). Gamers weren't approached to fund anti-nuke dump ads. *Las Vegas Review-Journal.*

Mowrer, O.H. (1960a). *Learning theory and behavior.* New York: John Wiley & Sons.

Mowrer, O.H. (1960b). *Learning theory and the symbolic processes.* New York: John Wiley & Sons.

Mundy, W. (1992). The impact of hazardous material on property value. *The Appraisal Journal, 60,* 155–162.

Mushkatel, A., Nigg, J., & Pijawka, D. (1989). *Urban risk survey: Public response, perception, and intended behavior of Las Vegas metro residents to the HLNW Repository.* Carson City:: NWPO.

Mushkatel, A.H., & Pijawka, K.D. (1992). *Institutional trust, information, and risk perceptions: Report of findings of the Las Vegas metropolitan area survey, June 29–July 1, 1992* (NWPO-SE055-92). Carson City, NV: Nevada NWPO.

Mushkatel, A.H., & Pijawka, K. (1994). *Nuclear waste transportation in Nevada: A case for stigma-induced economic vulnerability.* Carson City: Nuclear Waste Project Office (NWPO).

Mushkatel, A., Pijawka, D., & Glickman, T. (1993). The perceived risks of transporting hazardous materials and nuclear waste. *Proceedings of the international conference on the risks of transporting dangerous goods* (pp. 617-635). Canada: University of Waterloo.

National Highway Traffic Safety Administration (NHTSA). (1984). *Regulatory impact analysis of FMVSS 208: Occupant crash protection.* Washington, DC: U.S. Department of Transportation.

National Highway Traffic Safety Administration (NHTSA). (1996a, April). *Air bag warning labels: Focus group report* (Report prepared by Global Exchange Inc., Bethesda, MD). Washington, DC: U.S. Department of Transportation.

National Highway Traffic Safety Administration (NHSTA). (1996b, August). *Fatality reduction by airbags: Analysis of accident data through early 1996* (DOT-HS-808-470). Washington, DC: U.S. Department of Transportation.

National Highway Traffic Safety Administration (NHSTA). (1996c, November). *Air bag warning labels II: Focus group findings* (Report prepared by Global Exchange Inc., Bethesda, MD). Washington, DC: U.S. Department of Transportation.

National Highway Traffic Safety Administration (NHSTA). (1996d, December). *Third report to Congress: Effectiveness of occupant protection systems and their use.* Washington, DC: U.S. Department of Transportation.

National Highway Traffic Safety Administration (NHSTA). (1997, February). *Final regulatory evaluation: Actions to reduce the adverse effects of air bags: FMVSS No. 208: Depowering.* Washington, DC: Office of Regulatory Analysis, Plans and Policy.

National Research Council (NRC). (1989). *Improving risk communication.* Washington, DC: National Academy Press.

National Research Council (NRC). Board on Radioactive Waste Management. (1996). *Barriers to science: Technical management of the Department of Energy Environmental Remediation Program.* Washington, DC: National Academy Press.

National Research Council (NRC). National Academy of Science. Board on Radioactive Waste Management. (1990, July). *Rethinking high-level radioactive waste disposal: A position statement of the Board on Radioactive Waste Management.* Washington, DC: National Academy Press.

National Transportation Safety Board (NTSB). (1996, September). *Safety: The performance and use of child restraint systems, seatbelts, and airbags for children in passenger vehicles, Vol. 1.* Washington, DC: Author.

Naylor, S. (1997, August 15). Family's CJD anguish as daughter's life ebbs away. *PA News.*

Nealey, S.M., & Hebert, J.A. (1983). Public attitudes toward radioactive wastes. In C.A. Walker, L.C. Gould, & E.J. Woodhouse (Eds.), *Too hot to handle? Social and policy issues in the management of radioactive wastes* (pp. 94–111). New Haven, CT: Yale University Press.

Nelkin, D. (1995). Forms of intrusion: comparing resistance to information technology and biotechnology in the USA. In M. Bauer (Ed.), *Resistance to New Technology* (pp. 379–390). New York: Cambridge University Press.

Nelson, A., Genereaux, C., Genereaux, M. (1992). Price effects of landfills on house values. *Land Economics, 68*(4), 359–365.

Nelson, J.P. (1982). Highway noise and property values. *Journal of Transport Economics and Policy, XVI*(2), 117–138.

Nelson, T.F., Sussman, D., & Graham, J.D. (1999). Airbags: An exploratory survey of public knowledge and attitudes. *Accident Analysis and Prevention, 31*(4), 371–379.

Nemeroff, C., & Rozin, P. (1992). Sympathetic magical beliefs and kosher dietary practice: The interaction of rules and feelings. *Ethos: The Journal of Psychological Anthropology, 20,* 96–115.

Nemeroff, C., & Rozin, P. (1994). The contagion concept in adult thinking in the United States: Transmission of germs and interpersonal influence. *Ethos: The Journal of Psychological Anthropology, 22,* 158–186.

Nevada Agency for Nuclear Projects (NANP). (1995a). *Report on agency activities and oversight of the U.S. Department of Energy's high-level radioactive waste management program.* Carson City, NV: NWPO.

Nevada Agency for Nuclear Projects (NANP). (1995b). *State of Nevada socioeconomic studies biannual report, 1993–1995.* Carson City, NV: NWPO.

New evidence on hazards of drinking raw milk. (1984, August). *Consumer Reports, 49,* 425.

The New York Times. (1972, September 19).

The New York Times. (1974, August 28).

The New York Times. (1976, July 29).

The New York Times. (1980, June 27).

Nisbett, R.E, Borgida, E., Crandall, R., & Reed, H. (1982). Popular induction: Information is not necessarily informative. In D. Kahneman, P. Slovic, & A. Tversky (Eds.), *Judgment under uncertainty: Heuristics and biases* (pp. 102-116). New York: Cambridge University Press.

Nisbett, R., & Wilson, T.D. (1977). Telling more than we can know: Verbal reports on mental processes. *Psychological Review, 84,* 231–259.

Nomani, A.Q., & Stern, G. (1996, November 29). New campaign aims to save air bags' image. *Wall Street Journal,* p. B1.

Norton, B.G. (1991). *Toward unity among environmentalists.* New York: Oxford University Press.

Nuclear Waste Technical Review Board. (1992). *Sixth report to the U.S. Congress and the U.S. Secretary of Energy.* Washington, DC: U.S. Government Printing Office.

Nuke exec mad at DOE. (1992, March 11). *Nevada Appeal.*

O'Brien, C. (1996, March 29). Mad cow disease: Scant data cause widespread concern. *Science, 271,* p. 1798.

O'Callaghan, M. (1991, November 30). Hot response to nuke waste ads. *Las Vegas Sun.*

O'Donnell, J., & Healey, J.R. (1996, June 13). Air bags unsafe for under-12s, even with belts. *USA Today.*

Oakes, A.S., Donnelly, S., Garcia, M., & Karvia, N. (1993). *Community perceptions of risks associated with moving transuranic waste in southeastern Idaho.* Idaho State University.

Obmascik, M. (1994, March 1). DOE "knew" cleanup pacts doomed. *Denver Post,* pp. 1A, 7A.

Odwalla. (1996, November 8). *Odwalla expresses condolences to Denver family* [Press release]. Half Moon Bay, CA: Author.

Office of the Nuclear Waste Negotiator. (1993). *1992 annual report to Congress.* Boise, ID: Office of the United States Nuclear Waste Negotiator.

Oliver-Smith, A. (1996). Anthropological research on hazards and disasters. *Annual Review of Anthropology, 25,* 303.

Oram, K., & Allison, E., Eds. (1991). *The Nevada initiative: The long term program, an overview.* Proposal to ANEC.

Ottaway, D.B. (1996, October 27). A safety device with a fatal flaw: Designed to cushion body, air bags putting short people at risk. *Washington Post*, p. Al, A8.

Ottoboni, R. (1984). *The dose makes the poison: A plain language guide to toxicology.* Berkeley, CA: Vincente Books.

Page, G.W., & Rabinowitz, H. (1993). Groundwater contamination: Its effects on property values and cities. *Journal of the American Planning Association, 59*(4), 473–485.

Paivio, A. (1979). *Imagery and verbal processes.* Hillsdale, NJ: Erlbaum.

Patchin, P.J. (1991a). Contaminated properties—Stigma revisited. *The Appraisal Journal, 59*, 167–172.

Patchin, P.J. (1991b). The valuation of contaminated properties. *Real Estate Issues, 16*(2), 50–54.

Patino, M.M., Liu, J.J., Glover, J.R., & Lindquist, S. (1996, August 2). Support for the prion hypothesis for inheritance of a phenotypic trait in yeast. *Science*, 273, 622–626.

Patterson, P. (1989). Reporting Chernobyl: Cutting the government fog to clear the nuclear cloud. In L.M. Walters, L. Wilkins, & T. Walters (Eds.), *Bad tidings: Communication and catastrophe* (pp. 131–147). Hillsdale, NJ: Lawrence Erlbaum.

Paushkin, S.V., Kushnirov, V.V., Smirnov, V.N., Ter-Avanesyan, M.D. (1997). In vitro propagation of the prion-like state of yeast sup35 protein. *Science, 277*, 381.

Pbert, L.A., & Goetsch, V.L. (1988). A multifaceted behavioral intervention for pill-taking avoidance associated with Tylenol poisoning. *Journal of Behavior Therapy and Experimental Psychiatry, 19*(4), 311–315.

Perin, C. (1977). *Everything in its place.* NJ: Princeton University Press.

Perry, R., Lindell, M., & Greene, M. (1982). Threat perception and public response to volcano hazard. *The Journal of Social Psychology, 116*, 199-204.

Peterson, C., & Seligman, M. (1983). Learned helplessness and victimization. *Journal of Social Issues, 39*(2), 103-116.

Petterson, J.S. (1988). Perception vs. reality of radiological impact: The Goiânia model. *Nuclear News, 31*(14), 84–90.

Picard, A. (1995). *The gift of death: Confronting Canada's tainted blood tragedy.* Harper Collins.

Pickering, J.F., & Isherwood, B.C. (1974). Purchasing probabilities and consumer buying behavior. *Journal of the Market Research Society, 16*, 203–226.

Pitts, R., & Woodside, A. (1983). Personal value influences on consumer product class and brand preferences. *Journal of Social Psychology, 119*, 37-53.

Planning Information Corporation. (1995). *Project description* (Report). Carson City, NV: NWPO.

Pocock, D.C.D. (1974). *The nature of environmental perception.* UK: University of Durham, Department of Geography.

Pollycove, M. (1997, March). Invited commentary to the Heitzmann and Wilson article: Myron Pollycove. *Belle Newsletter, 6*(1), 13–18.

Powell, D.A., Harris, L.J., & Griffiths, M.W. (1997). *Professional and media warnings about the hazards of escherichia coli O157:H7 prior to and after the 1993 Jack In The Box outbreak.* Manuscript submitted for publication.

Powell, D.A., & Leiss, W. (1997). *Mad cows and mother's milk: The perils of poor risk communication.* Montréal, Quebec: McGill-Queen's University Press.

Pratto, F., & John, O.P. (1991). Automatic vigilance: The attention-grabbing power of negative social information. *Journal of Personality & Social Psychology, 61*, 380–391.

Pre-emptive strike at mad cows is rational response [Editorial]. (1997, January 4). *Chicago Tribune.*

Prusiner, S. (1995). The prion diseases. *Scientific American, 272*(1), 47–57.

Public Opinion Strategies. (1996). *Children and air bag safety.* Alexandria, VA: Author.

Purchase, I.F.H., & Auton, T. (1995). Thresholds in chemical carcinogenesis. *Regulatory Toxicology and Pharmacology, 22*, 199–205.

Purchase, I.F.H., & Slovic, P. (1999). Quantitative risk assessment breeds fear. *Human and Ecological Risk Assessment, 5*(3), 445–453.

Quam, M.D. (1990). The sick role, stigma, and pollution: The case of AIDS. In D.A. Feldman (Ed.), *Culture and AIDS* (pp. 9–28). Westport, CT: Praeger.

Radelfinger, S. (1965). Some effects of fear-arousing communications on preventive health behavior. *Health Education Monographs, 192*, 2-15.

Read, S. (1983).Once is enough: Causal reasoning from a single instance. *Journal of Personality and Social Psychology, 45*(2), 323-334.

Regens, J., Dietz, T., & Rycroft, R. (1983, March/April). Risk assessment in the policy-making process: Environmental health and safety protection. *Public Administrative Review*, pp. 137-145.

Reich, M. (1983, March). Environmental politics and science: The case of PBB contamination in Michigan. *American Journal of Public Health, 73*(3), 302-313.

Reinhold, R. (1989, April 3). Nevada draws the line: No Hitler in the casinos. *The New York Times*, p. 10A.

Reinhold, R. (1990, November 16). Arizona struggles anew with lingering racist image. *The New York Times*, p. 18A.

Reuters Library Report. (1990, May 10). Perrier puts cost of Benzene scare at 79 million dollars.

Rhodes, R. (1997a). *Deadly feasts: Tracking the secrets of a terrifying new plague*. Simon & Schuster.

Rhodes, R. (1997b, December 1). Pathological science. *The New Yorker*, pp. 53–62.

Richert, K. (1992, September 20). The Yucca Mountain experiment. *Post Register*, p. A5.

Riddell, P. (1997, April 8). Report and indictment of government—Labour. *PA News*.

Riley, B. (1992, February 7). Senator rips DOE on Yucca. *Las Vegas Sun*.

Roberts, L. (1987). Radiation accident grips Goiânia. *Science, 238*, 1028–1031.

Roberts, T. (1989). Human illness costs of food-borne bacteria. *American Journal of Agricultural Economics, 71*, 468–474.

Rocky Flats Future Site Use Working Group. (1995, July). *Rocky Flats Future Site Use Working Group recommendations for Rocky Flats environmental technology site*. Denver, CO: Author.

Roddewig, R.J. (1994, March). *Environmental risk and the real estate appraisal process*. Chicago: The Appraisal Institute.

Rogers, K. (1991, December 5). Ad funding angers watchdog group. *Las Vegas Review-Journal*.

Rogers, K. (1992, March 10). Nuclear industry exec bashes DOE, threatens to pull funds. *Las Vegas Review-Journal*.

Rogers, R. (1975). A protection motivation theory of fear appeals and attitude change. *The Journal of Psychology, 91*, 93-114.

Rosa, E., & Dunlap, R. (1994). Nuclear power: Three decades of public opinion. *Public Opinion Quarterly, 58*, 295–325.

Rosen, H.S., & Burke, J.F. (1987, November). *Preliminary report on the property value effects of the feed materials production center at Fernald, Ohio on single family residences only* [Prepared for Mr. Stanley M. Chesley, Attorney at Law].

Ross, L., & Anderson, C. (1982). Shortcomings in the attribution process: On the origins and maintenance of erroneous social assessments. In D. Kahneman, P. Slovic, & A. Tversky (Eds.), *Judgment under uncertainty: Heuristics and biases* (pp. 129-152). New York: Cambridge University Press.

Rothman, S., & Lichter, S. (1987). Elite ideology and risk perception in nuclear energy policy. *American Political Science Review, 81*, 388–404.

The Royal Society. (1983). Perception of risk. In *Risk assessment: A study group report* (pp. 94-148). London: Author.

Rozin, P. (1997). Moralization. In A. Brandt & P. Rozin (Eds.), *Morality and health* (pp. 379–401). New York: Routledge.
Rozin, P., Ashmore, M.B., & Markwith, M. (1996). Lay American conceptions of nutrition: Dose insensitivity, categorical thinking, contagion, and the monotonic mind. *Health Psychology, 15*, 438–447.
Rozin, P., & Fallon, A.E. (1987). A perspective on disgust. *Psychological Review, 94*, 23–41.
Rozin, P., Fallon, A.E., & Augustoni-Ziskind, M. (1985). The child's conception of food: The development of contamination sensitivity to "disgusting" substances. *Developmental Psychology, 21*, 1075–1079.
Rozin, P., Fallon, A.E., & Mandell, R. (1984). Family resemblance in attitudes to food. *Developmental Psychology, 20*, 309–314
Rozin, P., Haidt, J., McCauley, C.R., & Imada, S. (1997). The cultural evolution of disgust. In H. M. Macbeth (Ed.), *Food preferences and taste: Continuity and change* (pp. 65–82). Oxford, UK: Berghahn.
Rozin, P., Markwith, M., & McCauley, C.R. (1994). The nature of aversion to indirect contact with another person: AIDS aversion as a composite of aversion to strangers, infection, moral taint and misfortune. *Journal of Abnormal Psychology, 103*, 495–504.
Rozin, P., Markwith, M., & Nemeroff, C. (1992). Magical contagion beliefs and fear of AIDS. *Journal of Applied Social Psychology, 22*, 1081–1092.
Rozin, P., Markwith, M., & Ross, B. (1990). The sympathetic magical law of similarity, nominal realism and the neglect of negatives in response to negative labels. *Psychological Science, 1*, 383–384.
Rozin, P., Millman, L., & Nemeroff, C. (1986). Operation of the laws of sympathetic magic in disgust and other domains. *Journal of Personality and Social Psychology, 50*, 703–712.
Rozin, P., & Nemeroff, C.J. (1990). The laws of sympathetic magic: A psychological analysis of similarity and contagion. In J. Stigler, G. Herdt, & R. A. Shweder (Eds.), *Cultural psychology: Essays on comparative human development* (pp. 205–232). UK: Cambridge University Press.
Rozin, P., Nemeroff, C., Horowitz, M., Gordon, B., & Voet, W. (1995). The borders of the self: Contamination sensitivity and potency of the mouth, other apertures and body parts. *Journal of Research in Personality, 29*, 318–340.
Rozin, P., & Royzman, E. (2000). *Negativity bias, negativity dominance, and contagion.* Manuscript submitted for publication.
Rozin, P., & Singh, L. (1998). *The moralization of cigarette smoking in America.* Manuscript submitted for publication.
Rozin, P., & Weinberg, S. Unpublished results.
Saarinen, T.F., & Sell, J.L. (1980). Environmental perception. *Progress in Human Geography, 4*, 525–548.
Sahlins, M. (1976). *Culture and practical reason.* Chicago and London: The University of Chicago Press.
Satchell, M. (1996, January 8). Dances with nuclear waste. *U.S. News and World Report*, pp. 29–30.
The Saturday Review of Literature. (1936, November 28). *XV*(5), p. 4.
Schneider, K. (1991, November 13). Nuclear industry plans ads to counter critics. *The New York Times.*
School bans beef over BSE fears. (1995, November 17). *PA News.*
Schwab, J. (1994). *Deeper shades of green: The rise of blue collar and minority environmentalism in America.* San Francisco: Sierra Club Books.
Scott, J. C. (1990). *Domination and the arts of resistance.* New York: Yale University.
Several drug companies introduce new products in wake of Tylenol tragedy. (1983, January 10). *Advertising Age*, p. 1.
Severo, R. (1979, May 27). Two crippled lives mirror disputes on herbicides. *The New York Times.*

Shapiro, C. (1983, November). Premiums for high quality products as returns to reputations. *Quarterly Journal of Economics*, 659–679.
Shaw, D. (1996). An economic framework for analyzing facility siting procedures in Taiwan and Japan. In P. Kleindorfer, H. Kunreuther, & D. Hong (Eds.), *Energy, environment and the economy: Asian perspectives.* Cheltenham, UK: Edward Elgar.
Sheppard, B.H., Hartwick, J., & Warshaw, P.R. (1988). The theory of reasoned action: A meta-analysis of past research with recommendations for modifications and future research. *Journal of Consumer Research, 15*, 325–343.
Short, J.F., Jr. (1992). Defining, explaining, and managing risks. In J.F. Short & L. Clarke (Eds.), *Organizations, uncertainties, and risk.* Boulder, CO: Westview Press.
Siegal, M. (1988). Children's knowledge of contagion and contamination as causes of illness. *Child Development, 59*, 1353–1359.
Sigmon, E.B. (1987). Achieving a negotiated compensation agreement in siting: The MRS case. *Journal of Policy, Analysis and Management, 6*(2), 170–179.
Singer, E., & Endreny, P.M. (1993). *Reporting on risk: How the media portray accidents, diseases, disasters, and other hazards.* New York: Russell Sage Foundation.
Skowronski, J.J., & Carlston, D.E. (1989). Negativity and extremity biases in impression formation: A review of explanations. *Psychological Bulletin, 105*, 131–142.
Slack rules for a poison [Editorial]. (1981, August 13). *The Globe and Mail.*
Slovic, P. (1986). Informing and educating the public about risk. *Risk Analysis, 6*(4), 403–415.
Slovic, P. (1987). Perception of risk. *Science, 236*, 280–285).
Slovic, P. (1992). Perception of risk: Reflections on the psychometric paradigm. In S. Krimsky & D. Golding (Eds.), *Social theories of risk* (pp. 117–152). New York: Praeger.
Slovic, P. (1993). Perceived risk, trust, and democracy. *Risk Analysis, 13*, 675–682.
Slovic, P. (1997, June). *Perceived risk, stigma, and the vulnerable society.* Paper presented at the One-day Conference on Risk, London.
Slovic, P., Fischhoff, B., & Lichtenstein S. (1978). Accident probabilities and seat belt usage: A psychological perspective. *Accident Analysis and Prevention, 10*, 281–285.
Slovic, P., Fischhoff, B., & Lichtenstein, S. (1979). Rating the risks. *Environment, 21*, 14–20, 36–39.
Slovic, P., Fischoff, B., & Lichtenstein, S. (1982). Facts versus fears: Understanding perceived risk. In D. Kahneman, P. Slovic, & A. Tversky (Eds.), *Judgment under uncertainty: Heuristics and biases* (pp. 463-492). New York: Cambridge University Press.
Slovic, P., Flynn, J., & Gregory, R. (1994). Stigma happens: Social problems in the siting of nuclear waste facilities. *Risk Analysis, 14*(5), 773–777.
Slovic, P., Flynn, J., & Layman, M. (1991). Perceived risk, trust, and the politics of nuclear waste. *Science, 254*, 1603–1607.
Slovic, P., Kraus, N., & Covello, V. (1990). What *should* we know about making risk comparisons? *Risk Analysis 10*(3), 389–391.
Slovic, P., Layman, M., & Flynn, J. (1990a, September). *What comes to mind when you hear the words "nuclear waste repository"? A study of 10,000 images* (Report No. NWPO-SE-028-90). Carson City, NV: Nevada Agency for Nuclear Projects, NWPO.
Slovic, P., Layman, M., & Flynn, J. (1990b). *Images of a place and vacation preferences: Report of the 1989 surveys* (Report No. NWPO-SE-030-90). Carson City, NV: NWPO.
Slovic, P., Layman, M., & Flynn, J. (1991). Risk perception, trust, and nuclear waste: Lessons from Yucca Mountain. *Environment, 33*(3), 6–11, 28–30.
Slovic, P., Layman, M., Kraus, N., Flynn, J., Chalmers, J., & Gesell, G. (1991). Perceived risk, stigma, and potential economic impacts of a high-level nuclear waste repository in Nevada. *Risk Analysis, 11*(4), 683–696. Reprinted as chapter 6 of this volume.
Slovic, P., Lichtenstein, S., & Fischhoff, B. (1979). Images of disaster: Perception and acceptance of risks from nuclear power. In G. Goodman & W. Rowe (Eds.), *Energy risk management* (pp. 223–245). London: Academic Press.

Slovic, P., Malmfors, T., Mertz, C. K., Neil, N., & Purchase, I.F.H. (1997). Evaluating chemical risks: Results of a survey of the British Toxicology Society. *Human & Experimental Toxicology, 16*, 289–304.

Smith, R.J. (1982, July 9). Hawaiian milk contamination creates alarm. *Science, 217*, 137-8.

Smith, V.K., & Desvouges, W.H. (1986). The value of avoiding a LULU: Hazardous waste disposal sites. *Review of Economics and Statistics, 68*(2), 293–299.

Smolen, G., Moore, G., and Conway, L. (1992). Hazardous waste landfill impacts on local property values. *The Real Estate Appraiser, 58*(4), 4–11.

Sontag, Susan. (1978). *Illness as metaphor.* New York: Vintage.

Spence, R.K., et al. (1993). Current problems in surgery. In S.A. Wells, Jr. (Ed.), *Transfusion and surgery* (p. 1114). St. Louis, MO: Mosby.

Spence, R.K., et al. (1994). Preoperatively assessing and planning blood use for elective vascular surgery. *The American Journal of Surgery, 168*, 192.

Spranca, M. (1998). *How the naturalness of water affects its desirability.* Unpublished manuscript.

Srole, L. (1965). Social integration and certain corollaries. *American Sociological Review, 21*(6), 709ff.

Stallings, D.L. (1975). *Environmental cognition and land use controversy: An environmental image study of Seattle's Pike Place Market.* Doctoral dissertation, University of Washington, Seattle.

State of Nevada v. Watkins, 914 F. 2d 1545 (9th Cir. 1990).

Stein, R.L., & Nemeroff, C.J. (1995). Moral overtones of food: Judgments of others based on what they eat. *Personality & Social Psychology Bulletin, 21*, 480–490.

Steingraber, S. (1997). *Living downstream.* New York: Addison-Wesley.

Streisand, B. (1996, November 4). A new look at child-killing air bags. *U.S. News and World Report*, p.10.

Sunstein, K. (1996). Social norms and social roles. *Columbia Law Review, 96*, 903–968.

Szalay, L. B., & Deese, J. (1978). *Subjective meaning and culture: An assessment through word associations.* Hillsdale, NJ: Erlbaum.

Szasz, A. (1994). *Ecopopulism: Toxic waste and the movement for environmental justice: Vol. 1. Social movements, protest, and contention.* Minneapolis: University of Minnesota Press.

Talmey-Drake Research and Strategies. (1990). *Broomfield issues survey.* Broomfield, CO: Author.

Taubes, G. (1995). Epidemiology faces its limits. *Science, 269*, 164–169.

Taylor, S. (1982). The availability bias in social perception and interaction. In D. Kahneman, P. Slovic, & A. Tversky (Eds.), *Judgment under uncertainty: Heuristics and biases* (pp. 190-200). New York: Cambridge University Press.

Telling, G.C., Parchi, P., DeArmond, S.J., Cortelli, P., Montagna, P., Gabizon, R., Mastrianni, J., Lugaresi, E., Gambetti, P., & Prusiner, S.B. (1996). Evidence for the conformation of the pathologic isoform of the prion protein enciphering and propagating prion diversity. *Science, 274*, 2079–2082.

Telser, L.G. (1980, January). A theory of self-enforcing agreements. *Journal of Business*, pp. 27–44.

The bad news on Tylenol. (1982, November 15). *Drug Store News*, p. 1.

The fight to save Tylenol. (1982, November 29). *Fortune*, p. 44.

The race to grab up Tylenol's market. (1982, November 3). *Chemical Week*, p. 30.

Thompson, K. (1969). Insalubrious California: Perception and reality. *Annals of the Association of American Geographers, 59*, 50–64.

Thompson, M.R., Ellis, R., & Wildavsky, A. (1990). *Cultural theory.* Boulder, CO: Westview.

Tiepolo. (1997, January 27). St. Francis receiving the stigmata. *The New Yorker*, p. 57.

Todd, E.C.D. (1989). Preliminary estimates of costs of foodborne disease in Canada

and costs to reduce salmonellosis. *Journal of Food Protection, 52,* 586–594.
Tuan, Y. (1979). *Landscapes of fear.* New York: Pantheon.
Tunnelkommissionen (SOU). (1998a). *Kring Hallandsåsen: Delrapport från Tunnelkommissionen* (60). Stockholm: Fritzes.
Tunnelkommissionen (SOU). (1998b). *Miljö i grund och botten: Erfarenheter från Hallandsåsen* (137). Stockholm: Fritzes.
Tversky, A., & Kahneman, D. (1982). Judgment under uncertainty: Heuristics and biases. In D. Kahneman, P. Slovic, & A. Tversky (Eds.), *Judgment under uncertainty: Heuristics and biases* (pp. 3-22). New York: Cambridge University Press.
Tylenol competitors boom. (1982, October 18). *Advertising Age,* p. 1.
Tylenol maker must bear cost of 1982 recall. (1986, September 18). *Wall Street Journal,* p. 4.
Tylenol regains most of no. 1 market shares, amazing doomsayers. (1982, December 24). *Wall Street Journal,* p. 1.
Tylenol rivals hike ad activity. (1986, February 24). *Advertising Age,* p. 88.
Tylenol tablets lead rebound. (1982, December 13). *Advertising Age,* p. 1.
Tylor, E. B. (1974). *Primitive culture: Researches into the development of mythology, philosophy, religion, art and custom.* New York: Gordon. (Original work published 1871)
U.K. Institute of Food Science & Technology (IFST). (1997, April 14). *Bovine spongiform encephalopathy (BSE) position statement.* Author.
U.S. Department of Energy (DOE). (1999, July). *Draft environmental impact statement for a geologic repository for the disposal of spent nuclear fuel and high-level radioactive waste at Yucca Mountain, Nye County, Nevada: Volume I–Impact analyses* (DOE/EIS-0250D). Washington, DC: Author.
U.S. Department of Energy (DOE). Office of Environmental Management. (1995, March). *Estimating the cold war mortgage: The 1995 baseline environmental management report: Vol. II: Site summaries.* Springfield, VA: NTIS.
U.S. Environmental Protection Agency (EPA). (1991, September). *National priorities list sites: Colorado* (EPA/540/8-91-022). Washington, DC: Author.
U.S. Environmental Protection Agency (EPA). (1996). *Brownfields action agenda* (500/F-95/001). Washington, DC: Author.
U.S. General Accounting Office (GAO). (1983). *Siting of hazardous waste landfills and their correlation with the racial and socioeconomic status of surrounding communities* (Report No. GAO/RCED-83-168). Washington, DC: Author.
U.S. General Accounting Office (GAO). (1992, September). *Nuclear materials: Removing plutonium residues from Rocky Flats will be difficult and costly* (GAO/RCED-91-219). Washington, DC: GAO.
U.S. General Accounting Office (GAO). (1993). *Superfund: Cleanups nearing completion indicate future challenges* (GAO/RCED-93-188). Washington, DC: U.S. Government Printing Office.
U.S. Office of Technological Assessment. (1991). *Complex cleanup: The environmental legacy of nuclear weapons production.* Washington, DC: U.S. Government Printing Office.
U.S. Secretary of Energy Advisory Board. Task Force on Radioactive Waste Management. (1993, November). *Earning public trust and confidence: Requisites for managing radioactive waste* Washington, DC: Author.
United Church of Christ. Commission for Racial Justice. (1987). *Toxic wastes and race in the United States: A national report on the racial and socio-economic characteristics of communities with hazardous waste sites.* New York: Public Data Access.
Uzzell, D.L. (1982). Environmental pluralism and participation: A co-orientational perspective. In J.R. Gold & J. Burgess (Eds.), *Valued environments* (pp. 189–203). London: George Allen and Unwin.
Van Ginneken, J. (1998). *Understanding global news. A critical introduction.* London: Sage.
Vyner, H. (1988). *Invisible trauma: The psychosocial effects of invisible environmental contaminants.* Lexington, MA: Lexington Books.

Wald, M.L. (1996, September 18). More children are killed by air bags, and parents are blamed. *The New York Times*, p.A16.
Ward, K. (1991, November 24). All aboard the Yucca Mountain fantasy bus. *Las Vegas Sun*, p. 3K.
Weart, S.R. (1988). *Nuclear fear: A history of images*. Cambridge, MA: Harvard University Press.
Weber, K. (1995, April-June). Rear-facing restraint for small child passengers: A medical alert. *UMTRI Research Review*, pp.12–17.
Weber, M. (1947). *The theory of social and economic organization* (T. Parsons & A.M. Hendersen, Eds. & Trans.). New York: Oxford University Press.
Weinberger, M., Greene, J., Mamlin, J., & Jerin, M. (1981, November). Health beliefs and smoking behavior. *American Journal of Public Health, 71*(11), 1253-1255.
Weinstein, N. (1984). Why it won't happen to me: Perceptions of risk factors and susceptibility. *Health Psychology, 3*, 431–457.
Weinstein, N.D. (1988). *Attitudes of the public and the department of environmental protection toward environmental hazards* (Final report). New Jersey Department of Environmental Protection.
West, S.G., & Baker, E.J. (1983). Public reaction to nuclear power: The case of offshore nuclear power plants. In R.F. Kidd & M.J. Saks (Eds.), *Advances in applied social psychology* (Vol. 2, pp. 101–129). Hillsdale, NJ: Erlbaum.
White, A.L., Edwards, S., & Emani, S. (1990). *Risk perceptions of the Yucca Mountain repository: A comparative assessment of Caliente and other southern Nevada communities*. Worcester, MA: Clark University, CENTED.
Wickizer, T., Brilliant, L., Copeland, R., & Tilden, R. (1981, February). Polychlorinated biphenyl contamination of nursing mother's milk in Michigan. *American Journal of Public Health, 71*(2), 132-137.
Wildavsky, A. (1995). *But is it true?* Cambridge, MA: Harvard University Press.
Wildavsky, A, & Dake, K. (1990, Spring). Theories of risk perception: Who fears what and why? *Deadulus*, 41–60.
Wilkins, L. (1987). *Shared vulnerability*. New York: Greenwood.
Williams, H.B. (1964). Human factors in warning and response systems. In G.H. Grosser, H. Wechsler, & M. Greenblatt (Eds.), *The threat of impending disaster* (pp. 79-104). Cambridge, MA: The M.I.T. Press.
Wolf, C.P. (1977). Social impact assessments: The state of the art updated. *Social Impact Assessment, 20*, 3–22.
Wolfe, B. (1997, April). An apology to Dan Rather—But are Mr. Rather and his colleagues still dangerous? *Nuclear News*, 38–39.
Wood, W., & Eagley A. (1981). Stages in the analysis of persuasive messages: The role of causal attributions and message comprehension. *Journal of Personality and Social Psychology, 40*(2), 246-259.
Wundt, W. (1883). Uber psychologische methoden. *Philosophische Studien, 1*, 1–38.
Wurtele, S., Roberts, M., & Leeper, J. (1982). Health beliefs and intentions: Predictors of return compliance in a tuberculosis detection drive. *Journal of Applied Social Psychology, 12*(2), 128-136.
Yucca Mountain Socioeconomic Study Team. (1993). *The state of Nevada, Yucca Mountain socioeconomic studies* (Report No. NWPO-SE-056-93). Carson City, NV: NWPO.
Zimmerman, R. (1993). Social equity and environmental risk. *Risk Analysis, 13*(6), 649–666.
Zimmerman, R. (1994). Issues of classification in environmental equity—How we manage is how we measure. *Fordham Urban Law Journal, 21*, 633–669.

About the Authors

Gregory D. Adams is a Principal and Managing Director in the Environmental and Natural Resources practice of LECG, a unit of Navigant Consulting, Inc., and an Adjunct Assistant Professor at the University of Utah where he teaches environmental and natural resource economics. He has a B.A. in economics from Wake Forest University and a Ph.D. in environmental and natural resource economics from the University of California at Berkeley.

Sulemein Ashur is an Assistant Professor in the College of Engineering, University of Texas at El Paso.

John Blair is Research Associate, Center for Environmental Studies, Arizona State University, Tempe, Arizona.

Åsa Boholm (1953, FD, docent in social anthropology) has a position as researcher at CEFOS (Centre for Public Sector Research), Gothenburg University.

Robin Cantor is a Principal and Managing Director in the Environmental and Natural Resources practice of LECG, a unit of Navigant Consulting, Inc. Prior to joining LECG in 1996, she was Program Director for Decision, Risk, and Management Sciences, a multidisciplinary research program of the National Science Foundation. Dr. Cantor has a faculty appointment in the Part-time Program in Environmental Engineering of the Johns Hopkins University. She has a B.S. in mathematics from Indiana University of Pennsylvania and a Ph.D. in economics from Duke University.

James A. Chalmers is the National Coordinator of the PricewaterhouseCoopers LLP Environmental Damages Practice, which specializes in valuing the economic effects of environmental contamination. A Ph.D. economist, Dr. Chalmers has over 25 years of experience in real estate, environmental, and litigation consulting.

Penny Chan, at the time of writing, was scientific advisor to Justice Krever on the Commission of Inquiry on the Blood System in Canada. She is currently the scientific secretariat for the National Blood Safety Council (Canada) that advises the federal Minister of Health on issues of blood safety. She has been a member of the Risk Institute since her work in regulation of toxic substances and biohazards in the workplace.

Doug Easterling is the Associate Director of the Center for the Study of Social Issues at the University of North Carolina, Greensboro. In addition to conducting research on risk perception, facility siting, and community development, he works with foundations, nonprofit organizations, and researchers to improve the practice of program evaluation.

Michael R. Edelstein is Professor of Environmental Psychology at Ramapo College of New Jersey, where he heads the Environmental Studies Program. He is also President of a non-profit organization, Orange Environment, Inc. He has written *Contaminated Communities: The Social and Psychological Impacts of Residential Toxic Exposure* (1988 and 2nd Edition forthcoming) and *Radon's*

Deadly Daughters: Science, Environmental Policy, and the Politics of Risk (1998 with William Makofske).

Celio Ferreira (Ph.D. in Social Anthropology from Gothenburg University) is currently a consultant in the field of multimedia and an independent researcher. He is affiliated with CEFOS (Centre for Public Sector Research), Gothenburg University.

Baruch Fischhoff is University Professor in the Departments of Social and Decision Sciences and of Engineering and Public Policy, Carnegie Mellon University, Pittsburgh, Pennsylvania 15213 (baruch@cmu.edu).

James Flynn joined Decision Research in 1990 and is a Senior Research Associate. He is a graduate of Eastern Washington State University (1964) and the University of Washington (M.A., 1986; Ph.D., 1974).

Gail Gesell worked as a researcher with Mountain West Research in Phoenix, Arizona at the time this article was researched and written.

John D. Graham is Professor of Policy and Decision Sciences at the Harvard School of Public Health and founding Director of the Harvard Center for Risk Analysis.

Robin Gregory is a Senior Researcher with Decision Research in Vancouver, Canada. He works on problems of environmental and risk management, value elicitation, and community-based resource decision making.

Subhrajit Guhathakurta is Assistant Professor, School of Planning and Landscape Architecture, Arizona State University.

Wayne L. Hunsperger is President of Hunsperger & Weston, Ltd., a real estate consulting firm specializing in the valuation of conservation easements, the impacts of environmental damage, and eminent domain issues.

Hank C. Jenkins-Smith received his Ph.D. in political science from the University of Rochester, and is currently Professor of Political Science at the University of New Mexico. He is also the Director of the University of New Mexico's Institute for Public Policy. Hank would brew beer in his spare time, if he had any time to spare.

Nayna Jhaveri is Assistant Professor of Geography at the University of Washington, Seattle, Washington, USA.

Jeanne X. Kasperson is Research Associate Professor and Research Librarian at the George Perkins Marsh Institute, Clark University, Worcester, Massachusetts, USA.

Roger E. Kasperson is University Professor (Government and Geography) and Director of the George Perkins Marsh Institute, Clark University, Worcester, Massachusetts, USA.

Nancy Kraus. *See* Nancy Neil.

Howard Kunreuther is the Cecelia Yen Koo Professor of Decision Sciences and Public Policy, as well as Co-Director of the Wharton Risk Management and Decision Processes Center at the University of Pennsylvania. His current research examines the role of insurance compensation, incentive mechanisms, and regulation as policy tools for dealing with technological and natural hazards.

Mark Layman was a data analyst at Decision Research in 1991 when the original version of "Perceived Risk, Stigma, and Potential Economic Impacts of a High-Level Nuclear Waste Repository" was written.

Dr. Ragnar Löfstedt is a Reader in Social Geography at the Centre for Environmental Strategy, University of Surrey, United Kingdom and Visiting Associate Professor at both the Center for Public Sector Studies, University of Gothenburg, Sweden and the Harvard Center for Risk Analysis, Harvard School of Public Health. He is the Editor-in-Chief of *Journal of Risk Research* and on the Editorial Board of *Risk Analysis.*

Sarah Lebiednik is an Environmental Planner and completed her MEP degree at Arizona State University.

William Leiss is President of the Royal Society of Canada (1999-2001) and holds the Research Chair in Risk Communication and Public Policy in the Faculty of Management at the University of Calgary.

Richard K. Long worked at Dow Chemical for 24 years, including six as director of communications. He joined Weyerhaeuser Co. in 1991 as a vice president. In 1999, he was named professor of communications at Brigham Young University, where he earlier received BS and MA degrees in journalism.

C. K. Mertz is a data analyst with Decision Research in Eugene, Oregon.

Mark Mitchell is Associate Professor of Business Administration (in the Finance Area) at Harvard Business School. He was a member of the finance faculty at the University of Chicago during 1990-1999, and was a financial economist at the U.S. Securities and Exchange Commission during 1987-1990.

Nancy Neil was a research associate at Decision Research at the time this article was prepared. She is currently Health Economist at Virginia Mason Medical Center in Seattle, Washington, where her focus is on topics related to clinical judgment and decision making.

Ellen Peters is a research associate at Decision Research and an adjunct professor at the University of Oregon at Eugene, Oregon. Her research examines the affective and analytical processes underlying the decisions that people make in an increasingly complex world.

K. David Pijawka is Professor, School of Planning and Landscape Architecture, and Center Professor, Center for Environmental Studies, Arizona State University.

Douglas Powell completed a doctoral degree in the department of food science at the University of Guelph in 1996, and is currently an assistant professor in the department of plant agriculture at the University of Guelph. Dr. Powell continues as a freelance journalist, and his first book, *Mad Cows and Mother's Milk,* was published by McGill-Queen's University Press in 1997. His next book, *Reclaiming Dinner*, will be published later this year (2000).

Paul Rozin is the Edmund J. and Louise W. Kahn Professor for Faculty Excellence, in Psychology at the University of Pennsylvania. His main areas of scholarly work include determinants of human food choice, biocultural evolution of food and other traditions, magical thinking, and the emotion of disgust.

Theresa (Terre) Satterfield is a Research Associate with Decision Research, and teaches in the graduate \environmental studies program at the University of British Columbia. An anthropologist by training, her work focuses on environmental disuptes, the social impact of technological hazards, social theories of risk, and environmental ethics and values.

Paul Slovic is a Professor of Psychology at the University of Oregon and the President of Decision Research in Eugene, Oregon. His research interests include judgment, decision making, risk perception, and risk communication.

Vern R. Walker is Professor of Law at Hofstra University School of Law, in Hempstead, New York. His area of expertise is legal factfinding, especially involving scientific evidence, and he teaches courses in scientific evidence, health and safety regulation, products regulation and liability in the United States and the European Union, and torts.

Index